Microfluidic Applications in Biology

Edited by Niels Lion,
Joël S. Rossier, and
Hubert H. Girault

Related Titles

Omenn, G. S. (Ed.)

Human Plasma Proteomics

2007

ISBN 3-527-31757-0

Jungblut, P. R., Hecker, M. (Eds.)

Proteomics of Microbial Pathogens

2007

ISBN 3-527-31759-7

Kockmann, N. (Ed.)

Micro Process Engineering

Fundamentals, Devices, Fabrication, and Applications

2006

ISBN 3-527-31246-3

Klipp, E., Herwig, R., Kowald, A., Wierling, C., Lehrach, H.

Systems Biology in Practice

Concepts, Implementation and Application

2005

ISBN 3-527-31078-9

Niemeyer, C. M., Mirkin, C. A. (Eds.)

Nanobiotechnology

Concepts, Applications and Perspectives

2004

ISBN 3-527-30658-7

Geschke, O., Klank, H., Telleman, P. (Eds.)

Microsystem Engineering of Lab-on-a-Chip Devices

2004

ISBN 3-527-30733-8

Microfluidic Applications in Biology

From Technologies to Systems Biology

Edited by
Niels Lion, Joël S. Rossier, and Hubert H. Girault

WILEY-VCH
WILEY-VCH Verlag GmbH & Co. KGaA

The Editors

Dr. Niels Lion
Lab. d'Elec. Phy. et Anal.
EPFL
1015 Lausanne
Schweiz

Dr. Joël S. Rossier
DiagnoSwiss SA
CIMO SA
Route de l'Ile-aux-Bois 2
1870 Monthey
Schweiz

Prof. Dr. Hubert H. Girault
Lab. d'Elec. Phy. et Anal.
EPFL
1015 Lausanne
Schweiz

Library of Congress Card No.:
applied for

British Library Cataloguing-in-Publication Data
A catalogue record for this book is available from the British Library.

Bibliographic information published by the Deutsche Nationalbibliothek
The Deutsche Nationalbibliothek lists this publication in the Deutsche Nationalbibliografie; detailed bibliographic data are available in the Internet at <http://dnb.d-nb.de>.

Typesetting X Con Media AG, Bonn
Printing betz-druck GmbH, Darmstadt
Binding Litges & Dopf GmbH, Heppenheim

Printed in the Federal Republic of Germany
Printed on acid-free paper

ISBN-13: 978-3-527-31761-5
ISBN-10: 3-527-31761-9

Contents

Microfluidic Applications in Biology. Edited by Niels Lion, Joël S. Rossier, and Hubert H. Girault

ISBN-10: 3-527-31761-9

Preface

The incentive for this book is entirely due to Prof. Berthold Radola, who approached us during the HPCE 2004 conference in the lovely city of Salzburg. We would like to take this opportunity to testify of Prof. Radola's scientific leadership, kindness, and humanism. The basic idea of this book stems from the fact that two major trends are revolutionizing analytical biochemistry in a broad sense: first, all –omic approaches to life sciences put unprecedented constraints on analytical techniques, in terms of dynamic range, number of different molecules to be analysed at the same time, and time-to-result. In parallel, miniaturisation is pervading all branches of analytical chemistry, due to reduced analyte and solvent consumption, laboratory space saving, improved ease-of-use, and in some instances, better analytical results. The goal of this book is thus to provoke a discussion whether miniaturisation and microfluidics has the potential to address some of the analytical challenges raised by systems biology, and possibly cluster the analytical community around this problem.

Not surprisingly, there is no clear answer to this question yet, and only years will clarify what the most adequate analytical technology is for systems biology. Most probably, there will not be a unique platform, but more likely an alliance of generic systems, such as CE chips for fast analysis of DNA, RNA, and proteins, or the HPLC chip MS recently introduced by Agilent, with more dedicated systems, such as the HPLC chip MS that has been tailored by Ninonuevo *et al* for the separation of oligosaccharides. When speaking about microfluidics, we also have to keep in mind that the production of microfluidic devices requires dedicated facilities and equipments; in this perspective, the involvement of industrial suppliers becomes indispensable. Luckily, there are more and more microfluidic products available on the market (from DNA microarrays to CE chips or mass spectrometry interfaces, for example) that will help a lot convince biologists of the interest of microfluidics and miniaturisation for biological analyses, and also help delineate how to use them best.

As can be seen from the papers collected in this book, the field of microfluidics for systems biology is largely technology-driven, including the development of detection techniques in miniaturised systems, the hyphenation of microfluidic devices with both electrospray and matrix-assisted-laser-desorption-ionisation mass

Microfluidic Applications in Biology. Edited by Niels Lion, Joël S. Rossier, and Hubert H. Girault

ISBN-10: 3-527-31761-9

spectrometry, which remains a central analytical tool in systems biology, dielectrophoresis tools for cell manipulation... as well as more fundamental investigations of fluidic processes pertinent to the microscale that can ultimately be of relevance for analytical tools.

This is our hope that this book will stimulate the reflection about the design and use of microfluidic analytical systems in genomics, proteomics and metabolomics; beyond extremely successful proofs-of-principle of smart devices, we feel there is a real need for a deep reflection about what really needs to be developed, and what can be expected from microfluidics. We thus would like to warmly thank all authors who have contributed to this book and hope that it will be useful for all researchers involved in microfluidics.

Niels Lion
Joël S. Rossier
Hubert H. Girault
July 2006

List of Contributors

Professor Dr. Jicun Ren
College of Chemistry and
Chemical Engineering,
Shanghai Jiaotong University,
800 Dongchuan Road,
Shanghai 200240,
P. R. China

Professor Zhao-Lun Fang
Research Center for Analytical
Sciences,
Northeastern University,
Shenyang 110004,
P.R. China

Professor Peter C. Hauser
Department of Chemistry,
University of Basel,
Spitalstrasse 51,
CH-4004 Basel,
Switzerland

Professor Hong-Yuan Chen
The Key Laboratory of Analytical
Chemistry for Life Science,
Department of Chemistry,
Nanjing University,
Nanjing 210093,
P. R. China

Dr. Harold G. Craighead
School of Applied and Engineering
Physics,
Cornell University,
Ithaca, NY, USA

Professor Don L. DeVoe
Department of Mechanical
Engineering,
2129 Martin Hall,
University of Maryland,
College Park, MD 20742, USA

Professor Carlito Lebrilla
Department of Chemistry,
University of California,
One Shields Ave,
Davis, CA 95616, USA

Dr. Alina D. Zamfir
Institute for Medical Physics and
Biophysics,
University of Münster,
Robert-Koch-Strasse 31,
D-48149 Münster, Germany

Dr. Bill Wilson
Agilent Technologies,
Wilmington,
DE 19808, USA

Microfluidic Applications in Biology. Edited by Niels Lion, Joël S. Rossier, and Hubert H. Girault

ISBN-10: 3-527-31761-9

Professor Toshimasa Toyo'oka
Department of Analytical Chemistry,
School of Pharmaceutical Sciences
and COE Program in the 21st
Century,
University of Shizuoka,
52-1 Yada Suruga Shizuoka,
Shizuoka 422-8526, Japan

Dr. Alexandra Ros
Experimental Biophysics &
Applied Nanosciences,
Bielefeld University,
Universitätsstrasse 25,
D-33615 Bielefeld, Germany

Dr. Janko Auerswald
Centre Suisse d'Electronique et de
Microtechnique (CSEM S.A.),
Untere Gruendlistrasse 1,
CH-6055 Alpnach, Switzerland

Professor Jean-Louis Viovy
Laboratoire Physicochimie-Curie,
UMR/CNRS 168, Institut Curie,
Section de Recherche,
11 rue Pierre et Marie Curie,
F-75005 Paris, France

Dr. Howard A. Stone
Division of Engineering and
Applied Sciences, Pierce Hall,
Harvard University, 29 Oxford St,
Cambridge, MA 02138 USA

Professor Hsueh-Chia Chang
182 Fitzpatrick Hall,
Department of Chemical
Engineering,
University of Notre Dame,
Notre Dame, IN 46556, USA

Professor Nadine Aubry
New Jersey Institute of Technology,
Mechanical Engineering,
200 Central Avenue,
Newark, NJ 07102, USA

Professor Marc J. Madou
4200 Engineering Gateway Building,
Rm S3231,
University of California,
Irvine, Irvine, CA 92697–3975, USA

Dr. Young Sook Yoo
Bioanalysis and Biotransformation
Research Center,
Korea Institute of Science and
Technology,
P.O. BOX 131, Cheongryang,
Seoul 130-650, Korea

Dr. Johan Vanderhoeven
Department of Chemical
Engineering,
Vrije Universiteit Brussel,
Brussels, Belgium

Professor Bingcheng Lin
Department of Biotechnology,
Dalian Institute of Chemical Physics,
Chinese Academy of Sciences,
Zhongshan Road 457,
Dalian, 116023, China

Dr. Ildikó Kustos
Department of Medical
Microbiology and Immunology,
Faculty of Medicine,
University of Pécs,
Szigeti út. 12,
H-7643 Pécs, Hungary

1
On-line chemiluminescence detection for isoelectric focusing of heme proteins on microchips*

Xiangyi Huang, Jicun Ren

In this paper we present a sensitive chemiluminescence (CL) detection of heme proteins coupled with microchip IEF. The detection principle was based on the catalytic effects of the heme proteins on the CL reaction of luminol-H_2O_2 enhanced by *para*-iodophenol. The glass microchip and poly(dimethylsiloxane) (PDMS)/glass microchip for IEF were fabricated using micromachining technology in the laboratory. The modes of CL detection were investigated and two microchips (glass, PDMS/glass) were compared. Certain proteins, such as cytochrome *c*, myoglobin, and horseradish peroxidase, were focused by use of Pharmalyte pH 3–10 as ampholytes. Hydroxypropylmethylcellulose was added to the sample solution in order to easily reduce protein interactions with the channel wall as well as the EOF. The focused proteins were transported by salt mobilization to the CL detection window. Cytochrome *c*, myoglobin, and horseradish peroxidase were well separated within 10 min on a glass chip and the detection limits (S/N = 3) were 1.2×10^{-7}, 1.6×10^{-7}, and 1.0×10^{-10} M, respectively.

1.1
Introduction

Over the last 10 years, extensive work has been directed toward miniaturization of analytical methods. The integration of CE on a chip has been developed for biological and medical applications, such as DNA sizing and sequencing, PCR, and clinical diagnostics. The advantages of CE on chips over conventional separation techniques are short analysis time, high resolution, small reagent and sample requirements, portability, and lab on-a-chip versatility [1–7]. However, the extremely small sample size (usually at the picoliter level) makes it a challenge to achieve high detection sensitivity. It is very important to choose an appropriate

* Originally published in Electrophoresis 2005, 26, 3595–3601

Microfluidic Applications in Biology. Edited by Niels Lion, Joël S. Rossier, and Hubert H. Girault

ISBN-10: 3-527-31761-9

detector. Commonly used detection schemes for CE on chip mainly included LIF, electrochemical detections, and MS. Chemiluminescence (CL) detection is a simple, highly sensitive method, and it has been widely used in CE [8, 9], HPLC [10], and immunoassay [11]. Due to its simple optical devices, inexpensive price, and high sensitivity, CL is uniquely suited to on-line detection for microchips. Recently, some reports [12–16] have shown that CL is a promising alternative detection method for CE microchips.

IEF is a powerful tool for separation of proteins and other zwitterionic biomolecules based on their p*I*s. And CIEF continues to play an increasingly important role in the rapidly developing field of proteomics. There are three detection schemes in CIEF including single-point detection [17, 18], whole-column scanning detection [19], and whole-column imaging detection [20–22]. Single-point detection is commonly used, and it needs a mobilization step to electroosmotically [23–25], hydrodynamically [26, 27], or chemically [28, 29] migrate the focused sample bands past the detection point. Chemical mobilization can be achieved by replacing anolyte with a base (anodic mobilization), by replacing catholyte with an acid (cathodic mobilization), or by adding zwitterions to anolyte or catholyte. Compared with pressure mobilization, it provided an excellent resolution and simple procedure [28, 29].

IEF is a powerful separation technique for proteins, and miniaturized IEF (MIEF) has shown the superiority of high-speed analysis [30–35], coupling of multidimensional separations [36, 37], and high-throughput analysis [19, 38]. Due to the short length of the separation channel, MIEF is substantially faster than that of conventional capillary [30–35]. However, the smaller sample size makes it a challenge to achieve high detection sensitivity. It is, therefore, desirable to apply more sensitive detection methods to MIEF. LIF and whole-column imaging detection coupling with MIEF have been used for the analysis of labeled peptides and proteins with detection limits of $10^{-6}-10^{-8}$ M [32, 33, 37].

Although CL is a very sensitive technique and uniquely suited for on-line detection for microfluidic chips, few reports [12–16] have been found on application of CL in CE microchips compared to other methods. In this paper, we developed a sensitive and on-line CL detection for microchip IEF of proteins, and evaluated the feasibility of the on-line CL detection of proteins, such as cytochrome *c*, myoglobin, and horseradish peroxidase. The modes of CL detection and two microchips (glass, poly(dimethylsiloxane) (PDMS)/glass)) were compared.

1.2
Materials and methods

1.2.1
Chemicals

The borosilicate wafers were purchased from Shaoguang Microelectronics (Changsha, China). PDMS prepolymer and curing agents were generously gifted by Rhodia Silicon (Shanghai). Pharmalyte 3–10 (Beckman, Fullerton, CA, USA)

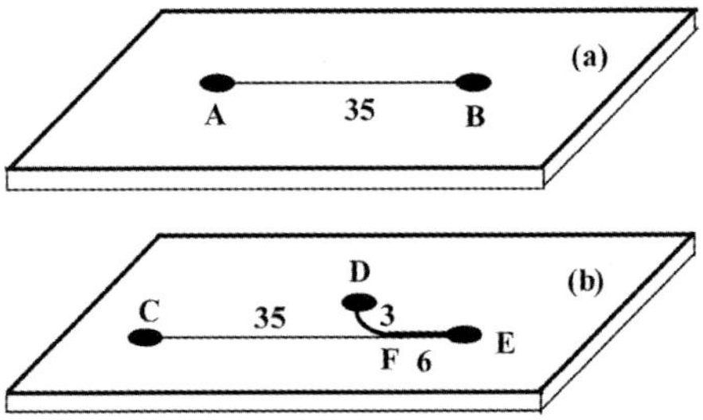

Fig. 1 Schematic layouts of microchips. (a) Chip pattern of a line. (b) Chip pattern of a line combining with Y. Channel dimensions of the microchips are shown in Tab. 1. Scales of the channel length in the figure were in millimeters.

was used as carrier ampholyte. Hydroxypropylmethylcellulose (HPMC), horse heart cytochrome *c* (p*I* 9.6), horse heart myoglobin (p*I* 7.2 and 6.8), and ammonium persulfate were products of Sigma (St. Louis, MO, USA). TEMED and luminol were purchased from Fluka (Buchs, Switzerland). Horseradish peroxidase was purchased from Dongfeng Biochemical (Shanghai, 300 U/mg). *Para*-iodophenol was a product of J&K (Shanghai, China). Ultrapure water (18.2 MΩ), double-distilled and purified on Millipore Simplicity (Millipore, Bedford, MA, USA), was used for preparation of all aqueous solutions.

1.2.2
Fabrication of microchips

The PDMS/glass microchips were fabricated in the laboratory as described in Fig. 1 following the method published in a previous paper [39]. Briefly, different masters were made on silicon wafers using an SU-8 photoresist and photolithographic procedures. The prepolymers of PDMS and curing agent were mixed at a 10:1 w/w ratio, stirred thoroughly, and then degassed under vacuum for 20 min. The viscous mixture was smoothly decanted onto the silicon master. The reservoirs were defined by placing the glass posts on the silicon master. The wafer was placed in an oven at 65°C and cured for 2 h. After curing, the PDMS replica containing negative relief of channels was easily peeled away from the master, and the glass posts were removed. Then the PDMS replica was placed under a 6 W mercury lamp (wavelength 365 nm) and irradiated for 3 h. The distance between the substrate and the lamp was about 5 cm and the estimated power of the exposure was about 15 mJ/cm^2. The microchip was composed of PDMS (upper) and glass substrates (bottom), and irreversible sealing was formed by UV irradiation.

Tab. 1 Channel dimensions of microchips

	Length, mm	Width, μm	Depth[a], μm
AB	35	70	24
CF	35	70	24
DF	3	500	24
EF	6	500	24

a) The depth of the PDMS/glass microchips is 40 μm.

This borosilicate glass microchip (35.0 mm × 62.0 mm) was fabricated through standard photolithography, wet chemical etching, and heat bonding technology [40]. The chip patterns of a line and a line combining with "Y" shape were designed (Fig. 1). The dimensions of the channels are shown in Tab. 1.

1.2.3
Instrumental setup

The MIEF-CL system was built in our laboratory and is depicted in Fig. 2. The schematic layouts of the microchips are given in Fig. 1. A high-voltage supply (Institute of Nuclear Science, Shanghai, China) was employed to offer potentials for IEF and electrophoresis separation. A vacuum was created to pull the reaction reagents through the detection window by the microinfusion pump modified (10 μL/min) (WZ-50C2, Zhejiang University Medical instrument, China). The microchip and a photomultiplier tube (PMT) (Model H5784-04, Hamamatsu, Japan) were shut up in a light-tight box. The back of the microchip was packed with black tape, on which a rectangular window (2.0 mm × 3.0 mm) was made. The window was just located on the joint of the Y. The microchip was fixed above the PMT, and the centers of the detection cell and PMT window were close to each other as near as possible. The PMT integrated with an amplifier closely situated in the detection window and transferred CL signal into electric signal. The output from the amplifier was not further amplified, and directly fed to a computer. We used Caesar software (version 4.0) for data collection from Prince Technologies (Emmen, The Netherlands).

1.2.4
IEF on microchips

For preparing sample solutions, heme proteins were dissolved in a solution of 2.0% carrier ampholytes pH 3–10 containing 0.2% HPMC and 0.5% TEMED. Herewith, HPMC was used for the purpose of minimizing the EOF as well as protein interactions with the capillary wall [41]. The channel was filled with the sample solution. Focusing was performed under an applied electric field strength of 200 or 300 V/cm for 2–4 min till the current dropped down to zero, using 10 mM phosphoric acid as anolyte and 20 mM sodium hydroxide as catholyte. Salt mobilization was subse-

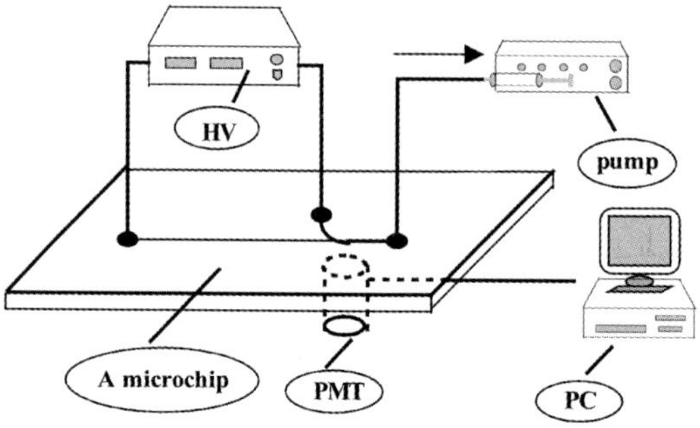

Fig. 2 Schematic of instrumental setup for MIEF coupled with CL detection.

quently carried out by replacing the catholyte with CL reagents consisting of 1.0 mM luminol, 10 mM H_2O_2, 1.5×10^{-4} M EDTA, 5.0×10^{-4} M *para*-iodophenol, 1.0×10^{-4} M ammonium persulfate, and 5.0 mM NaCl in 10 mM Tris solution. EDTA was used for elimination of the interferences with metal ion (possibly coexisting) such as cobalt (II) and copper (II). In addition, *para*-iodophenol and ammonium persulfate served as enhancers in the CL reactions. The emission wavelength range of the CL reaction of luminol was 425–435 nm. Applied electric field strength for mobilization process was 100 or 150 V/cm.

1.3 Results and discussion

1.3.1 Two CL detection modes for IEF on microchips

The aim of this work was to develop an effective on-line CL detector for MIEF and improve the detection sensitivity. Consequently, the CL reaction of luminol-H_2O_2 catalyzed by heme proteins was used to evaluate different on-chip configurations with CL detection in terms of sensitivity and peak symmetry.

The present CL detection modes included end-column and on-line detection mode. Optical detection modes usually involve the following problem. Some analytes may focus ahead of the detection window, and others behind. Consequently, these analytes that were focused between the detection window and the channel end were never detected. In contrast, the present end-column CL detection mode enables all analytes in the microchip channel to be located prior to the detection point. In order to reduce the loss of samples in the on-line detection mode, we placed the detection window close to the outlet of the channel, and TEMED was used to extend the pH gradient to pH 12 and serve as a cathodic blocker [42].

When salt mobilization is adopted for transporting focused proteins, an electrolyte solution must be exchanged before the process. The cathodal reservoir of the present system could be used to conveniently fill the catholyte at the focusing process and the CL reagent containing NaCl at the mobilization process.

Initially, a chip with a line channel pattern as described in Fig. 1a was investigated. The CL detection in the microchip was similar to the end-column scheme used in CE. An asymmetric peak (tailing) was obtained with low detection sensitivity as given in Fig. 3a. This phenomenon also occurred in CE with end-column CL detection. A reasonable explanation was that matrix in the separation channel continuously mobilized out to reduce the local concentration of CL reagents on the detection window. Additionally, the diffusion of analytes would lead to band-broadening and lowering of CL sensitivity.

Another channel pattern, a line combining with Y shape as shown in Fig. 1b, was employed. We used a syringe pump to continuously transport fresh chemiluminescent reagent from D point in Fig. 1 to E, instead of the molecular diffusion on end-column detection. In the on-line detection mode, in order to reduce the loss of

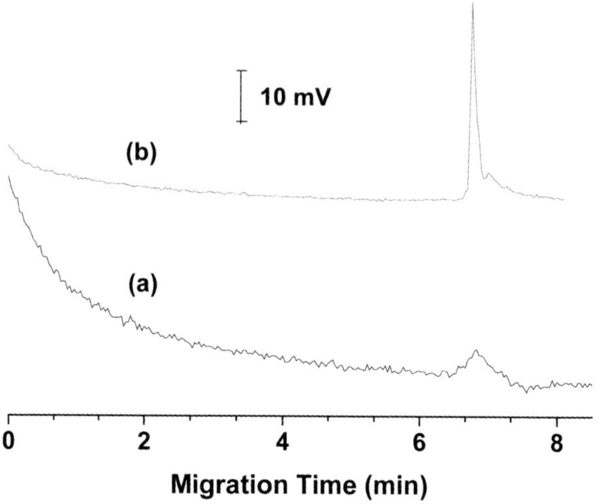

Fig. 3 Electropherograms of cytochrome *c* on two chips with CL detection. Glass chips: effective length, 35 mm. Sample: 1.0×10^{-6} M cytochrome *c*, 2.0% ampholyte containing 0.5% TEMED and 0.2% HPMC. Focusing: anolyte, 10 mM H_3PO_4, catholyte, 20 mM NaOH, $E = 300$ V/cm^{-1} for 2 min. Mobilization: anolyte, 10 mM H_3PO_4; catholyte, CL reagent, $E = 150$ V/cm^{-1}. (a) End-column detection by a line chip, (b) on-line detection by a line combining with the Y shape chip.

the sample focused between F and D, TEMED was used to extend the pH gradient to pH 12 as a cathodic blocker [42]. A vacuum was created to pull the chemiluminescent reagents through D to E by the syringe pump modified, and the chemiluminescent reagent was mixed with heme proteins from CF and then light was generated at EF (Fig. 1). As shown in Fig. 3b, a symmetric peak was achieved with about eightfold improvement in sensitivity. We also tried to push chemiluminescent reagent through D to E by a syringe pump [16], however, mostly we did not get the detection signals of proteins (data not shown). The reason is that the EOF was reduced by adding HPMC in the sample solution, the elecroosmotic pressure was not high enough to decrease the effect of the back pressure of pump flow, and the analytes did not move to the detection window under the push pressure.

1.3.2
Optimization of CL reaction conditions

Compared with IEF in capillary (~1 μL) with CL detection [29], the sample volume (~40 nL) was only about 1/25 on the microchip used in this paper. Thus, enhancing sensitivity of CL detection is of paramount importance to successful IEF assay in the microchips. Besides the structure of the CL detector, the CL sensitivity is mainly dependent on the CL reaction and electrophoresis conditions, which needs to be optimized systematically.

We optimized systematically the concentrations of H_2O_2, luminol, and EDTA, and found that their optimal concentrations were 1.0×10^{-2}, 1.0×10^{-3}, and 1.5×10^{-4} M, respectively.

It has been found that some phenol derivatives greatly enhanced the CL detection in luminol-H_2O_2 reaction catalyzed by peroxidase [43]. *Para*-iodophenol is one of the broadly used CL enhancers. We systematically investigated the influences of *para*-iodophenol concentrations on the CL detection and observed that *para*-iodo-

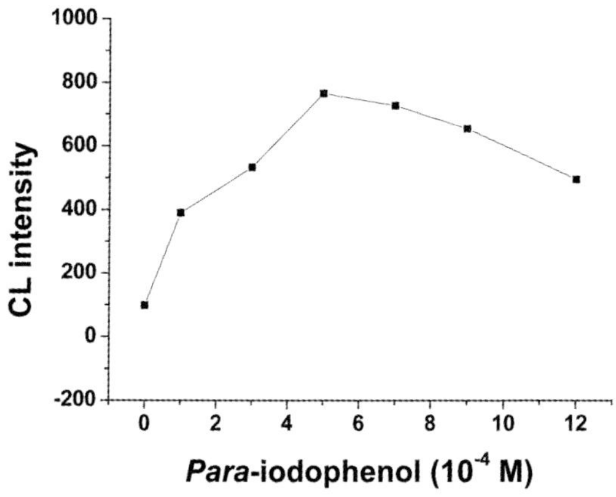

Fig. 4 Effects of *para*-iodophenol. Glass chip pattern of a line combining with Y as described in Fig. 1b. Sample: 1.25×10^{-6} M cytochrome *c*. Other conditions are the same as in Fig. 3.

phenol dramatically increased the CL intensity, and its optimum concentration was 5.0×10^{-4} M (as shown in Fig. 4). Under the optimal condition, the enhanced CL intensity was about seven times compared to CL detection without an enhancer, and the results are shown in Fig. 5. The enhancement factor is similar for all tested proteins (data not shown). However, the mechanism of the ECL is not completely clear so far. According to the deduction from [44–47], in the presence of *para*-iodophenol, the luminol-H_2O_2 reaction catalyzed by heme proteins could take place as the following processes:

$B + H_2O_2 \rightarrow B\text{-I} + H_2O \quad (1)$

$B\text{-I} + EH \rightarrow B\text{-II} + EH^* + H_2O \quad (2)$

$B\text{-II} + EH \rightarrow B + EH^* \quad (3)$

$EH^* + L^- \rightarrow EH + L^* \quad (4)$

Emission of light

$2L^* \rightarrow L + L^- \quad (5)$

$L + H_2O_2 \rightarrow LO_2^{\,2-} \quad (6)$

$LO_2^{\,2-} \rightarrow AP^{2-*} \quad (7)$

$AP^{2-*} \rightarrow AP^{2-} + h\nu \quad (8)$

In the above equations, B expresses heme proteins; B-I and B-II represent heme proteins intermediate-I and intermediate-II, respectively; L^- and L^* express luminol anion and its radical, respectively; EH and EH^* express enhancer and its radical, respectively; $LO_2^{\,2-}$ indicates luminol endoperoxide dianion; AP^{2-*} and AP^{2-} represent 3-aminophthalate dianion excited and 3-aminophthalate dianion, respectively. However, this mechanism needs to be confirmed further in future.

Furthermore, it has been reported that ammonium persulfate can sensitize the CL reaction of luminol and hydrogen peroxide [48, 49], especially when the concentration of ammonium persulfate was 1.0×10^{-4} M (data not shown). In the presence of ammonium persulfate and *para*-iodophenol, the S/N increased about ten times compared with no enhancer, and the results are shown in Fig. 5. The mechanism of the enhanced CL reaction is not yet completely clear. This may be explained by the fact that the free radicals from decomposition of ammonium per-

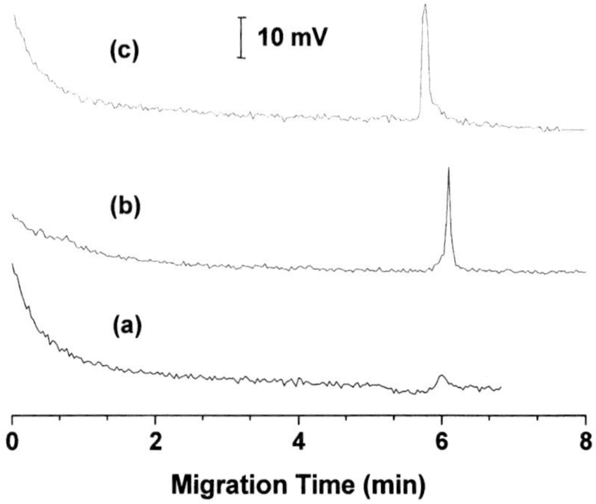

Fig. 5 Electropherograms of cytochrome *c* in the presence of *para*-iodophenol and ammonium peroxydisulfate. Glass chip pattern of a line combining with Y as described in Fig. 1b. Sample, 5.0×10^{-7} M cytochrome *c*. CL reagent contains: (a) no *para*-iodophenol, (b) 5.0×10^{-4} M *para*-iodophenol, and (c) 5.0×10^{-4} M *para*-iodophenol and 1.0×10^{-4} M ammonium peroxydisulfate. Other conditions are the same as in Fig. 3.

sulfate enhance the CL signal. These free radicals could oxidize heme proteins to form new free radicals, and then these free radicals oxidized from the proteins could catalyze the CL reaction of luminol and hydrogen peroxide.

1.3.3
Comparison of IEF on the PDMS/glass chip and glass chip

During the mobilization, the current gradually increased with migration of Cl^- to the anodic electrode. The current suddenly increased up to 50–60 μA, and then the current dropped down to zero slowly in a few minutes. This phenomenon implied that the chemical mobilization process was finished. Fig. 6 shows the electropherogram of heme proteins using CL detection. At optimized conditions, cytochrome *c*, myoglobin, and horseradish peroxidase were separated within 13 min using IEF on a PDMS/glass chip with CL detection except myoglobin (p*I* 6.8 and 7.2). Two peaks of myoglobin are observed with two p*I* values because of two horse heart myoglobin variants.

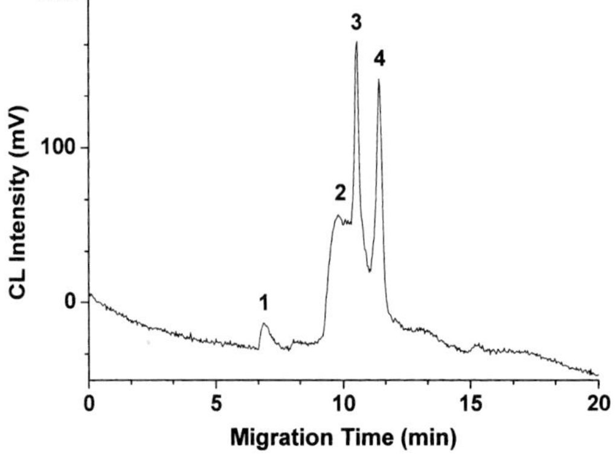

Fig. 6 IEF on a glass/PDMS chip with CL detection. Chip effective length, 35 mm. Sample consisted of 1.0×10^{-6} M cytochrome *c*, 5.0×10^{-6} M myoglobin, and 1.0×10^{-8} M peroxidase containing 2.0% carrier ampholyte containing 0.5% TEMED and 0.2% HPMC. Peak identification: (1) cytochrome *c* (p*I* 9.6), (2) myoglobin (p*I* 7.2), (3) myoglobin (p*I* 6.8), (4) peroxidase (p*I* 3.5). Focusing: $E = 200$ V/cm, mobilization: $E = 100$ V/cm. Other conditions are the same as in Fig. 3.

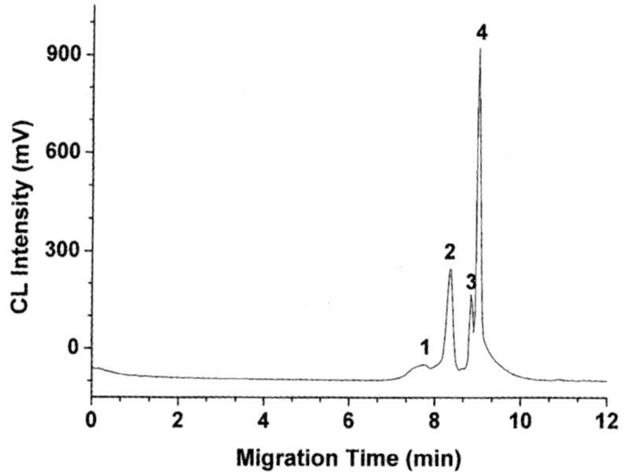

Fig. 7 IEF on a glass chip with CL detection. Chip effective length, 35 mm. Sample consisted of 2.5×10^{-7} M cytochrome *c*, 5.0×10^{-6} M myoglobin, and 1.0×10^{-8} M peroxidase including 2.0% carrier ampholyte containing 0.5% TEMED and 0.2% HPMC. Peak identification: (1) cytochrome *c* (p*I* 9.6), (2) myoglobin (p*I* 7.2), (3) myoglobin (p*I* 6.8), (4) peroxidase (p*I* 3.5). Focusing: $E = 200$ V/cm^{-1}, mobilization: $E = 100$ V/cm^{-1}. Other conditions are as described in the legend to Fig. 3.

And under the same condition, cytochrome *c*, myoglobin, and horseradish peroxidase were well separated within 10 min using IEF on a glass chip with CL detection, as shown in Fig. 7. It should be noticed that the peaks 3 and 4 did not resolve completely despite the big p*I* difference of these species (p*I* 6.8 and 3.5). This may be attributed to the following reasons. First, it is hard to avoid the disadvantages associated with the mobilization process, such as distortion of pH gradient, which led to loss in resolution. Second, there may be still certain adsorption of proteins on inner surface of channel although HPMC was used as dynamic coating reagent. Third, the effective separation length on the microchips was only 3.5 cm, so the separation capability of the microchips was less than the capillary. This phenomenon was also observed in [29]. Compared with Figs. 6, 7, the PDMS/glass chip has longer migration time and lower separation efficiencies. The reasons may be due to the different EOF and surfaces on two chips. The surface of the glass microchip is more hydrophilic, and the residual EOF is bigger on the glass chip than on the PDMS/glass chip, though HPMC has been used as a dynamic coating reagent.

1.3.4
Detection limits and linearity

We measured the detection limits for heme proteins using microchips IEF with CL detection and the results obtained are shown in Tab. 2. A detection limit down to 10^{-10} M for horseradish peroxidase was realized, and Tsukagoshi *et al.* [50] reported

Tab. 2 LODs for heme proteins (S/N = 3) on the PDMS/glass chip and the glass chip

Samples	PDMS/glass chip LOD, M	Glass chip LOD, M
Cytochrome *c*	2.5×10^{-7}	1.2×10^{-7}
Myoglobin	1.0×10^{-6}	1.6×10^{-7}
Horseradish peroxidase	2.4×10^{-9}	1.0×10^{-10}

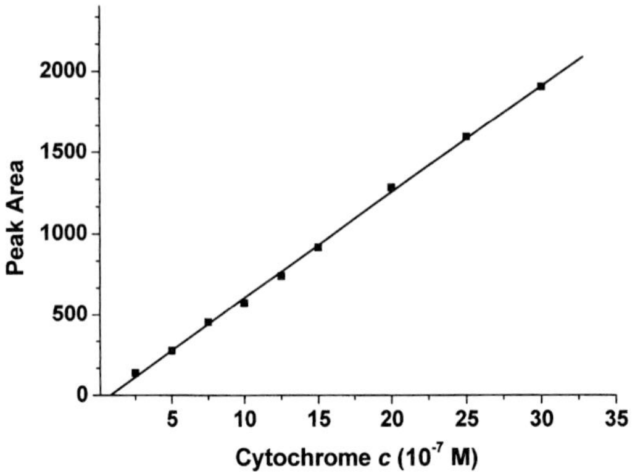

Fig. 8 Calibration curves of cytochrome *c* using a glass microchip. Experiments were carried out under the same conditions as described in Fig. 3.

that the detection limit of horseradish peroxidase was 2.0×10^{-9} M. However, the separation efficiency of our system was a little inferior compared with some whole-column imaging detections [34, 36] because of the mobilization step. The detection limits of heme proteins were lower in the glass microchip than in the PDMS/glass microchip. Despite the sample volume (~100 nL) on the PDMS/glass microchip (70 μm × 40 μm × 35 mm) being more than on the glass microchip (~40 nL) (0.5 × (70 μm + 30 μm) × 24 μm × 35 mm), the detection limits of heme proteins on the glass microchip were lower than on the PDMS/glass microchip. The reasons may be from two aspects: First, the residual EOF was smaller on the PDMS/glass chip than on the glass chip, and the migration time of the proteins was longer on the PDMS/glass chip, which would lead to band-broadening and lowering of CL sensitivity due to the diffusion of analytes. Second, HPMC was added to the sample solution to serve as a dynamic coating reagent. It formed a stable coating on PDMS and glass surfaces, eliminating the inhomogeneous EOF in channels on PDMS/glass microchips, and improving the hydrophilicity of PDMS surfaces. However, the inner surface of the PDMS channels probably had certain affinity for proteins [51], which caused a stronger adsorption onto the channel wall.

We quantitatively examined the linearity of cytochrome *c* and horseradish peroxidase since they were detected as a single peak on the electropherograms, respectively. Fig. 8 shows the calibration curve for cytochrome *c* using a glass chip as shown in Fig. 1b. The linear range of the calibration curve was from 2.5×10^{-7} to 3.0×10^{-6} M ($R = 0.999$), the detection limit (S/N = 3) of cytochrome *c* was 1.2×10^{-7} M. The linear range of the calibration curve for horseradish peroxidase was from 3.0×10^{-10} to 4.0×10^{-9} M ($R = 0.994$), the detection limit (S/N = 3) was 1.0×10^{-10} M. The RSDs of migration times and peak areas for cytochrome *c* and horseradish peroxidase were less than 5.0 and 10.0% ($n = 5$), respectively.

1.4 Concluding remarks

Our preliminary results demonstrated that the microchip IEF-CL detection was an effective method for the separation and determination of heme proteins. Compared to the end-detection, the on-line detection has good sensitivity and symmetric peak shapes of proteins. Certain proteins such as cytochrome *c*, myoglobin, and horseradish peroxidase were satisfactorily separated, and a detection limit down to 10^{-10} M for horseradish peroxidase was realized. This method is simple, rapid, and easy to operate, and exhibits advantages from the hybrid of MIEF with a high separation efficiency and CL with a high sensitivity. We believe that MIEF-CL system has potential applications in biochemical analyses and clinical diagnosis such as hemoglobin variants in blood samples.

This work was financially supported by the National Natural Science Foundation of China (Nos. 20271033, 90408014), the key project of the National Natural Science Foundation of China (No. 20335020), the Nano-Science Foundation of Shanghai (0452NM052), and mega-projects of Science Research for the 10th Five-Year-Plan of MSTPRC (2004BA720A04).

1.5 References

[1] Manz, A., Graber, N., Widmer, H., *Sens. Actuators B* 1990, *1*, 244–248.

[2] Harrison, D. J., Fluri, K., Seiler, K., Fan, Z., Effenhauser, C. S., Manz, A., *Science* 1993, *261*, 895–897.

[3] Jacobson, S., Hergenröder, R., Koutny, L., Warmack, R., Ramsey, J., *Anal. Chem.* 1994, *66*, 1107–1113.

[4] Raymond, D. E., Manz, A., Widmer, H. M., *Anal. Chem.* 1996, *68*, 2515–2522.

[5] Li, J., Kelly, J. F., Chernushevich, I., Harrison, D. J., Thibault, P., *Anal. Chem.* 2000, *72*, 599–609.

[6] Figeys, D., Pinto, D., *Electrophoresis* 2001, *22*, 208–216.

[7] Hebert, N. E., Brazill, S. A., *Lab Chip* 2003, *3*, 241–247.

[8] Tsukagoshi, K., Nakahama, K., Nakajima, R., *Anal. Chem.* 2004, *76*, 4410–4415.

[9] Ren, J. C., Huang, X. Y., *Anal. Chem.* 2001, *73*, 2663–2668.

[10] Yamaguchi, M., Yoshida, H., Nohta, H., *J. Chromatogr. A* 2002, *950*, 1–19.

[11] Hage, D. S., *Anal. Chem.* 1999, *71*, 294R–304R.

[12] Hashimoto, M., Tsukagoshi, K., Nakajima, R., Kondo, K., Arai, A., *J. Chromatogr. A* 2000, *867*, 271–279.

[13] Mangru, S. D., Harrison, D. J., *Electrophoresis* 1998, *19*, 2301–2307.

[14] Liu, B. F., Ozaki, M., Utsumi, Y., Hattori, T., Terabe, S., *Anal. Chem.* 2003, *75*, 36–41.

[15] Liu, B. F., Ozaki, M., Hisamoto, H., Luo, Q., Utsumi, Y., Hattori, T., Terabe, S., *Anal. Chem.* 2005, *77*, 573–578.

[16] Su, R. G., Lin, J. M., Qu, F., Chen, Z. F., Gao, Y. H., Yamada, M., *Anal. Chim. Acta* 2004, *508*, 11–15.

[17] Hjertén, S., Zhu, M. D., *J. Chromatogr.* 1985, *346*, 265–270.

[18] Cruickshank, K. A., Olvera, J., Müller, U. R., *J. Chromatogr. A* 1998, *817*, 41–47.

[19] Sanders, J. C., Huang, Z., Landers, J. P., *Lab Chip* 2001, *1*, 167–172.

[20] Mao, Q., Pawliszyn, J., Thormann, W., *Anal. Chem.* 2000, *72*, 5493–5502.

[21] Mao, Q., Pawliszyn, J., *J. Biochem. Biophys. Methods* 1999, *39*, 93–110.

[22] Liu, Z., Pawliszyn, J., *Anal. Chem.* 2003, *75*, 4887–4894.

[23] Hiraoka, A., Tominaga, I., Hori, K., *J. Chromatogr. A* 2002, *961*, 147–153.

[24] Hagmann, M. L., Kionka, C., Schreiner, M., Schwer, C., *J. Chromatogr. A* 1998, *816*, 49–58.

[25] Lillard, S. J., Yeung, E. S., *J. Chromatogr. B* 1996, *687*, 363–369.

[26] Huang, T. L., Richards, M., *J. Chromatogr. A* 1997, *757*, 247–253.

[27] Thorne, J. M., Goetzinger, W. K., Chen, A. B., Moorhouse, K. G., Karger, B. L., *J. Chromatogr. A* 1996, *744*, 155–165.

[28] Jenkins, M. A., Ratnaike, S., *Clin. Chim. Acta* 1999, *289*, 121–132.

[29] Hashimoto, M., Tsukagoshi, K., Nakajima, R., Kondo, K., *J. Chromatogr. A* 1999, *852*, 597–601.

[30] Xu, Y., Zhang, C. X., Janasek, D., Manz, A., *Lab Chip* 2003, *3*, 224–227.

[31] Tan, W., Fan, Z. H., Qiu, C. X., Ricco, A. J., Gibbons, I., *Electrophoresis* 2002, *23*, 3638–3645.

[32] Hofmann, O., Che, D., Cruickshank, K. A., Müller, U. R., *Anal. Chem.* 1999, *71*, 678–686.

[33] Macounová, K., Cabrera, C. R., Holl, M. R., Yager, P., *Anal. Chem.* 2000, *72*, 3745–3751.

[34] Macounová, K., Cabrera, C. R., Yager, P., *Anal. Chem.* 2001, *73*, 1627–1633.

[35] Wen, J., Lin, Y., Xiang, F., Matson, D. W., Udseth, H. R., Smith, R. D., *Electrophoresis* 2000, *21*, 191–197.

[36] Wang, Y. C., Choi, M. H., Han, J., *Anal. Chem.* 2004, *76*, 4426–4431.

[37] Herr, A. E., Molho, J. I., Drouvalakis, K. A., Mikkelsen, J. C., Utz, P. J., Santiago, J. G., Kenny, T. W., *Anal. Chem.* 2003, *75*, 1180–1187.

[38] Zilberstein, G., Korol, L., Bukshpan, S., Baskin, E., *Proteomics* 2004, *4*, 2533–2540.

[39] Chen, L., Ren, J. C., Bi, R., Chen, D., *Electrophoresis* 2004, *25*, 914–921.

[40] Wooley, A. T., Mathies, R. A., *Proc. Natl. Acad. Sci. USA* 1994, *91*, 11348–11352.

[41] Kubach, J., Grimm, R., *J. Chromatogr. A* 1996, *737*, 281–289.

[42] Guo, Y. J., Bishop, R., *J. Chromatogr.* 1982, *234*, 459–462.

[43] Thorpe, G. H., Kricka, L. J., Moseley, S. B., Whitehead, T. P., *Clin. Chem.* 1985, *31*, 1335–1341.

[44] Navas Díaz, A., García Sánchez, F., González Garcia, J. A., *J. Biolumin. Chemilumin.* 1998, *13*, 75–84.

[45] Candy, T. E. G., Jones, P., *J. Biolumin. Chemilumin.* 1991, *6*, 239–243.

[46] Vlasenko, S. B., Arefyev, A. A., Klimov, A. D., Kim, B. B., Gorovits, E. L., Osipov, A. P., Gavrilova, E. M., Yegorov, A. M., *J. Biolumin. Chemilumin.* 1989, *4*, 164–176.

[47] Hodgson, M., Jones, P., *J. Biolumin. Chemilumin.* 1989, *3*, 21–25.

[48] Li, J., Dasgupta, P. K., *Anal. Chim. Acta* 1999, *398*, 33–39.

[49] Huang, G. M., Yang, J. O., Delanghe, J. R., Baeyens, W. R. G., Dai, Z. X., *Anal. Chem.* 2004, *76*, 2997–3004.

[50] Tsukagoshi, K., Nakamura, K., Nakajima, R., *Anal. Chem.* 2002, *74*, 4109–4116.

[51] Duffy, D. C., McDonald, J. C., Schueller, O. J. A., Whitesides, G. M., *Anal. Chem.* 1998, *70*, 4974–4984.

2
A simple microfluidic system for efficient capillary electrophoretic separation and sensitive fluorimetric detection of DNA fragments using light-emitting diode and liquid-core waveguide techniques*

Shi-Li Wang, Xiao-Feng Fan, Zhang-Run Xu, Zhao-Lun Fang

A miniaturized CE system has been developed for fast DNA separations with sensitive fluorimetric detection using a rectangle type light-emitting diode (LED). High sensitivity was achieved by combining liquid-core waveguide (LCW) and lock-in amplification techniques. A Teflon AF-coated silica capillary on a compact 6×3 cm baseplate served as both the separation channel for CE separation and as an LCW for light transmission of fluorescence emission to the detector. An electronically modulated LED illuminated transversely through a 0.2 mm aperture, the detection point on the LCW capillary without focusing, and fluorescence light was transmitted to the capillary outlet. To simplify the optics and enhance collection of light from the capillary outlet, an outlet reservoir was designed, with a light transmission window, positioned directly in front of a photomultiplier tube (PMT), separated only by a high pass filter. Automated sample introduction was achieved using a sequential injection system through a split-flow interface that allowed effective release of gas bubbles. In the separation of a ϕX174 *Hae*III DNA digest sample, using ethidium bromide as labeling dye, all 11 fragments of the sample were effectively resolved in 400 s, with an S/N ratio comparable to that of a CE system with more sophisticated LIF.

2.1
Introduction

LIF is currently employed most broadly as detection system in microfluidic chip-based CE systems, owing to its high sensitivity and noninvasiveness in many bioapplications when dealing with subnanoliter detection. Miniaturization of CE-LIF systems is highly desirable for the production of portable systems suitable for field applications or point-of-care testing. However, conventional lasers are relatively expensive, bulky, and have rather limited lifetime, and nonsuitable for integration in chip-based micro-

* Originally published in Electrophoresis 2005, 26, 3602–3608

Microfluidic Applications in Biology. Edited by Niels Lion, Joël S. Rossier, and Hubert H. Girault

ISBN-10: 3-527-31761-9

devices. The availability of solid-state laser diodes have somewhat alleviated these limitations, but still insufficient for achieving these goals. Recently, light-emitting diodes (LEDs) are being exploited to achieve fluorescence detection in CE and micro-analytical systems, while providing excellent prospects for detector integration and miniaturization [1–12]. As a light source, LEDs are small, cheap, easily operated, available in a wide range of wavelengths from violet to near infrared, and exceptionally stable. They also have long lifetimes, and provide high emission intensities. Despite their multiple advantages, the emission from LED sources is spatially incoherent and not monochromatic. When using LEDs as the excitation source for fluorimetric detection, considerable power loss might be expected owing to failure in achieving proper focusing and obtaining the required spectral output. To overcome such limitations, a complex and bulky optical system might be required [6]. Dasgupta *et al.* [7], and later our group [8], combined the use of a liquid-core waveguide (LCW) technique with an LED source to produce an extremely simple optical system for LED-excited fluorescence detection in CE without requiring a focusing system. A fused-silica capillary coated with Teflon AF was used as CE separation channel, which also functioned as an LCW when filled with an aqueous solution. When the capillary was illuminated transversely at the detection point by an LED, only fluorescence light excited by the LED traveled to the terminal of the capillary, where it could be collected with high efficiency. Satisfactory isolation of fluorescence radiation from the excitation light was demonstrated merely using a single broadband filter with such LCW-CE devices. The LCW-CE system was used successfully for the separation and determination of FITC-labeled amino acids [8]. Further related works using LCW techniques were reviewed by Dallas and Dasgupta [13] in a recent article.

Despite the simplicity, compactness, low cost, and high efficiency of this approach, the detection limits achieved were significantly worse than those often reported for chip-based CE, employing more sophisticated LIF optical detection systems. For example, the detection limit of FITC-amino acids was a factor of 1–2 worse than those obtained with a chip-based CE-LIF system.

In order to overcome such deficiencies, numerous methods for enhancing the S/N ratio of LED-based CE systems have been proposed, mostly exploiting the favorable property of LEDs of being electrically modulable with any desired pattern. An often-adopted approach is to modulate the LED driving voltage, followed by detection of the modulated signals employing a lock-in amplifier [9–11].

Presumably, in order to compensate for the above-mentioned deficiencies in the LED-based fluorescence detection, high fluorescence efficiency dyes, such as SYBR Green [10, 11], SYTO [12] or YOYO-1 [6], were used almost exclusively for such applications, when dealing with the separation of DNA fragments.

In this work, an extremely simple and sensitive microfluidic CE system (μCE) with LED-excited fluorescence detection for DNA fragment separations was developed, based on a combination of LCW and lock-in amplifier techniques, with a performance comparable to that of chip-based CE-LIF. Ethidium bromide (EB), which is cheap and most widely used as a DNA dye in practical applications, but much less sensitive than the previously mentioned labeling dyes, was used in this work to demonstrate the favorable performance of the optical/detecting system.

2.2
Materials and methods

2.2.1
Chemicals

All reagents were of analytical grade. Deionized water was used throughout and was obtained from a Water Pro PS system (LabConco, Kansas, MO, USA). ϕX174-*Hae*III DNA digest sample (50 ng/μL, containing 11 fragments) was purchased from TaKaRa Biotechnology (Dalian) (Dalian, Liaoning, P.R. China). Hydroxypropylmethylcellulose (HPMC, average MW = 86 000) was purchased from Aldrich Chemical (Milwaukee, WI, USA). 0.5 × Tris-borate-EDTA (TBE) was used as working electrolyte for CE separation and carrier for the sequential injection (SI) system, and obtained by diluting from a 1 × TBE (89 mM Tris, 89 mM borate, and 2 mM EDTA) solution. Working DNA sample solutions were prepared daily by diluting DNA stock solutions with deionized water.

2.2.2
Miniaturized CE (μCE) system

The experimental setup of the LCW-LED based CE system was similar to that described previously [8] with minor modifications as shown in Fig. 1. The entire system was built on a 6 × 3 cm glass plate. The central part of the CE device was a 5 cm long, 75 μm ID, 375 μm OD Teflon AF coated fused-silica capillary (TSU075375; Polymicro Technologies, Phoenix, AZ, USA), which served as both the separation channel and LCW for transferring fluorescence emission light. The inlet terminal of the capillary was connected to a split-flow sample injection interface (left end in Fig. 1) that was in turn connected to an SI sampling device, and its outlet was connected to a windowed reservoir (right end in Fig. 1).

The split-flow interface was produced from a 20 mm lower section of a plastic pipet tip, cut lengthwise from its wider end to produce a 15 mm long opening (see inset in Fig. 1). This formed a trough at the wider end for guiding waste flows, while the thinner end was used as the inlet for continuous sample/carrier (running buffer) introduction. A small hole was drilled at 5 mm from the thinner end of the interface to accommodate the capillary inlet. The thinner end of the interface was connected to the SI sample/carrier transportation line. The trough was then fixed on the glass baseplate with epoxy.

The windowed reservoir at the capillary outlet was produced from a 10 mm section of a 6 mm ID plastic microcentrifuge tube that was cut lengthwise into half, and painted black outside. A 0.4 mm hole was drilled at one end of the cut-tube for later accommodating the capillary outlet. A 0.1 mm thick cover glass was epoxyed to the cut-end of the tube to form a window, and the entire setup was epoxyed on the baseplate forming a reservoir.

The inlet of the Teflon AF-coated silica capillary was then inserted 1 mm into the trough of the split-flow interface, and epoxyed to it, while the other end was inserted all way through the hole on the outlet reservoir and butted against the wall of

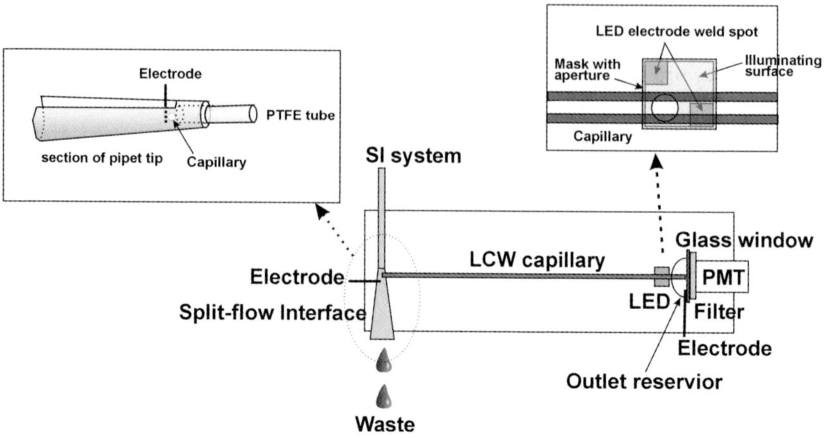

Fig. 1 Schematic diagram of CE system (not to scale).

the glass window. The small space between the capillary and the hole was then filled with epoxy both to avoid leakage and to fix the capillary. A photomultiplier tube (PMT, Model CR114; Beijing Hamamatsu, Beijing, P.R. China) was installed directly outside the glass window with a longpass filter (560 nm cutoff wavelength, CB4 model; Haiguang Optical Components, Shanghai, P.R. China) inserted between the reservoir and PMT windows. The position of the outlet of the LCW capillary and the PMT window were aligned for optimum fluorescence light collection. A rectangle type green InGaN LED (HT-S91NG5; Harvateck, Taiwan, P.R. China) with maximum emission at 520 nm with a full-width at half-maximum (FWHM) of 40 nm, which served as the excitation light source, was fixed beneath the LCW capillary, 10 mm from its outlet. Light from the LED illuminating the excitation point was restricted using a diaphragm with an orifice of 200 μm fixed on the LED surface (see inset in Fig. 1).

2.2.3
Sequential injection sample introduction system

The sample introduction system used for the CE system was as described previously [8], and consisted of a syringe pump with a two-way valve (Model P/N 50300; Kloehn, Las Vegas, NV, USA), and a multiposition selector valve (Model P/N 50120; Kloehn). Connections were made as shown in Fig. 2. PTFE tubing (0.5 mm ID) was used for connecting all components of the SI system, and an 11 cm length of PTFE tubing (0.25 mm ID) was used for connecting the SI system to the split-flow interface. The SI system was controlled by software that was programmed by LabVIEW6i (National Instruments, Austin, TX, USA).

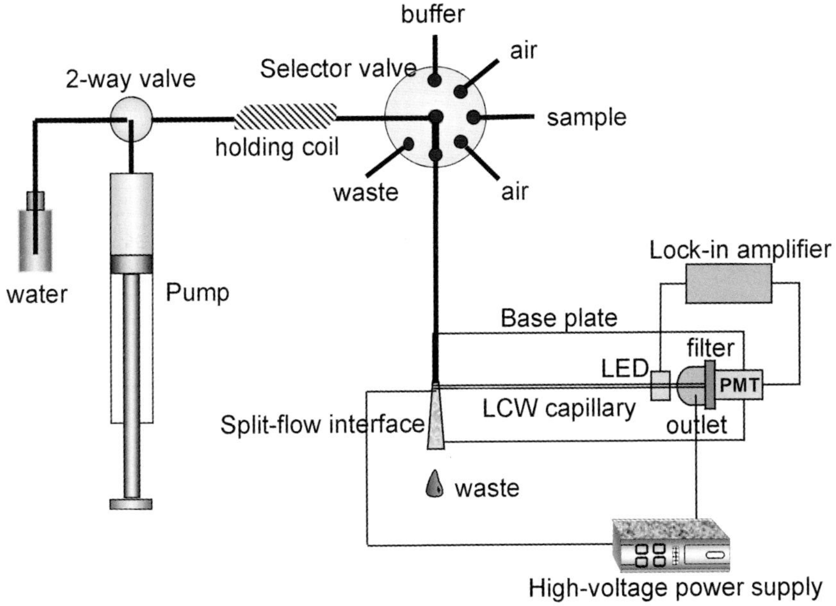

Fig. 2 Schematic diagram of combined SI sample introduction μCE system.

2.2.4
Procedures

Before use, the LCW capillary was sequentially flushed with 0.1 M NaOH, water, and 0.5 × TBE buffer, each for 15 min. The capillary was then filled with 2.5% HPMC in 0.5 × TBE solution, containing 3 μg/mL EB using a syringe. The outlet reservoir on the chip was filled with the same solution, and one of the platinum electrodes was immersed into the solution of the outlet reservoir along its side, while the other was inserted into the trough of the interface from its upper opening, touching the tip side of the capillary inlet.

A home-built programmable high-voltage (HV) power supply, with four electrode terminals (only two used in this work), variable in the 0–1800 V range, was used for CE separation. Voltages were then applied to the electrodes, with +1.00 kV applied to the outlet reservoir during the separation and sample injection stages, and with the split-flow interface electrode set at 0 V.

During sample loading, 3.0 μL samples, intercalated between two 5.0 μL air segments, were aspirated into the holding coil of the SI system, and during injection stage the intercalated sample was transported by the carrier (working electrolyte) solution through the split-flow interface at 75 μL/min in 5 s. Following sample injection, the SI system delivered the working electrolyte through the interface at the same flow rate for another 10 s to rinse the interface, then the flow rate was reduced to 10 μL/min. During the separation stage, a small fraction of the injected sample electrokinetically (depending on the split-flow ratio) entered the LCW capillary inlet when passing through the interface, and its constituents were separated in the capillary under the applied voltage.

The LED used for excitation was driven by a home-built pulsed power supply at 900 Hz. The pulsed signal also served as reference signal for the lock-in amplifier. The emitted light transmitted through the LCW was detected at the capillary outlet by the PMT. The signal output from the detector was amplified by a lock-in amplifier (ND-201; Nanjing University, Nanjing, P.R China) and recorded by a PC using a PCL-1711 DAQ card (Advantech, Taiwan, P.R. China). The entire software used for controlling the SI, LED, HV, and PMT were programmed using LabVIEW6i. In studies for comparison, when no lock-in amplifier was used, the LED was driven by a home-made constant-current supply and the detection signals were recorded directly by the PC.

2.3
Results and discussion

2.3.1
Considerations in the μCE system design

The ultimate objective of our project is to develop a cheap, portable, and compact μCE system for DNA fragment and other separations, suitable for continuous analysis of a series of samples, and achieving detection power comparable to LIF detection. In the present work, we mainly focused on, first, simplification of the optical system by modifying the LCW approach; and second, substitution of the laser as an excitation source with a much cheaper and more compact LED.

Our previous experiences working with LCW using Teflon AF-coated silica capillaries showed that excellent isolation of the fluorescence emission from excitation light is feasible with no optical components other than a broadband filter, and a short length of optical fiber. In this work, we further simplified the system by positioning the PMT directly in front of a specially designed reservoir window (Fig. 1), less than 1 mm distance from the LCW capillary outlet housed within the reservoir. Thus, light transmission using an optical fiber was avoided without sacrificing the efficiency of light collection.

As discussed in Section 1, in many respects, LEDs are ideal for developing cheap and compact systems with fluorimetric detection. Their main deficiency lies in their lower light energies and the difficulties in focusing the available energy into a tight focus point. In this work we attempted to overcome such deficiencies by: (i) modulating the driving voltage of the LED combined with lock-in amplification of the modulated PMT signal; (ii) by using a rectangle-type LED with higher emission efficiency and planar light-emitting surface that can be positioned closer to the LCW capillary than a dome-topped LED; and (iii) by shielding the light-emitting surface of the LED using a diaphragm with a small aperture that restricted illumination of the capillary to about 0.2 mm. The combined effects of these measures ensured high separation efficiencies while achieving detection power comparable to that of CE-LIF systems.

Our previous works [14–17] have shown the efficiency of combining flow injection (FI) and SI sample introduction with CE. Although the SI-μCE system was used in this work mainly for the sake of convenience, some modifications were made in the split-flow interface to achieve a more stable flow. The half-open nature of the interface (see Fig. 1) facilitated release of gas segments in the flow introduced by the SI system for limiting dispersion of samples during transport to the interface. The design minimized fluctuations that could occur in the falling-drop approach owing to discontinuity between drops.

2.3.2
Conditions for split-flow sampling

In the present system, samples were introduced electrokinetically when being transported through the split-flow interface. The amount of sample introduced into the separation channel was decided by the sample volume injected by the SI system as well as the split-flow ratio. With a fixed sample volume, larger volumes are introduced with a higher ratio. Both the applied sampling voltages and sampling time (*i.e.*, carrier flow rate) are important for determining the ratio, which should be optimized both in relation to separation efficiency and sensitivity. Larger ratios favor the introduction of larger sample volumes, and therefore, higher sensitivity, but lower separation efficiencies.

Fig. 3 shows the theoretical plate number of the 603 bp peak in CE separation of DNA samples within the carrier flow rate range of 25.0–100.0 μL/min and field strength range of 150–250 V/cm during sample introduction. With a fixed sample volume of 3.0 μL, a separation field strength of 200 V/cm and 75 μL/min carrier flow rate were chosen in this study, under which conditions the split-flow ratio was about 1/25 000, and approximately 130 pL of sample were introduced into the separation channel, which is similar to the amount injected into the separation channel of microfabricated chips.

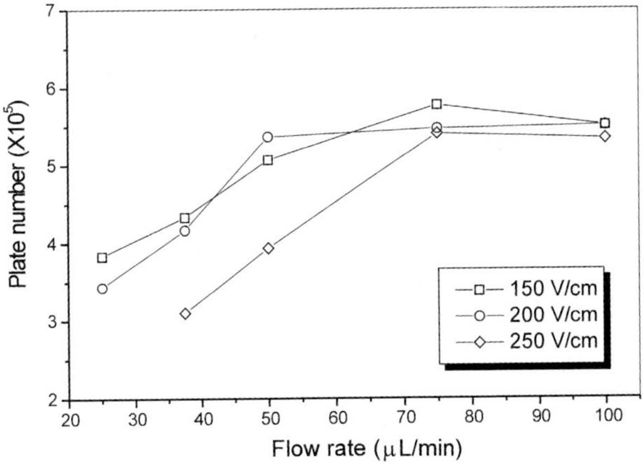

Fig. 3 Effect of flow rate and electrical field strength on sample introduction. Sample was ϕX174-*Hae*III DNA digest (25 ng/μL), injected volume 3 μL. Effective CE separation length 4 cm, separated in 2.5% HPMC containing 3 μg/mL EB in 0.5 × TBE. Separation voltage, 1 kV.

2.3.3
CE separation medium for DNA fragments

Various sieving matrices, consisting of a polymer dissolved in a BGE, are being widely employed for the separation of DNA fragments by CE [18]. A number of polymers, including hydroxyethylcellulose (HEC), HPMC, hydroxypropylcellulose (HPC), linear polyacrylamide (LPA), and poly(ethylene oxide) (PEO), have all been employed successfully for that purpose. One of the most frequently used polymers, HPMC (average MW 86 000), was used in this study. For chip-based CE separation of DNA, Xu *et al.* [19] recommended the use of HPMC with a lower molecular weight of average 11 500, which was less viscous, and more easily filled into the microfabricated chip channels; however, polyhydroxy additives were included in the separation media to achieve satisfactory resolution. In our application, a capillary with ID similar to those used in conventional CE, but much shorter, was used for separation, and no difficulties were encountered in hand-filling the 5 cm capillary with HPMC (MW 86 000) using a syringe, even at a higher concentration than usually reported. In this work 2.5% HPMC was employed instead of the more frequently recommended lower concentration of 0.5% to ensure better separation efficiency within the short separation length.

The concentration of the EB fluorescence dye was reported by Liang *et al.* [20] to affect the performance of CE separation of DNA samples and an EB concentration of about 3 μg/mL was recommended for good performance. EB concentrations in the range 2–4 μg/mL were tested in our system and no significant differences were observed in this range. A concentration of 3 g/mL EB was adopted in further experiments.

2.3.4
Effect of excitation light sources

At a preliminary stage of this work, the performance of a nonmodulated LED as an excitation light source for fluorimetric detection in the LCW-CE system was compared with the same LCW-CE system but with LIF detection. This is readily achievable on a single experimental setup, using a laser beam and an LED to illuminate the same detection point on the capillary from radially opposite directions, one source working at a time. The LED was substituted by a green diode laser (532 nm, 18.7 mW), the light of which was transferred to the LCW capillary through a 200 μm ID optical fiber.

Fig. 4 shows electropherograms gained during eight continuous injections of the same DNA sample (25 ng/μL), using different excitation sources, with all other conditions kept constant while switching from LED to laser, and *vice versa*. Higher baseline levels and lower peak heights were observed with the LED source. In Fig. 4b, no marked differences in peak width and noise levels are observed between the two approaches; however, higher stability with lower drift of the LED source signal is evident. Employing a laser source, the estimated theoretical plate number for the 603 bp peak was 7.5×10^5/m and the S/N ratio was 18 at the DNA concentration used. With

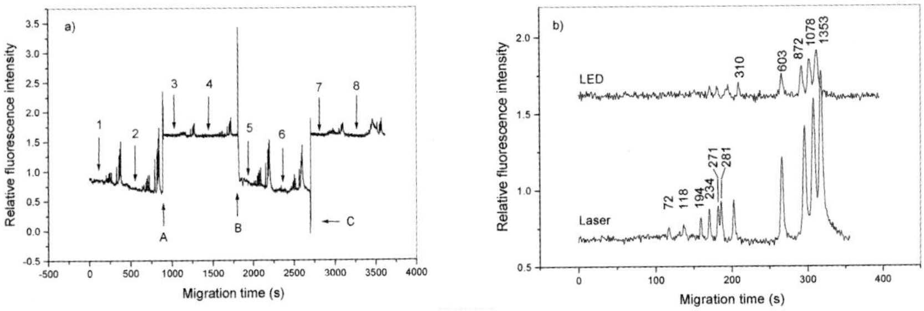

Fig. 4 Effect of excitation light sources in separation of ϕX174-*Hae*III DNA digest sample (25 ng/μL). (a) Electropherograms recorded continuously using different excitation sources: 1, 2, 5, and 6 using laser and 3, 4, 7, and 8 using LED as exciting light source; 1–8 are injection points, A–C are points where light source were changed. (b) Enlarged electropherogram of samples. Effective CE separation length 4 cm, separated in 2.5% HPMC containing 3 μg/mL EB in 0.5 × TBE. Separation and sampling voltage 1 kV. Injection volume, 3 μL, and SI carrier flow rate, 75 μL/min.

the LED source, the separation efficiency was not significantly degraded (theoretical plate number 6.5×10^5 /m) but the S/N ratio was reduced to 4. Such results showed favorable prospects for maintaining high separation efficiency using LED sources, while at the same time stressed the need for further improving S/N levels.

The higher background signal for the LED is a consequence of a lack of focusing system for this source, leading to more excitation light being transmitted through the capillary walls to the PMT. In preliminary studies, we tried to minimize light transmission through the capillary walls by covering the capillary end with silver mirror, but later abandoned the attempt because the silver gradually peeled off when soaked in the buffer solution, and in the worst case, caused blockages of the capillary. A secondary reason for the higher background is the much broader emission spectrum of the LED source, which caused a larger fraction of the excitation stray light to leak through the cutoff filter.

The S/N level of the fluorescence detection system was significantly improved by applying lock-in amplification, with the LED source driving current being subjected to square-wave modulation at a specified frequency. The lock-in amplifier functions as a phase-sensitive detector, where only signals with a specified frequency and phase identical to the reference waveform of the lock-in amplifier are extracted and amplified, while other frequency and phase combinations are rejected. By tuning the phase and frequency of the amplifier reference wave to that of the modulated excitation signal, highest output could be obtained, and following demodulation and low-pass filtering by the lock-in amplifier, the noise level could be significantly suppressed. In preliminary studies during this work, different frequencies were tested. Best S/Ns were obtained in the proximity of 1 kHz, and 900 kHz square-wave modulation was used in all later studies. The lock-in amplification achieved at least two orders of magnitude improvement in S/N level, compared to a nonmodulated LED source.

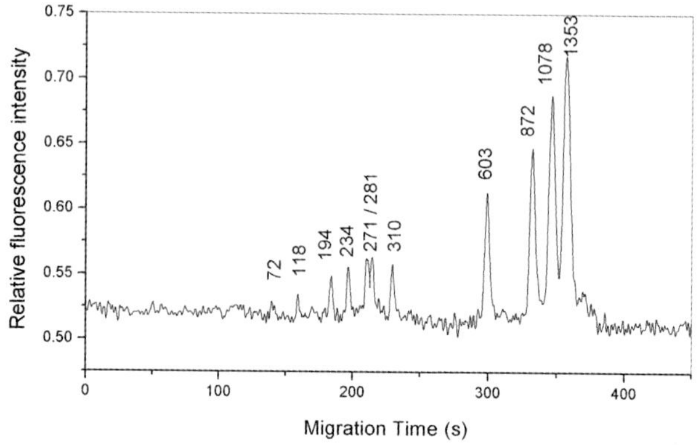

Fig. 5 Electropherogram of ϕX174-*Hae*III DNA digest separation. Experimental condition same as for Fig. 4, but using LED as exciting light source with lock-in amplifier of the PMT signal.

2.3.5
Performance of the μCE system

The performance of the LCW-LED-based CE system employing lock-in amplification was demonstrated in the separation and determination of a 25 ng/μL ϕX174 *Hae*III DNA digest sample (Fig. 5). Samples were aspirated and injected automatically through the preprogrammed SI system, and a complete analytical cycle lasted about 400 s. The electropherogram shows effective separation of all 11 fragments in the sample, with a theoretical plate number of 7.6×10^5/m, and S/N ratio of 16 for the 603 bp peak. These results are comparable to those obtained using the laser excitation source shown in Fig. 4a. The LOD (3 × S/N) was about 5 ng/L DNA, corresponding to about 200 fg absolute DNA mass introduced into the capillary.

2.4
Concluding remarks

The present work demonstrated the feasibility of combining extremely simple optics with LED excitation and lock-in amplification to produce an efficient, simple, and cheap microfluidic system for achieving CE separation and fluorimetric detection of DNA fragments, with a performance comparable to those of more sophisticated CE-LIF systems.

The sensitivity of the system could be doubled by employing dual LED sources positioned on opposite sides of the LCW capillary; and noise level could be further reduced by depositing a platinum layer at the outlet tip of the capillary to improve shielding of stray light. Further improvement of the detection power of the system is also possible by employing DNA labeling dyes with higher fluorescence efficiencies.

The present μCE system, including the separation and integrated detection system, measures 10 × 2 × 2 cm in size. Admittedly, currently the lock-in amplifier and the SI sample introduction system used in this work are still too large to be compatible with the separation and optical components. Our future aim is to miniaturize these components by developing a compact lock-in amplification device, dedicated to this application, and by adopting a recently reported slotted-vial array sample introduction system [21] for achieving continuous sample introduction.

We gratefully acknowledge support from Natural Science Foundations of China under Project No. 20299030 for this work.

2.5
References

[1] Mogensen, K. B., Klank, H., Kutter, J. P., *Electrophoresis* 2004, *25,* 3498–3512.

[2] Dang, F. Q., Zhang, L. H., Hagiwara, H., Mishina, Y., Baba, Y., *Electrophoresis* 2003, *24,* 714–721.

[3] Willauer, H. D., Collins, G. E., *Electrophoresis* 2003, *24,* 2193–2207.

[4] Tsai, C. H., Huang, H. M., Lin, C. H., *Electrophoresis* 2003, *24,* 3083–3088.

[5] Johns, C., Macka, M., Haddad, P. R., *Electrophoresis* 2004, *25,* 3145–3152.

[6] Kuo, J. S., Kuyper, C. L., Allen, P. B., Fiorini, G. S., Chiu, D. T., *Electrophoresis* 2004, *25,* 3796–3804.

[7] Dasgupta, P. K., Genfa, Z., Li, J., Boring, C. B., Jambunathan, S., Al-Horr, R., *Anal. Chem.* 1999, *71,* 1400–1407.

[8] Wang, S.-L., Huang, X.-J., Fang, Z.-L., Dasgupta, P. K., *Anal. Chem.* 2001, *73,* 4545–4549.

[9] Chabinyc, M. L., Chiu, D. T., McDonald, J. C., Stroock, A. D., *Anal. Chem.* 2001, *73,* 4491–4498.

[10] Burns, M. A., Johnson, B. N., Brahmasandra, S. N., Handique, K., *Science* 1998, *282,* 484–487.

[11] Webster, J. R., Burns, M. A., Burke, D. T., Mastrangelo, C. H., *Anal. Chem.* 2001, *73,* 1622–1626.

[12] Yu, L., Yuan, L., Feng, H., Li, S. F. Y., *Electrophoresis* 2004, *25,* 3139–3144.

[13] Dallas, T., Dasgupta, P. K., *Trends Anal. Chem.* 2004, *23,* 385–392.

[14] Fang, Z.-L., Chen, H.-W., Fang, Q., Pu, Q.-S., *Anal. Sci.* 2000, *16,* 197–203.

[15] Fang, Z.-L., Liu, Z.-S., Shen, Q., *Anal. Chim. Acta* 1997, *346,* 135–143.

[16] Fang, Q., Wang, F.-R., Wang, S.-L., Liu, S.-S., *Anal. Chim. Acta* 1999, *390,* 27–37.

[17] Huang, X. J., Pu, Q. S., Fang, Z. L., *Analyst* 2001, *126,* 281–284.

[18] Albarghouthi, M. N., Barron, A. E., *Electrophoresis* 2000, *21,* 4096–4111.

[19] Xu, F., Jabasini, N., Baba, Y., *Electrophoresis* 2002, *23,* 3608–3614.

[20] Liang, D., Zhang, J., Chu, B., *Electrophoresis* 2003, *24,* 3348–3355.

[21] Du, W.-B., Fang, Q., He, Q.-H., Fang, Z.-L., *Anal. Chem.* 2005, *77,* 1330–1337.

3
Determination of biochemical species on electrophoresis chips with an external contactless conductivity detector*

Eva M. Abad-Villar, Pavel Kubáň, Peter C. Hauser

Contactless conductivity measurements were found to be suitable for the direct detection, *i.e.*, without needing any labels, of a range of biochemically relevant species, namely amino acids, peptides, proteins, immunoglobulin, and DNA. It was also possible to monitor the products of the enzymatic digestion of HSA with pepsin. Detection was carried out on bare electrophoresis chips made from poly(methyl methacrylate) by probing the conductivity in the channel with a pair of external electrodes, which are fixed on the chip holder. Separation efficiencies up to 15 000 plates could be obtained and LODs are in the low μM-range, except for immunoglobulin G (IgG) which could be determined down to 0.4 nM. Linear dynamic ranges of two to three orders of magnitude were obtained for the peptides as examples.

3.1
Introduction

The current trend to systems biology as an approach to the understanding of biochemistry poses increased demands on analytical sciences to provide the large number of data required for the desired complete quantitative modeling of interactions. Planar electrophoresis, introduced in the 1950s, is still widely used in proteomics and other applications for its high separation efficiency when used in a 2-D approach. However, the staining and blotting procedures required lend themselves poorly to automation and quantitation. The latter limitations are, on the other hand, readily overcome with CE. In particular microfluidic electrophoresis devices have gained a lot of attention in recent years for biochemical analysis because of their fast analysis times and the possibility to integrate several sample processing steps (*e.g.*, see review [1]). Electrophoresis chips have for example been successfully used in the separation of amino acids [2, 3], carbohydrates [4], lipids [5], peptides [6], proteins [7,

* Originally published in Electrophoresis 2005, 26, 3609–3614

Microfluidic Applications in Biology. Edited by Niels Lion, Joël S. Rossier, and Hubert H. Girault

ISBN-10: 3-527-31761-9

8], and DNA [9, 10]. It has also been found possible to carry out immunoassays on this platform [11] as well as to monitor enzymatic reactions [12–14].

For the high sample throughputs required in systems biology, simplicity, robustness, and low costs are desirable. The latter implies the use of polymeric chips, which can be manufactured by embossing, rather than the microlithographically produced glass chips used in the early days of the development of these devices. Poly(methyl methacrylate) (PMMA) is commonly used. Detection is most often carried out by fluorescence measurement [15] as UV-absorption is not sufficiently sensitive for the narrow channels used on chips. Fluorescence detection allows high sensitivity in the small detection volumes but requires the labeling of the biochemical analytes with fluorescent compounds. Electrochemical detection methods are in principle simpler, as labeling is not needed and the detector consists of electronic components only.

Amperometric detection as well as the nearly universal conductometric method have indeed successfully been used by many researchers (see, for example, the following reviews [16–18]). On the other hand, the use of electrodes for detection generally requires the careful alignment of separation channel and electrodes, either by using micromanipulators or by embedding the electrodes during manufacture. As has recently been shown, these limitations can be overcome by using contactless conductivity detection with external detached electrodes [19, 20]. In this arrangement the electrodes are made a permanent part of a holder onto which the chips are placed during the analysis, rather than the separation devices themselves. This allows the application of bare chips, as used for optical detection, because the conductivity is probed remotely from underneath the chip through the substrate material below the separation channel. It has been shown so far that this approach is suitable for the detection of small organic ions of pharmaceutical relevance [21, 22], enzymatic reactions [23], and immunoassays [11]. Contactless conductivity detection has also often been used for conventional CE and relies on coupling an excitation signal into and out of the cell by making use of the capacitance formed between external electrodes and the solution inside the capillary. Hence it has been termed capacitively coupled contactless conductivity detection (C^4D). Detailed fundamental studies for capillaries [24, 25] and microchips [26], and a recent review [27] of this relatively new technique are available.

In this publication, an exploration of the scope of application of external conductivity probing for detection on electrophoresis chips is reported. A range of biomolecules of different size, from amino acids to a large immunoglobulin, were tested.

3.2
Materials and methods

3.2.1
Instrumentation

An electrophoresis chip (90 × 16 mm) in the double-T-configuration was used in this work (Microfluidic Chip Shop, Jena, Germany). For this device, the separation and injection channels with lengths of 85 and 8 mm, respectively, (effective

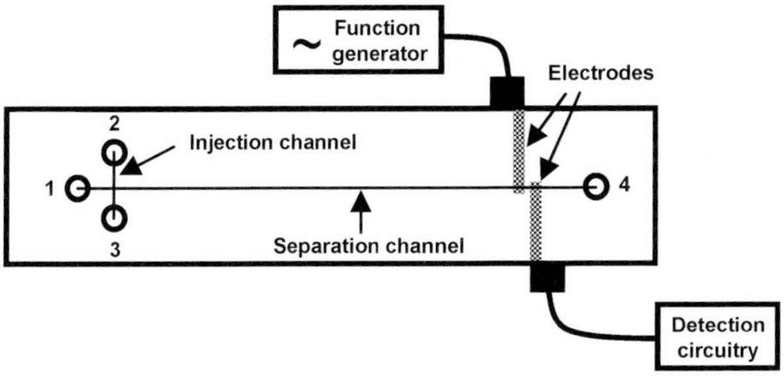

Fig. 1 Schematic drawing of the electrophoresis chip and the external detector electrodes used for C^4D. Detector electrodes lie underneath the chip except for the connection pads. (1, 3, 4) Buffer reservoirs; (2) sample reservoir.

separation length: 75 mm) and cross-sectional dimensions of 50 × 50 μm are located at the bottom of a PMMA plate of about 1 mm thickness and are sealed underneath with a PMMA foil of 175 μm thickness. Access holes are provided on top. Small perspex blocks with appropriate holes were glued with UV-epoxy to the top of the chip in order to provide reservoirs for the solutions. The chip was fixed onto a purpose-made holder with the help of a perspex clip. This holder carries a pair of detection electrodes of 1 mm width, 14 mm length, in the antiparallel configuration, and a detection gap of 0.5 mm, which come to lie underneath the separation chip as shown in Fig. 1. The electrodes were prepared by lithographic etching of the copper layer (35 μm) on a flexible printed circuit base material of 100 μm thickness (Lamitec Czech s.r.o., Pardubice, Czech Republic). Note that the holder also has a Faradaic shield in vertical orientation to minimize direct capacitive coupling between the two electrodes, which is not shown in the drawing for clarity.

Detection was carried out by passing a sinusoidal excitation signal of 20 V (peak to peak) and 600 kHz from a function generator (GFG-8019G; Goodwill Instruments, Tapei, Taiwan) to the first electrode and picking up the resulting cell current at the second electrode. The detector current was transformed into a voltage with an operational amplifier (OPA655; Texas Instruments, Dallas, TX, USA) in the current follower configuration (feedback resistor: 2.2 MΩ). The signal was then rectified and further amplified. The circuitry used is a slight modification of a design reported earlier [28]. For injection and electrophoretic separations two high-voltage power supplies (model CZE 1000R; Start-Spellman, Pulborough, England) were used. They were controlled by a purpose-built interface connected to a multifunctional I/O-card (model PCI-MIO-16XE-50; National Instruments, Austin, TX, USA) located in a PC. Injection and separation parameters were set in a program written in LabVIEW (National Instruments) and data acquisition was carried out with a MacLab/4s system (AD Instruments, Hastings, UK). Injection was always carried out by applying 3 kV for 3 s along the injection channel.

3.2.2
Reagents

Acetic acid and Tween 20 (polyoxyethylene sorbitan monolaurate, 70% in water) were supplied by Fluka (Buchs, Switzerland) and Sigma (Buchs, Switzerland), respectively. The amino acids, namely lysine (Lys), arginine (Arg), histidine (His), glycine (Gly), alanine, (Ala), valine (Val), isoleucine (Ile), serine (Ser), threonine (Thr), methionine (Met), glutamine (Gln), phenylalanine (Phe), and proline (Pro), were purchased from Fluka as was the peptide angiotensin II human (acetate). The following peptides were provided by Sigma: leucine enkephalin, bradykinin (triacetate salt), thyrotropin releasing hormone (TRH), and oxytocin (acetate salt hydrate). The proteins cytochrome *c* and myoglobin were purchased from Sigma and HSA and the enzyme pepsin were obtained from Fluka. Monoclonal immunoglobulin G (IgG) fraction of mouse ascites fluid (clone MB-11, 2 mg/mL) was purchased from Sigma. A purified fragment of bacterial DNA (1652 bps) was donated by the Division of Molecular Microbiology at the University of Basel. It had been obtained by standard procedures [29].

3.2.3 Procedures

Solutions were prepared daily with water purified with a Milli-Q plus 185 system from Millipore (Bedford, MA, USA), degassed in an ultrasonic bath for 5 min and filtered with 0.2 m syringe filters supplied by BGB-Analytik (Anwil, Switzerland) prior to use. Dilutions were made using the running buffer consisting of 2.3 M acetic acid (pH 2.1) containing 0.05% Tween 20 (to prevent wall interactions of the larger species [30]), and aliquots of 300 μL were pipetted into the reservoir on the chip used for injection. HSA, pepsin, and stock solutions of individual amino acids were stored at 4°C. Peptides, myoglobin, cytochrome *c*, and working aliquots of IgG and DNA fragment were stored at −20°C. The microchip was rinsed thoroughly with water and subsequently flushed with the buffer solution prior to the electrophoretic separation. Washing steps with buffer between runs ensured reproducibility and maintained a stable baseline. Experiments were performed at a constant laboratory temperature of 24 ± 1°C. Detection limits were determined as the concentrations giving peak heights corresponding to three times the baseline noise and the separation efficiencies (N) were determined from the peak half-width according to the following equation:

$$N = 5.54\left(\frac{t_m}{w_{1/2}}\right)^2$$

3.3 Results and discussion

3.3.1 Amino acids and peptides

An electropherogram obtained for 12 underivatized amino acids is shown in Fig. 2. For the separation of amino acids, the choice of the pH-value of the buffer is critical. As underivatized amino acids are zwitterionic over a wide pH-range, and in most cases they have no net charge, zone electrophoretic separation must be

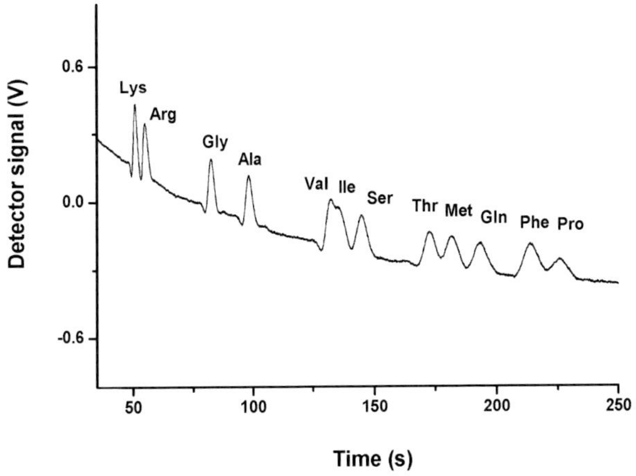

Fig. 2 Electrophoretic separation of essential amino acids at 500 μM, except for Lys, Arg, and His which were at 250 μM, in 2.3 M acetic acid (pH 2.1) containing 0.05% Tween 20. Separation voltage: 6 kV.

carried out at extreme pH-values. Both acidic and basic conditions have been used satisfactorily in conventional CE [31, 32]. However, strongly alkaline conditions (pH 10–11) were found to be not suitable for use with the PMMA chips and therefore an acidic buffer was used. The buffer adopted, acetic acid at a concentration of 2.3 M has a pH-value of 2.1, which renders all amino acids cationic [31]. As can be seen from the figure, all 12 amino acids, namely Lys, Arg, Gly, Ala, Val, Ile, Ser, Thr, Met, Gln, Phe, and Pro, could be separated, except Val and Ile which were not baseline resolved. The baseline drift notable in particular earlier in the electropherogram was found to be more pronounced for higher applied separation voltages and is thought to be due to Joule heating. Conductometric measurements always show a temperature coefficient. If sloping baselines are deemed undesirable, the effect can be suppressed at the cost of the analysis time. Typical detection limits are exemplified by 50 and 12 μM obtained for Gln and Arg, respectively. Plate numbers were 4500, 5200, and 6400 for Lys, Ala, and Ser, respectively. Note that the peaks are wider for longer migration times, presumably due to diffusional band broadening.

In contrast to amino acids, peptides may be determined at more moderate pH-values, as they show net charges at any value not corresponding to their individual p*I*s. However, the low-pH buffer was deemed useful for the determination of these larger species as well, as it assures a high positive charge of all compounds and therefore fast migration and sensitive detectability by conductometry. Furthermore, it allows concurrent determination of different peptides with very different p*I*s. The p*I*-values for the five peptides shown in Fig. 3, namely leucine enkephalin, bradykinin, angiotensin II, TRH, and oxytocin, are 5.2, 11.7, 11.0, 11.0, and 6.0 [33], respectively. The detection limits for these compounds were determined as 42, 10, 12, 50, and 5 μM in the respective order, which are comparable to the values for the amino acids. The linear dynamic range extends up to approximately 1 mM for angiotensin, TRH, and oxytocin taken as examples. The RSD of the peak areas for three successive injections was between 0.6 and 0.8% for the three compounds (for concentrations of 0.6, 1, and 0.1 mM, respectively). The migrations times for the three compounds showed RSD values between 0.4 and 0.6%. The plate numbers

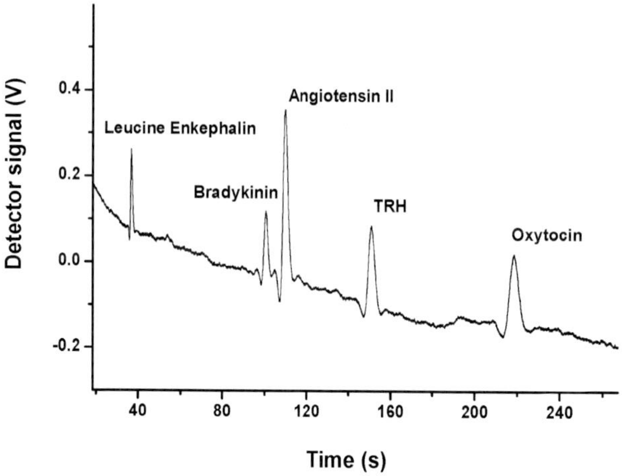

Fig. 3 Electropherogram obtained for a mixture of peptides in 2.3 M acetic acid containing 0.05% Tween 20: leucine enkephalin (1000 μM), bradykinin (190 μM), angiotensin II (640 μM), TRH (1186 μM), and oxytocin (109 μM). Separation voltage: 4 kV.

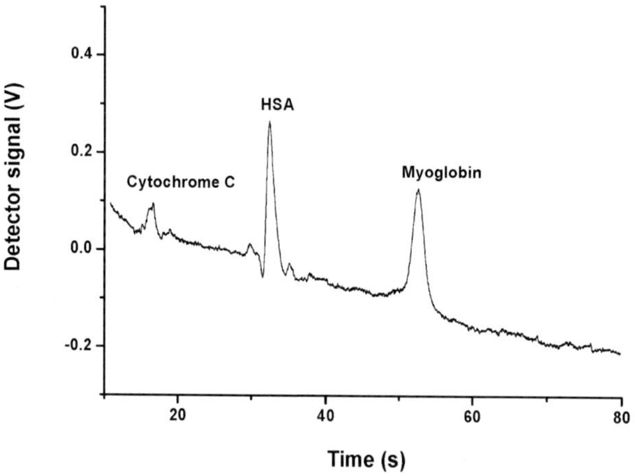

Fig. 4 Electrophoretic separation of the proteins cytochrome *c* (40 μM), HSA (15 μM), and myoglobin (30 μM) in 2.3 M acetic acid containing 0.05% Tween 20. Separation voltage: 6 kV.

varied between 7500 and 15 000. The latter values are somewhat better for the oligomers of the amino acids than the amino acids themselves, presumably because diffusional band-broadening is less pronounced for the larger species.

3.3.2
Proteins, IgG, and DNA-fragment

The acetic acid buffer was retained for the investigation of the separation of a mixture of proteins for the same reason as given above for the peptides. The three proteins used have p*I*-values of 10.7, 4.9, and 7.0 and large mass (13 000, 68 500, and 16 890 g/mol). Good separation could be achieved within less than 1 min, as shown in Fig. 4, and the plate numbers *N* are 1000, 4500, and 5400 for cytochrome *c*, HSA, and myoglobin, respectively. The detection limits were determined as 6, 0.6, and 1.4 μM for the three compounds in the respective order.

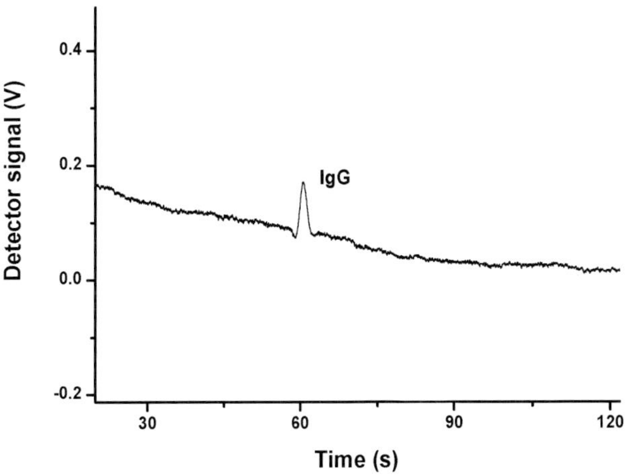

Fig. 5 Electropherogram for 3.3 nM IgG in 2.3 M acetic acid containing 0.05% Tween 20. Separation voltage: 3 kV.

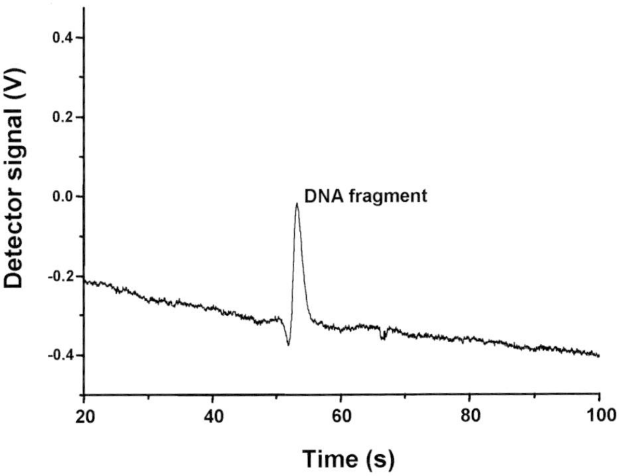

Fig. 6 Electropherogram for a DNA-fragment (1652 bps) in 2.3 M acetic acid containing 0.05% Tween 20. Separation voltage: 3 kV.

The IgG could be readily determined at the very low concentration of 3.3 nM injected for the electropherogram of Fig. 5. The three times S/N-detection limit for this compound is 0.4 nM and the plate numbers *N* are 7600. Presumably the high sensitivity found for this large compound (150 000 g/mol) is due to a favorable charge to size ratio. This is comparable to a value of 0.04 nM for IgM reported previously for contactless conductivity detection on a microchip using 20 mM TAPS/AMPD containing 0.01% Tween 20 [11].

Direct determination of a DNA fragment (1652 bps) in free solution is also possible using contactless conductivity detection as illustrated in Fig. 6. Note, however, that it is not possible to separate different DNA-fragments by zone electrophoresis as these show identical charge densities.

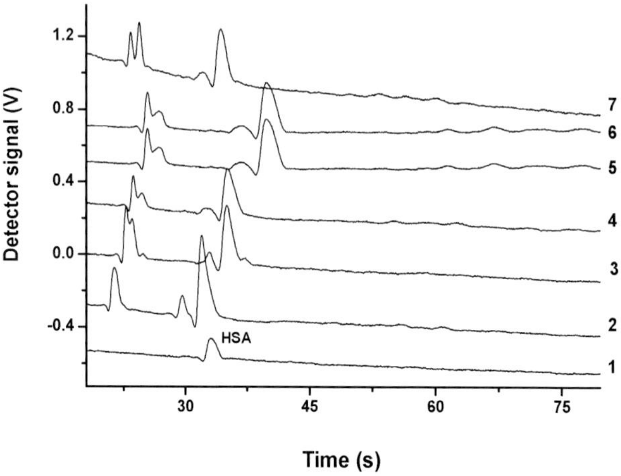

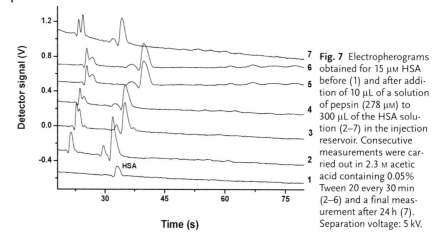

Fig. 7 Electropherograms obtained for 15 μM HSA before (1) and after addition of 10 μL of a solution of pepsin (278 μM) to 300 μL of the HSA solution (2–7) in the injection reservoir. Consecutive measurements were carried out in 2.3 M acetic acid containing 0.05% Tween 20 every 30 min (2–6) and a final measurement after 24 h (7). Separation voltage: 5 kV.

3.3.3
Enzymatic digestion

In enzymatic digestions of proteins these are enzymatically cleaved into specific fragments. Trace 1 of Fig. 7 shows a single peak for HSA. After acquisition of this electropherogram, pepsin was added to the sample reservoir. Pepsin is an enzyme that hydrolyzes ingested proteins in the stomach at peptide bonds on the amino-terminal side of the aromatic amino acids residues Phe, Trp, and Tyr, cleaving long polypeptide chains into a mixture of smaller peptides. Its activity has a pH-optimum of about 1.6–2 and it is thus suitable for use with the acetic acid buffer employed in this study. The first electropherogram obtained 30 min after addition of the enzyme shows pronounced peaks due to the peptide fragments obtained. As the enzymatic digestion proceeds, the amount of different fragments varies, resulting in changes in the peak heights and shapes (electropherograms 3–7 of Fig. 7). No effort was made to identify the individual peaks obtained, but the result clearly shows that it is possible to directly monitor the products of the enzymatic reaction with our approach.

3.4
Concluding remarks

This survey on the direct detectability of different biochemical species by external contactless conductivity measurement clearly showed that the technique has a high potential. The sensitivity was found to show a dependence on the size of the molecule, with a tendency to lower LODs for larger species. More studies are needed to further elucidate this effect. The sensitivity cannot match that of the established fluorescence detection, but the method does not require chemical derivatization for labeling and has the advantage of instrumental simplicity. The possibility to detect the products of the enzymatic cleavage of a protein is thought not only to be of general interest for enzymatic assays but, in particular, also for so-called peptide mapping, a method widely used in proteomics (and in systems biology) for the identification of proteins by their enzymatic digestion products.

Partial funding of this work was provided by the Swiss National Science Foundation (Grant No. 200020–105176/1) and the Ministerio de Educación y Ciencia of Spain under a postdoctoral fellowship. The authors would also like to thank T. A. Rhomberg from the Division of Molecular Microbiology of the Biocentre, at the University of Basel, for the donation of the DNA-fragment.

3.5 References

[1] Bilitewski, U., Genrich, M., Kadow, S., Mersal, G., *Anal. Bioanal. Chem.* 2003, *377*, 556–569.

[2] Dou, Y. H., Bao, N., Xu, J. J., Chen, H. Y., *Electrophoresis* 2002, *23*, 3558–3566.

[3] Kato, M., Gyoten, Y., Sakai-Kato, K., Nakajima, T., Toyo'oka, T., *Anal. Chem.* 2004, *76*, 6792–6796.

[4] Monahan, J., Gewirth, A. A., Nuzzo, R. G., *Electrophoresis* 2002, *23*, 2347–2354.

[5] Lin, S. S., Fischl, A. S., Bi, X. H., Parce, W., *Anal. Biochem.* 2003, *314*, 97–107.

[6] Liu, J., Pan, T., Woolley, A. T., Lee, M. L., *Anal. Chem.* 2004, *76*, 6948–6955.

[7] Xiao, D., Le, T. V., Wirth, J., *Anal. Chem.* 2004, *76*, 2055–2061.

[8] Dou, Y. H., Bao, N., Xu, J. J., Meng, F., Chen, H. Y., *Electrophoresis* 2004, *25*, 3024–3031.

[9] Liu, S., Guttman, A., *Trends Anal. Chem.* 2004, *23*, 422–431.

[10] Fu, L. M., Lin, C. H., *Electrophoresis* 2004, *25*, 3652–3659.

[11] Abad-Villar, E. M., Tanyanyiwa, J., Fernández-Abedul, M. T., Costa-García, A., Hauser, P. C., *Anal. Chem.* 2004, *76*, 1282–1288.

[12] Arenkov, P., Kukhtin, A., Gemmell, A., Voloshchuk, S., Chupeeva, V., Mirzabekov, A., *Anal. Biochem.* 2000, *278*, 123–131.

[13] Wang, J., *Electrophoresis* 2002, *23*, 713–718.

[14] Xu, H. W., Ewing, A. G., *Anal. Bioanal. Chem.* 2004, *378*, 1710–1715.

[15] Schwarz, M. A., Hauser, P. C., *Lab Chip* 2001, *1*, 1–6.

[16] Lacher, N. A., Garrison, K. E., Martin, R. S., Lunte, S. M., *Electrophoresis* 2001, *22*, 2526–2536.

[17] Tanyanyiwa, J., Leuthardt, S., Hauser, P. C., *Electrophoresis* 2002, *23*, 3659–3666.

[18] Wang, J., *Talanta* 2002, *56*, 223–231.

[19] Tanyanyiwa, J., Abad-Villar, E. M., Fernández-Abedul, M. T., Costa-García, A., Hoffmann, W., Guber, A. E., Herrmann, D., *et al.*, *Analyst* 2003, *128*, 1019–1022.

[20] Wang, J., Chen, G., Muck, A. J., *Anal. Chem* 2003, *75*, 4475–4479.

[21] Tanyanyiwa, J., Hauser, P. C., *Electrophoresis* 2004, *25*, 3010–3016.

[22] Tanyanyiwa, J., Abad-Villar, E. M., Hauser, P. C., *Electrophoresis* 2004, *25*, 903–908.

[23] Wang, J., Chen, G., Muck, A. J., Chatrathi, M. P., Mulchandani, A., Chen, W., *Anal. Chim. Acta* 2004, *505*, 183–187.

[24] Kubáň, P., Hauser, P. C., *Electrophoresis* 2004, *25*, 3398–3405.

[25] Kubáň, P., Hauser, P. C., *Electrophoresis* 2004, *25*, 3387–3397.

[26] Kubáň, P., Hauser, P. C., *Lab Chip* 2005, *5*, 407–415.

[27] Kubáň, P., Hauser, P. C., *Electroanalysis* 2004, *16*, 2009–2021.

[28] Tanyanyiwa, J., Galliker, B., Schwarz, M. A., Hauser, P. C., *Analyst* 2002, *127*, 214–218.

[29] Sambrook, J., Russell, D. W., *Molecular Cloning: A Laboratory Manual*, Cold Spring Harbor Laboratory Press, New York 2001.

[30] Castelletti, L., Verzola, B., Gelfi, C., Stoyanov, A., Righetti, P. G., *J. Chromatogr. A* 2000, *894*, 281–289.

[31] Coufal, P., Zuska, J., van de Goor, T., Smith, V., Gas, B., *Electrophoresis* 2003, *24*, 671–677.

[32] Tanyanyiwa, J., Schweizer, K., Hauser, P. C., *Electrophoresis* 2003, *24*, 2119–2124.

[33] Nelson, D. L., Cox, M. M., *Lehninger Principles of Biochemistry*, Worth Publishers, New York 2000, pp. 115–158.

4
In-channel indirect amperometric detection of nonelectroactive anions for electrophoresis on a poly(dimethylsiloxane) microchip*

Jing-Juan Xu, Ying Peng, Ning Bao, Xing-Hua Xia, Hong-Yuan Chen

In the present paper, we describe a microfluidics-based sensing system for non-electroactive anions under negative separation electric field by mounting a single carbon fiber disk working electrode (WE) in the end part of a poly(dimethylsiloxane) microchannel. In contrast to work in a positive separation electric field described in our previous paper (*Anal. Chem.* 2004, *76*, 6902—6907), here the electrochemical reduction reaction at the WE is not coupled with the separation high-voltage (HV) system, whereas the electrochemical oxidation reaction at the WE is coupled with the separation HV system. The electroactive indicator is the carbon fiber WE itself but not dissolved oxygen. This provides a convenient and sensitive means for the determination of nonelectroactive anions by amperometry. The influences of separation voltage, detection potential, and the distance between the WE and the separation channel outlet on the response of the detector have been investigated. The present detection mode is successfully used to electrochemically detect F^-, Cl^-, SO_4^{2-}, CH_3COO^-, H_2PO^{4-}. Based on the preliminary results, a detection limit of 2 μM and a dynamic range up to three orders of magnitude for Cl^- could be achieved.

4.1
Introduction

With the rapid development of CE microchips, there is an urgent need to develop compatible detection modes. LIF is the dominating detection mode for microfluidic devices because of its high sensitivity [1, 2]. The drawbacks of LIF are its limited compatibility with miniaturization chips and the requirement for labeling of most analytes. As alternative and effective detection methods, electrochemical methods have been widely developed in recent years [3–24]. Electrochemistry (EC) can offer a mass sensitivity level that is comparable to fluorescence, but usually without deriva-

* Originally published in Electrophoresis 2005, 26, 3615–3621

Microfluidic Applications in Biology. Edited by Niels Lion, Joël S. Rossier, and Hubert H. Girault

ISBN-10: 3-527-31761-9

tization. The analytical performance, unlike optical detection methods, is not compromised by miniaturization and the associated EC instrumentation is less expensive and less complex compared to LIF. However, for amperometric detection, the target material must be intrinsically electroactive, which strongly limits the applicability of the EC detection system. Moreover, amperometric detection generally suffers from interferences due to the presence of the CE separation voltage driving fluids to flow. Conductivity detection is another electrochemical technique with the ability to detect any analyte irrespective of whether it contains an electroactive species or not. The only requirement is that the migrating analyte zones possess a conductivity that is different from that of the carrier electrolyte [25]. Such detection mode can be accomplished either by a direct contact of the solution with electrodes or by a contactless method which requires inductively or capacitively coupled devices using high frequencies. However, the high electric field inside the column and the difficulty in positioning electrodes are obvious drawbacks of the contact conductivity detection mode for CE, while contactless conductivity measurements achieved with a pair of electrodes have a reduced sensitivity due to the presence of the insulating layer [26]. Compared to LIF, these electrochemical detection methods cannot truly realize direct detection in a separation channel.

As the field of chip-based separation microsystems continues its rapid growth, there are urgent needs for developing compatible detection modes with excellent performance including in-channel mode for high separation efficiency, a wide variety of applications, ease of miniaturization and integration, high sensitivity, and so on [27]. Recently, some valuable methods have been reported. An approach for employing electrochemical detection in a microchip system was performed by Lunte *et al.* [28] who used an electrically isolated potentiostat with the working electrode (WE) placed just within the separation channel from the cathodic reservoir. Manz *et al.* [29] first reported an in-channel wireless electrochemiluminescence detection mode using the potential difference induced by the CE separation electric field. Recently, Girault *et al.* [30] reported a passive electrochemical detection mode for CE in which the detection electrodes are directly placed in the separation channel along the flow; the separation electrical field is used to generate the detection signal. In our previous work [31], we reported a new in-channel indirect amperometric detection mode for microchip CE with positive separation electric field. In this detection mode, dissolved oxygen is used as an electroactive indicator. The potential difference induced by the CE separation electric field between the WE and the reference electrode brings about the change of the reduction potential of dissolved oxygen, which can be used to determine nonelectroactive analytes. The detection method has been successfully used to detect K^+, Na^+, Li^+, and biomolecules such as dopamine (DA) and epinephrine (EP), as well as EOF rates.

Here, we report a new microfluidics-based in-channel sensing system for nonelectroactive anions under negative separation electric field that relies on amperometric response of the oxidation of a carbon fiber electrode. The important new result is that the electrochemical reduction reaction at the WE is decoupled from the separation high-voltage (HV) system, while the electrochemical oxidation reaction at WE is coupled from the separation electric field. This approach enables us to carry out detection of nonelectroactive analytes *via* the oxidation of carbon fiber. That is, nonelectroactive analytes do not participate directly in the electrochemical process, but because the detection system and separation HV system are electrically coupled, the potential of the WE changes with the change of analytes' resistances, resulting in the change of oxidation current of the carbon fiber. This also provides a convenient and sensitive means for direct current indication of anions that do not directly participate in an electrochemical reaction and thus broadens the detection range of analytes with amperometry.

4.2
Materials and methods

4.2.1
Reagents

Histidine (His), MES, sodium chloride, sodium fluoride, potassium sulfate, sodium acetate, and potassium dihydrogen phosphate were purchased from Nanjing Chemical Reagents Factory (Nanjing, China). All reagents were of analytical grade. The MES/His buffer (20 mM each, pH 6.1) was prepared by dissolving MES and His in doubly distilled water and passed through a 0.22 μm cellulose acetate filter (Xinya Purification Factory, Shanghai, China). Stock solutions of the target anions (chloride, fluoride, sulfate, acetate, and dihydrogen phosphate 100 mM) were prepared by dissolving the corresponding salts in running buffer. Sylgard 184 was purchased from Dow Corning (Midland, MI, USA). Cathode electrophoresis paint (a kind of acrylic-based cathodic electrodeposition coating) was donated by Jiangsu Hongye Coatings Science-Technology Industry, P.R. China.

4.2.2
Apparatus

The integrated CE-EC poly(dimethylsiloxane) (PDMS) chip microsystem was similar to that described previously [31]. Briefly, a home-made HV power supply had an adjustable voltage range between 0 and -5000 V. Parameters such as sampling voltage, sampling time, separation voltage, and separation time could be setup and automatically switched *via* an RS232 communication port of PC through a home-made computer program. The separation current could be monitored graphically in real time. The simple-cross single-separation channel PDMS microchip (as schematically shown in Fig. 1) was made based on a master composed of a positive relief structure of GaAs made in No. 55 Electronic Institute (Nanjing, China) by using standard microphotolithographic technology and reaction ion etching. The chip had a 40 mm long separation channel (from injection cross to the channel outlet) and a 5 mm long injection channel (between the sample reservoir and injection cross). The channels had a maximum depth of 18 μm and a width of 50 μm. The microchip had a four-way injection cross that was connected to the three reservoirs and the separa-

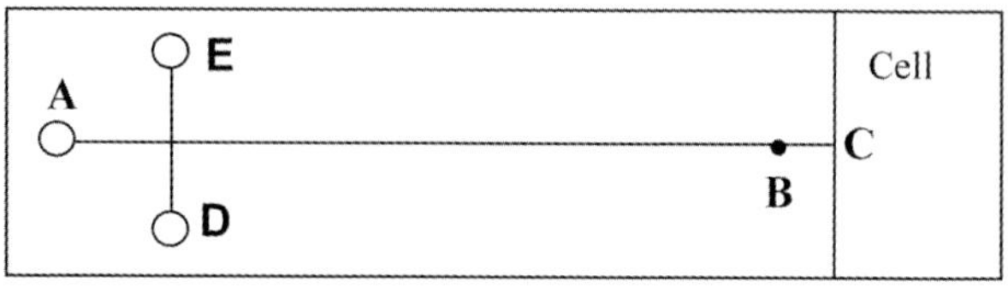

Fig. 1 Chip layout (see text for dimensions) and in-channel electrode alignment used in the studies: A, D and E represent fluid reservoirs, C represents the outlet of separation channel, B represents the relative location of the WE in the separation channel, the distance between B and C is denoted as L_{BC}. Buffer is placed in reservoir A and sample is placed in reservoir D; E and Cell correspond to the sample waste and detection reservoirs, respectively.

tion channel. A Plexiglas holder that integrated a precise 3-D system (Shanghai Lian Yi Instrument Factory of Optical Fibre and Laser, China) with the precision of ± 1 μm in each direction was fabricated for housing the separation chip and the detector and allowing their convenient replacement and reproducible positioning. A clip of optical fiber that can be fastened in the 3-D system was used to closely clip the carbon fiber microdisk electrode. Platinum wires, inserted into the individual reservoirs on the holder, served as contacts to the HV power supply.

4.2.3 Electrode fabrication

A single carbon fiber disk microelectrode was prepared according to the previous reports [31, 32]. Briefly, a glass capillary with an inner diameter of 0.5 mm was pulled under a multifunctional glass microelectrode puller (Shanghai Biological Institute, China) to form a fine tip. Then a single carbon fiber with diameter of 8 μm was carefully mounted onto the tip and fixed with epoxy. A copper wire was connected with the carbon fiber through carbon powder on the other end of capillary and then fastened with epoxy. This semifinished microelectrode protruded about 2 mm length was insulated with cathode electrophoresis paint according to the method reported previously [31, 32]. The electrodeposition was carried out at a constant voltage of 5 V between the anode (platinum wire) and the cathode (the carbon fiber electrode) for several minutes, to induce electrophoretic deposition of the paint at the cathode surface. After that, this painted microelectrode was heated at 170°C for 20 min. Before use, the tip of the painted carbon fiber was cut with a clean scalpel to form a microdisk electrode under a microscope.

4.2.4 In-channel electrochemical detection

The WE was placed into the end part of the separation channel (shown in Fig. 1, point B). The distance between the electrode surface and the channel outlet was controlled by a plastic screw of a 3-D system. Linear sweep voltammograms (LSVs) and amperometric detection were performed on a CHI 832 Electrochemical Analyzer (CHI, Shanghai, China) connected to an Intel Celeron 667 MHz PC with 128 MB RAM using the "linear sweep voltammetry" "amperometric *i–t* curve" modes. The electropherograms were recorded using a time resolution of 0.1 s while applying a fixed detection potential, and then smoothed with the least square point of 17. The WE was treated at +1.5 V for 200 s and then at −1.0 V for 200 s before use and once the baseline current changed. All experiments were performed at room temperature.

4.2.5
Electrophoretic procedure

Before use, the PDMS layer with microchannel and the PDMS flat were ultrasonically cleaned subsequently with water, methanol, and water for 10 min each and were then dried under infrared lamp. Then they were sealed together to form a reversible PDMS microchip. After the microchip was held on the holder, a WE was inserted into the electrode hole on the platform. Silicon grease was used to prevent leaking of the detection cell. The running buffer was 20 mM MES/His buffer (pH 6.1). The "running buffer" reservoir was filled with the buffer and the "sample" reservoir with the sample mixture. The injections were performed by applying a desired potential for 5 s to the sample reservoir with the "detection" reservoir grounded while all other reservoirs floating. Separations were performed by switching the HV contacts and applying the corresponding separation voltages to the running buffer reservoir with the detection reservoir grounded and all other reservoirs floating.

4.3
Results and discussion

4.3.1
Electrochemical behavior of the WE under negative separation electric field

To illustrate the effect of the negative separation HV field on the detector, LSVs were carried out under various separation HVs. Fig. 2 depicts the typical LSVs for the oxidation of a carbon fiber electrode under the conditions of 0 V (a), −1000 V (b), −1200 V (c), −1400 V (d), 1600 V (e), and −1800 V (f) separation voltages. These voltammetric profiles indicated that the separation voltage greatly influenced the voltammetric behavior of the detector. Without separation electric field, the onset oxidation potential of a carbon fiber electrode occurred at 0.8 V, while the reduction reaction of oxygen occurred at potentials more negative than −0.3 V. In the presence of negative separation electric field, the onset potential for the oxidation of carbon fiber shifted to negative value, while the reduction potential of oxygen was not changed. More negative HV induced a more negative onset potential for the oxidation of carbon fiber. It indicates that the electrochemical reduction reaction at the WE is not coupled from the separation HV system, while the electrochemical oxidation reaction at the WE is coupled from the separation HV

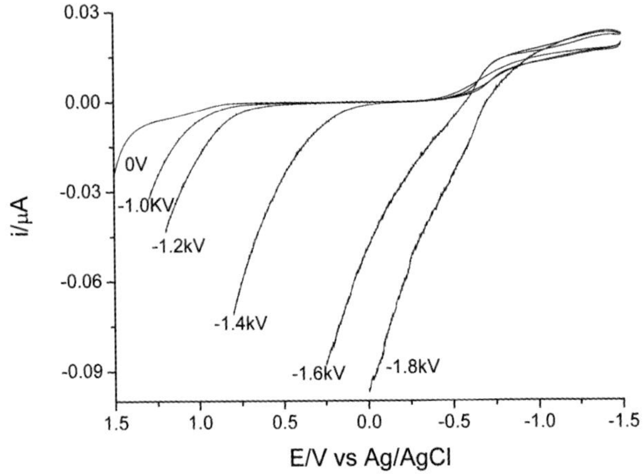

Fig. 2 Linear scanning voltammograms of a carbon fiber disc electrode in the separation channel with the presence of different negative separation voltages at a scan rate of 20 mV/s. Experimental parameters: running buffer solution, 20 mM MES+His; $L_{BC} = 30$ μm.

system. This phenomenon is opposite to that with positive separation voltage, which is due to the opposite direction of separation voltage. Under a positive separation electric field, the electrochemical reduction reaction at the WE is coupled from the separation HV, while the electrochemical oxidation reaction at the WE is not coupled. Thus, the separation electric field drives the oxygen reduction at a more negative electrode potential, producing a greater reduction current. Similarly, under negative separation electric field, the separation electric field induces the oxidation of a carbon fiber electrode at a more positive electrode potential, producing a greater anodic oxidation current. The higher the negative separation electric field is, the more positive the electrode potential for the oxidation of carbon fiber will be, and the greater the anodic oxidation current for the reaction is. This behavior can then be used to indirectly measure nonelectroactive anions by amperometry.

4.3.2
Effect of the separation potential

Since the electrochemical detection principle is based on the coupling of separation electric field on the electrochemical detector, the influence of separation voltage on detection is a major concern with respect to improving detection sensitivity. Fig. 3 exhibits the influence of separation voltage on the detection of 0.50 mM Cl^-, F^-, SO_4^{2-}, CH_3COO^-, and $H_2PO_4^-$. It is shown that Cl^-, F^-, and SO_4^{2-} could be separated in less than 30 s. By enhancing the separation electric field, the migration times of the five anions have no obvious change due to the opposite EOF and electrophoretic flow, which simultaneously increase with the increase of separation electric field. However, a relatively fast ascent of the amperometric signal with the increase of voltage was observed, because the coupling effect on amperometric detection potential is also strengthened with the enhancement of the separation electric field, thus resulting in larger peak current corresponding to accelerated electrochemical oxidation of the carbon fiber electrode. The separation voltage has a negligible effect on the peak-to-peak background noise level for voltages ranging from −1000 to −2000 V.

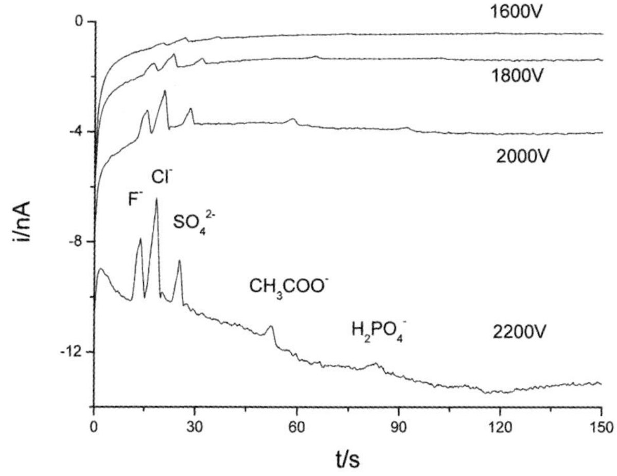

Fig. 3 Electropherograms of 0.50 mM Cl^-, F^-, SO_4^{2-}, CH_3COO^- and $H_2PO_4^-$ with different separation voltages. Experimental parameters: running buffer solution, 20 mM Tris; sampling voltage, 800 V; sampling time, 5 s; detection potential, 0.4 V; L_{BC} = 20 μm.

Separation electric fields higher than −2000 V result in higher background and noise levels (attributed to Joule heating effects). Moreover, a too high separation electric field will induce the formation of oxygen bubble at the carbon fiber electrode, then resulting in a large noise level or clogging of the channel.

4.3.3 Effect of the detection potential

The effect of the potential applied to the WE was analyzed between +0.3 and +1.1 V for Cl^-, F^-, SO_4^{2-}, CH_3COO^-, and $H_2PO_4^-$ anions at different separation voltages. Fig. 4 shows the typical electropherograms of the five unlabeled anions at different detection potentials with a separation voltage of −1800 V. Baseline separation of these anions was obtained. As expected, the peak currents increased with the increase of detection potential from 0.3 to 0.6 V. With further increase of the detection potential, the peak currents also increase; however, the baseline separation could not be obtained for Cl^-, F^-, SO_4^{2-} (not shown). Moreover, once the potential was moved up to the potential for the formation of oxygen bubble on the WE, a large noise is also produced. Further studies indicated that the effect of the detection potential on the response of the detector is also determined by the separation voltage. The plot of the peak currents as a function of detection potential at different separation voltage for Cl^- is shown in Fig. 5A. The separation voltage and the detection potential simultaneously determine the response of the detector. There is a similar trend for all anions. As shown in Fig. 5B, there exists a near-linear relationship between the detection potential and the separation voltage for the five anions at the detection sensitivity of 2 nA/mM (Cl^-, F^-, SO_4^{2-}) and 0.6 nA/mM (CH_3COO^- and $H_2PO_4^-$). Increase of 200 V separation voltage, a detection potential of *ca.* 130 mV for all these anions should be shifted negatively in order to obtain the same detection sensitivity. Although the exact reason for this relationship is not clear and needs further investigation, it may provide a way to select the detection potential at certain separation electric field.

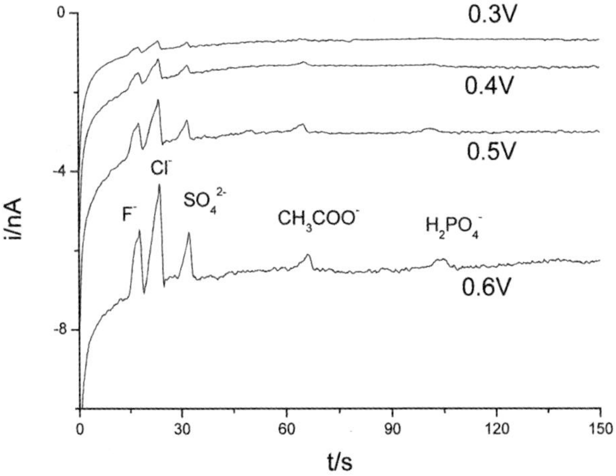

Fig. 4 Electropherograms of 0.50 mM Cl^-, F^-, SO_4^{2-}, CH_3COO^- and $H_2PO_4^-$ with different detection potentials. Experimental parameters: running buffer, 20 mM, pH 6.1 MES+His; sampling voltage, 800 V; sampling time, 5 s; separation voltage, −1800V, $L_{BC} = 20$ μm.

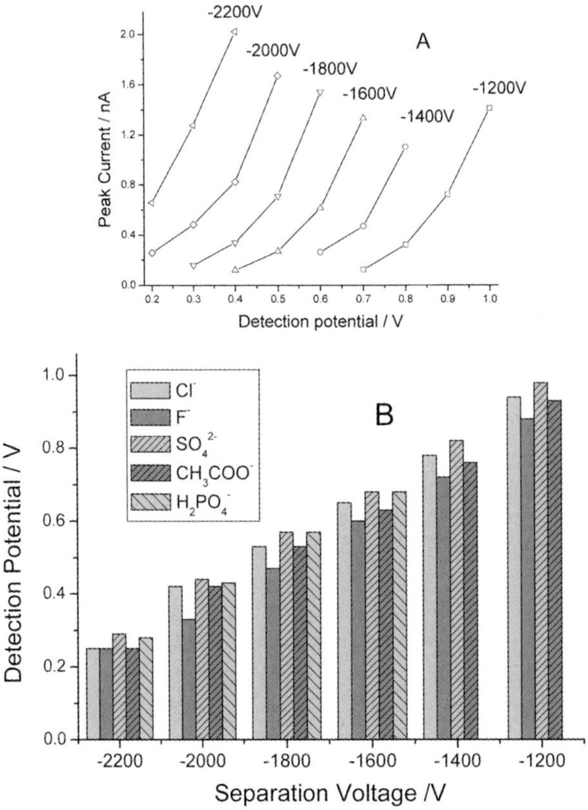

Fig. 5 (A) Plot of the peak current as a function of detection potential at different separation voltage for 0.50 mM Cl^-. (B) Relationships between the separation voltage and the detection potential for Cl^-, F^-, SO_4^{2-}, CH_3COO^- and $H_2PO_4^-$ at the detection sensitivity of 2nA/mM (Cl^-, F^-, SO_4^{2-}), and 0.6nA/mM (CH_3COO^- and $H_2PO_4^-$). Experimental parameters: running buffer, 20 mM pH 6.1, MES+His; sampling voltage, 800 V; sampling time, 5 s; $L_{BC} = 20$ μm.

4.3.4
Effect of the WE position

Electrode position in the channel has been shown to be a crucial parameter when optimizing the separation efficiency and detection sensitivity in CE-EC analysis because the coupled degree is also related to the potential of the position of the WE in-channel induced by the separation electric field. The distance between the carbon fiber disk WE and the separation channel outlet is visualized with a light microscope. For conventional amperometric detection mode, the closer the electrode is positioned away from the capillary outlet, the higher the sensitivity is because of the reduced dispersion of the analyte zone when ejected from the end of the capillary to the electrode surface. However, a very small capillary-to-electrode distance will result in relatively intense interference of separation voltage on detection, decreasing the sensitivity. Here, because the indirect amperometric detection mode is based on the coupling effect of the high separation electric field, the results are thus different from the conventional amperometric mode. Fig. 6 shows the effect of the distance (L_{BC}) from the electrode surface (Fig. 1, point B) to the channel outlet (Fig. 1, point C) on the detection signal. From Fig. 6, it is clear that with the increase of L_{BC}, the peak current increases. This can be explained as follows. At a certain high separation voltage, the larger the distance is, the more the coupling extent will be exerted on the electrode, and thus, the more positive the true oxidation potential of the carbon fiber is. At a certain detection potential, the large the distance, the higher the oxidation rate of the carbon fiber will be, and hence, the larger the change of the peak current induced by solution resistance is. If the distance was less than 20 μm, a significant loss in signal intensity was observed. Alternatively, if the distance was more than 60 μm, the background current levels became unsatisfactory and a loss of the S/N occurred. The optimal placement of the electrode was determined by the detection potential and separation voltage.

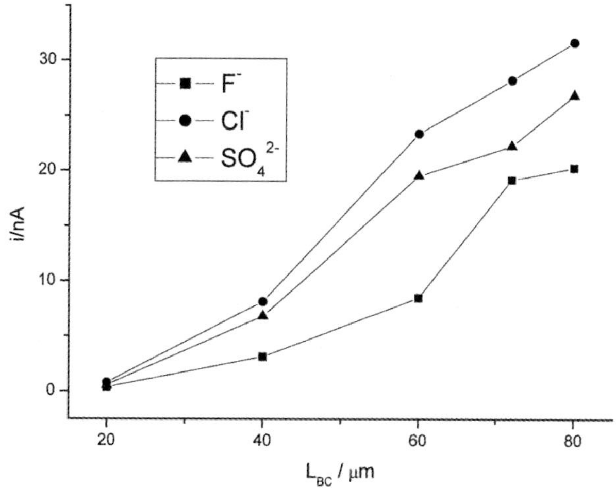

Fig. 6 Effect of L_{BC} on the detector response to 0.50 mM Cl^-, F^-, SO_4^{2-}. Experimental parameters: running buffer solution, 20 mM, pH 6.1 MES+His; sampling voltage, 800 V; sampling time, 5 s; separation voltage, −1800 V; detection potential, 0.4 V.

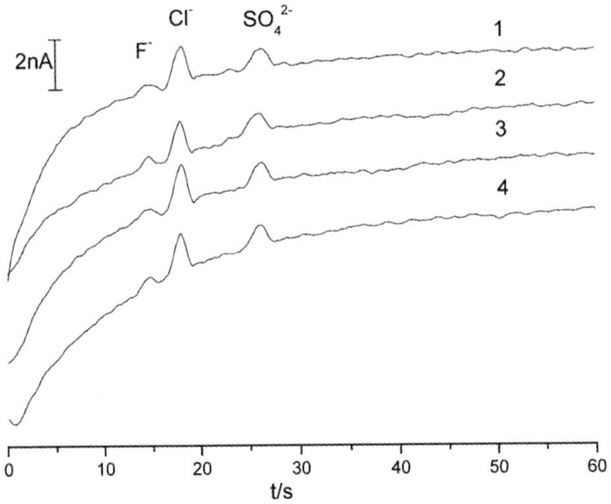

Fig. 7 Electropherograms of 10 μM Cl^-, F-, SO_4^{2-} for four successive samplings. Experimental parameters: running buffer solution, 20 mM, pH 6.1 MES+His; sampling voltage, 800 V; sampling time, 5 s; separation voltage, −1800 V; detection potential, 0.4 V; L_{BC} = 50 μm.

4.3.5
Linear range and detection limit

The in-channel indirect amperometric microchip detector displays a well-defined concentration dependence for the anionic species. Linear ranges of three orders of magnitude (peak area) (from 10 μM to 10 mM; correlation coefficients, >0.996) and two orders of magnitude (peak current) (from 10 μM to 1 mM; correlation coefficients, >0.996) were obtained for Cl^- under the separation voltage of −1800 V, detection potential of 0.4 V, and L_{BC} of 50 μm. LODs of 10.0, 2.0, and 4.0 μM were estimated for fluoride, chloride, and sulfate, respectively (based on an S/N of 2). Meanwhile, the microchip indirect amperometric detection system showed a very good reproducibility and stability. For example, a series of four repetitive injections of a mixture containing 10 μM fluoride, chloride, and sulfate ions yielded SDs of 4.4, 2.5, and 5.2%, respectively (shown in Fig. 7).

4.4
Concluding remarks

In conclusion, the results presented here combined with our previous report [31] clearly demonstrate that the combination of a carbon fiber electrode as in-channel detector with a microchip CE system would result in a versatile analytical device. As the coupling of separation electric field with detection circuit is used to generate the detection signal, decoupling between CE and electrochemical detection can be avoided. This simplification of the electrical design can be very useful for complex microsystems and compact microdevices that incorporate the CE technique. Such CE-EC microsystem is expected to find a wide range of applications.

We gratefully thank the National Natural Science Foundation of China for financial support of this research (Nos. 20475025, 20205007, 90206037, 20299030).

4.5 References

[1] Harrison, D. J., Manz, A., Fan, Z. H., Ludi, H., Widmer, H. M., *Anal. Chem.* 1992, *64*, 1926–1932.

[2] Schwarz, M. A., Hauser, P. C., *Lab Chip* 2001, *1*, 1–6.

[3] Gavin, P. F., Ewing, A. G., *Anal. Chem.* 1997, *69*, 3838–3845.

[4] Woolley, A. T., Lao, K., Glazer, A. N., Mathies, R. A., *Anal. Chem.* 1998, *70*, 684–688.

[5] Martin, R. S., Gawron, A. J., Fogarty, B. A., Regan, F. B., Dempsey, E., Lunte, S. M., *Analyst* 2001, *126*, 277–280.

[6] Gawron, A. J., Martin, R. S., Lunte, S. M., *Electrophoresis* 2001, *22*, 242–248.

[7] Martin, R. S., Gawron, A. J., Lunte, S. M., Henry, C. S., *Anal. Chem.* 2000, *72*, 3196–3202.

[8] Henry, C. S., Zhong, M., Lunte, S. M., Moon, K., Bau, H., Santiago, J. J., *Anal. Commun.* 1999, *36*, 305–308.

[9] Zeng, Y., Chen, H., Pang, D.-W., Wang, Z.-L., Cheng, J.-K., *Anal. Chem.* 2002, *74*, 2441–2445.

[10] Wang, J., Chatrathi, M. P., Ibanez, A., *Anal. Chem.* 2001, *73*, 1296–1300.

[11] Wang, J., Chatrathi, M. P., Tian, B., *Anal. Chem.* 2000, *72*, 5774–5778.

[12] Wang, J., Tian, B., Sahlin, E., *Anal. Chem.* 1999, *71*, 5436–5440.

[13] Rossier, J. S., Ferrigno, R., Girault, H. H., *J. Electroanal. Chem.* 2000, *492*, 15–22.

[14] Chen, D.-C., Hsu, F.-L., Zhan, D.-Z., Chen, C.-H., *Anal. Chem.* 2001, *73*, 758–762.

[15] Hilmi, A., Luong, J. H. T., *Anal. Chem.* 2000, *72*, 4677–4682.

[16] Hilmi, A., Luong, J. H. T., *Environ. Sci. Technol.* 2000, *34*, 3046–3050.

[17] Schwarz, M. A., Galliker, B., Fluri, K., Kappes, T., Hauser, P. C., *Analyst* 2001, *126*, 147–151.

[18] Liu, Y., Fanguy, J. C., Bledsoe, J. M., Henry, C. S., *Anal. Chem.* 2000, *72*, 5939–5944.

[19] Lacher, N. A., Garrison, K. E., Martin, R. S., Lunte, S. M., *Electrophoresis* 2001, *22*, 2526–2536.

[20] Wang, J., *Talanta* 2002, *56*, 223–231.

[21] Tanyanyiwa, J., Leuthardt, S., Hauser, P. C., *Electrophoresis* 2002, *23*, 3659–3666.

[22] Baldwin, R. P., Roussel, T. J., Jr., Crain, M. M., Bathlagunda, V., Jackson, D. J., Gullapalli, J., Conklin, J. A., *et al.*, *Anal. Chem.* 2002, *74*, 3690–3697.

[23] Hebert, N. E., Kuhr, W. G., Brazill, S. A., *Anal. Chem.* 2003, *75*, 3301–3307.

[24] Yan, J., Du, Y., Liu, J., Cao, W., Sun, X., Zhou, W., Yang, X., Wang, E., *Anal. Chem.* 2003, *75*, 5406–5412.

[25] Zemann, A. J., *Trends Anal. Chem.* 2001, *20*, 346–354.

[26] Laugere, F., Guijt, R. M., Bastemeijer, J., Van der Steen, G., Berthold, A., Baltussen, E., Sarro, P., *et al.*, *Anal. Chem.* 2003, *75*, 306–312.

[27] Landers, J. P., *Handbook of Capillary Electrophoresis*, 2nd edn., CRC Press, Boca Raton, FL 1997, pp. 380–382.

[28] Martin, R. S., Ratzlaff, K. L., Huynh, B. H., Lunte, S. M., *Anal. Chem.* 2002, *74*, 1136–1143.

[29] Arora, A., Eijkel, J. C. T., Morf, W. E., Manz, A., *Anal. Chem.* 2001, *73*, 3282–3288.

[30] Bai, X., Wu, Z., Josserand, J., Jensen, H., Schafer, H., Girault, H. H., *Anal. Chem.* 2004, *76*, 3126–3131.

[31] Xu, J. J., Bao, N., Xia, X. H., Peng, Y., Chen, H. Y., *Anal. Chem.* 2004, *76*, 6902–6907.

[32] Zhang, X. J., Ogorevc, B., Rupnik, M., Kreft, M., Zorec, R., *Anal. Chim. Acta* 1999, *378*, 135–143.

5
Coupling on-chip solid-phase extraction to electrospray mass spectrometry through an integrated electrospray tip*

Yanou Yang, Chen Li, Kelvin H. Lee, Harold G. Craighead

We report the integration of solid-phase extraction (SPE) with mass spectrometry (MS) through an on-chip electrospray tip for sample precleaning and preconcentration. An *in situ* polymerized alkylacrylate-based monolithic column was used as the stationary phase for the on-chip SPE. Each microchip consists of two sets of microchannels and their respective integrated electrospray tips, with a common gold electrode. After the microchip was fabricated from cycloolefin polymer by hot embossing, thermal bonding, and annealing steps, a mixture of monomers and porogenic solvents was pumped into the microchannels and certain areas of the main microchannels were exposed to UV irradiation through a mask. The resulting porous monolithic beds that were polymerized from different compositions of the mixture were characterized by scanning electron microscopy. The microchip containing the monolithic column was then interfaced to an ion trap (IT) mass spectrometer by modifying a commercially available interfacing system. Makeup solution from the side channel was infused concurrently with the solution flowing into the main channel, and the mixture of these two solutions was sprayed into the MS orifice. Both the adsorption and elution of a pharmaceutical test compound, imipramine, to and from the on-chip SPE columns were monitored by MS. The potential application of this device for sample cleanup was demonstrated by pretreatment of urine samples spiked with imipramine.

5.1
Introduction

Microfluidic systems have been investigated extensively to address biological and chemical analysis because miniaturization can offer improved performance, smaller amounts of sample consumption, lower cost, and higher throughput [1]. How-

* Originally published in Electrophoresis 2005, 26, 3622–3630

Microfluidic Applications in Biology. Edited by Niels Lion, Joël S. Rossier, and Hubert H. Girault

ISBN-10: 3-527-31761-9

ever, the ability of microfluidic devices to efficiently handle complex samples and subsequently perform the required analytical operations on the same chip will be essential to the final successful applications of microfluidic systems. Although most of the samples used in microchips have been manually prepared off-chip primarily to avoid the clogging problem due to the small features associated with microchips, the integration of sample pretreatment within microfluidic devices has recently become an important area of emphasis in the field of micrototal analysis systems (μTAS) [2, 3].

Sample preconcentration is one of the most common sample preparation steps in analyses because the concentrations of the analytes of interest in the raw samples are usually very low. In addition, the small dimensions of microchips result in ultrasmall fluid volumes manipulated on microchips, which makes sensitive detection methods a prerequisite for most on-chip analyses. This fact explains the predominance of the fluorescence detection method in chip-based analysis. Despite their extremely low mass and concentration detection limits, fluorescence detection methods are restricted to analytes containing either intrinsic or extrinsic fluorophores. Thus, coupling sample preconcentration into microfluidic devices allows the utilization of other general detection methods for chip-based analysis. SPE is a preconcentration as well as a precleaning process in which analytes are retained on a stationary phase, while solvents and undesired compounds are washed out, and retained analytes can thus be eluted in a more concentrated form. Due to its simplicity, speed, and effectiveness, SPE has been routinely used for concentration of selected analytes prior to chromatographic or electrophoretic analysis [4]. Further, SPE is especially well suited for integration into microfluidic devices because of the small elution volumes [5].

Microfluidic SPE was first achieved by coating channel walls in a glass microchip with octadecyltrimethylsilane to bear the desired C18 function [6]. To increase the specific surface area, alternative stationary phases for SPE have been investigated, including fabricated microstructures [7], hydrophobic polymer membranes [8–10], RP microparticles [11–15], and an *in situ* polymerized monolith [16, 17]. While having the same benefits of high surface area and easily controlled surface chemistry as packed microparticles, *in situ* polymerized monoliths have their distinct advantages. They can be prepared easily and rapidly because a monomer solution with low viscosity can be introduced by vacuum or pressure into the microchannel before initiation. The continuous polymer bed is attached to the channel walls, which makes frits and other retaining structures redundant. In addition, the porosity, pore size, and surface area of the monolith can be easily controlled by changing the composition of the initial monomer solution and the polymerization condition [18]. In addition to be used as stationary phase for SPE, monolithic stationary phases have been used to perform separation [19–21] and mixing [22] in microfluidic devices.

Since the first examples of coupling microfluidic devices to ESI-MS in 1997 [23–25], there has been growing interest in this research area, especially on the coupling of on-chip sample manipulation to MS. Considerable effort has been devoted to coupling on-chip separation to ESI-MS. The separation methods that have been investigated for integration with ESI-MS include IEF [26], CEC [27], CE of small molecules [28, 29], and CE of peptides and protein digests [30–33]. On-line SPE techniques for preconcentration or desalting before separation would reduce sample handling and analysis time and provide improved detection limits. But few studies have coupled on-chip sample preparation to ESI-MS [8, 10]. Tan *et al.* [17] recently reported on coupling on-chip SPE to MS through an off-chip micro-ion sprayer. The SPE was achieved by an *in situ* polymerized monolithic column in a microchannel on a cycloolefin polymer-based device with a diameter of 360 μm. The concentrated sample was eluted at a flow rate of 2–4 μL/min with a makeup liquid pumped to the sprayer at 4–8 μL/min. More investigation is necessary especially on integrating on-line sample preparation to ESI-MS by on-chip electrosprayer tips.

Following our recent activity on fabricating a cycloolefin polymer-based microchip with integrated tips and *in situ* polymerized monolith for ESI-MS applications [34], we report the first example of coupling on-chip SPE by *in situ* polymerized monolith to ESI-MS using an integrated on-chip electrospray tip. Both the adsorption and elution processes of a small molecule drug, imipramine, were monitored by ESI-MS. In addition, the monolithic polymer bed was applied to cleanup urine samples spiked with imipramine followed by ESI-MS.

5.2
Materials and methods

5.2.1
Materials

Zeonor polymer plates (ZEONOR1020R, 2 × 100 × 150 mm) were obtained from Zeon Chemicals (Louisville KY, USA). Imipramine hydrochloride, ammonium acetate, methyl methacrylate (MMA), ethylene dimethacrylate (EDMA), butyl methacrylate (BMA), 2,2-dimethoxy-2-phenylacetophenone (DPA), and benzophenone were acquired from Sigma-Aldrich (St. Louis, MO, USA). Ethanol, methanol, and acetonitrile (ACN) were from J. T. Baker (Phillipsburg, NJ, USA). Formic acid was from EM science (Gibbstown, NJ, USA). All aqueous solutions were prepared using deionized (DI) water purified by Nanopure® Infinity UV/UF system (Barnstead Thermolyne, Dubuque, IA, USA). Stock solutions of imipramine (4 mg/mL) were prepared in methanol and stored in a refrigerator.

5.2.2
Device fabrication

The device fabrication procedure (Fig. 1) was described in detail elsewhere [34]. An MTP Press (Tetrahedron, San Diego, CA, USA) was used to fabricate the zeonor chip. Briefly, microchannels (150 μm wide and 150 μm deep) were fabricated on one zeonor substrate by hot embossing against a premade silicon master (138°C,

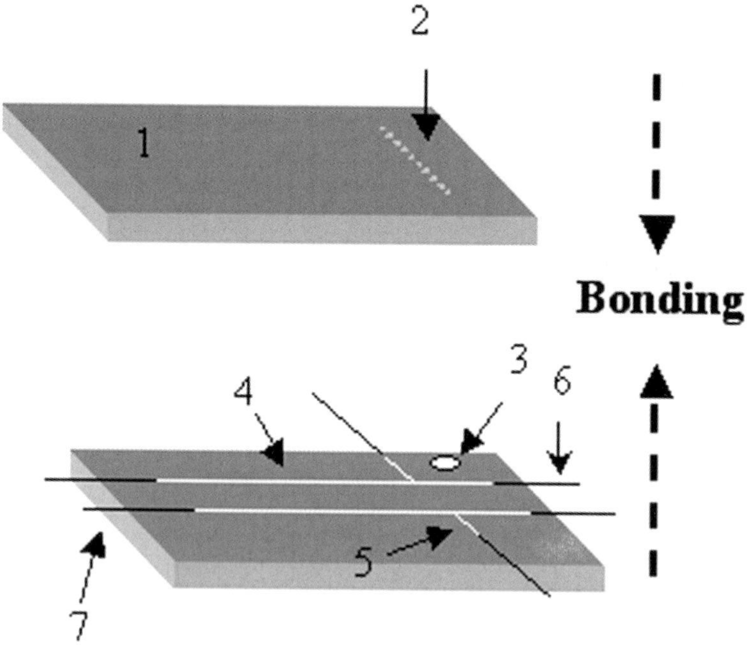

Fig. 1 Microchip fabrication procedures. (1) Zeonor substrate; (2) gold electrode; (3) access hole for gold electrode; (4) main channel; (5) side channel; (6) electrospray tip; (7) fused-silica capillary.

667 N, 10 min). Four fused-silica capillaries (150 μm OD, 50 μm ID, 30 cm length; Polymicro Technologies, Phoenix, AZ, USA) were inserted into the channels at one end and both sides of the microchip, and two commercial uncoated electrospray tips (150 μm OD, 50 μm ID, 50 μm tip; New Objective, Woburn, MA, USA) were inserted into the other end. A gold electrode with a width of 500 μm was fabricated onto another zeonor substrate by standard photolithography and lift-off techniques as described previously. After a hole for gold electrode access was drilled through the substrate containing the microfluidic channels, the two chips were cut into the desired size and bound together (107°C, 448 N, 5 min) followed by an annealing step (115°C, 0 N, 10 min). A copper screw was glued with silver paint into the hole to make contact with the electrode.

5.2.3
SPE monolithic column preparation

For the initial experiments, we modified the microchannel surface using a recently described photografting method [35, 36] to form covalent attachment between the channel wall and the monolith. The photografting solution was prepared as follows: a 30% stock solution of benzophenone was first prepared in ethanol, then the inhibitor molecules in EDMA and MMA were removed by mixing with fresh alumina power, finally, 225 μL EDMA, 225 μL MMA, and 50 μL 30% benzophenone stock solution were mixed together and purged with nitrogen gas for 5 min. After washing the microchannels with methanol extensively, the prepared photografting solution was pumped into the channel and exposed to two UV light sources for 20 min. One light source (Model UVGL-55, UVP, Upland, CA, USA; wavelength 254 nm) was held 4 cm above the device top, and the other (Model VWR LM-20E, VWR Scientific, West Chester, PA, USA; wavelength 302 nm) was held 6 cm below the device bottom. After grafting reaction, the channels were washed for 1 h with methanol at a flow rate of 3.0 μL/min before putting in the monomer solution for monolith preparation.

The SPE polymer bed was prepared as described previously [34]. The monomer mixture was prepared by mixing 2.5 mg DPA, 250 μL methanol, 125 μL ethanol, 75 μL BMA, and 50 μL EDMA. Prior to mixing with porogenic solvent, EDMA and BMA were mixed with fresh basic alumina powder (mesh 60–325; Fisher, Pittsburgh, PA, USA) to remove the inhibitor. The alumina powder was then removed by centrifugation at 1000 × g for 5 min. After purging with nitrogen for 3 min, the mixture was pumped into the microchannel prewashed extensively with methanol. The end of each capillary was closed by a union with a stop nut at one end (Upchurch Scientific, Oak Harbor, WA, USA). The location of the polymer bed inside the main channel was controlled by a low-resolution mask with a 5 mm long open window made from aluminum foil attached to a 4-in. glass wafer. A portable 365-nm lamp (Model UVGL-55, UVP) was used to provide UV light from the bottom through the mask to the microchip. The distance between the light and the device was 5 cm. After polymerization for 10 min, the polymer bed was washed for 30 min using methanol at 3.0 μL/min. A Nikon (Rochester, NY, USA) SMZ-1500 stereozoom fluorescence microscope attached with a SPOT Insight® CCD camera (Diagnostic Instruments, Sterling Heights, MI, USA) was used to take the whole view of the polymer bed in the microchannel. The microchips containing the polymer bed were cut perpendicular to the microchannels with a saw and the edges were cleaned with a sharp knife. The cross sections of the microchannels filled with monolith were imaged using a Leo 982 SEM (Leo, Thornwood, NY) after coating the surface with gold/platinum for 60 s using an SCD 500 Balzers sputter coater (Boeckeler Instruments, Tuscon, AZ, USA).

5.2.4 Interfacing the microchip to MS

The microchip containing the polymer bed was interfaced to an LCQ Deca IT mass spectrometer (ThermoFinnigan, San José, CA, USA) as shown in Fig. 2. A commercial PicoView® system (New Objective, Cambridge, MA, USA) was modified for this purpose. A plastic plate was placed onto the *X, Y, Z* transition stage in order to align the tip on the microchip to the MS orifice. A mirror was taped to the MS interface plate in order to visualize the electrospray tip by the camera and video monitor. The microchip was positioned onto the plastic plate on the *X, Y, Z* stage with ~ 7 mm distance from the MS orifice. Each chip contained two sets of channels integrated with their respective capillary electrospray tips, but only one tip was used at a time. Two syringe infusion pumps (Harvard Apparatus, Holliston, MA, USA) were used to generate the desired flow rates of the samples. A 25 μL syringe was used to pump solutions to the main channel containing the monolithic polymer bed at a flow rate of 0.15 μL/min, and a 100 μL syringe was used to infuse makeup solution to the side channel at a flow rate of 0.6 μL/min. This flow rate was chosen experimentally for the best performance of the integrated electrospray tip. An external power supply (Bertan, Franklin Park, IL, USA) was used to apply the high voltage (1.9 kV) between the MS orifice and the gold microelectrode. The MS heated inlet capillary was maintained at 180°C. The ESI-MS was performed in the positive ion mode. For full range MS scan, the maximum injection time was 200 ms. For MS/MS experiment, the maximum injection time was 400 ms, and the collision energy used in the MS/MS mode was 30%, using argon for the collision gas. Other scan parameters are noted in the caption of each figure.

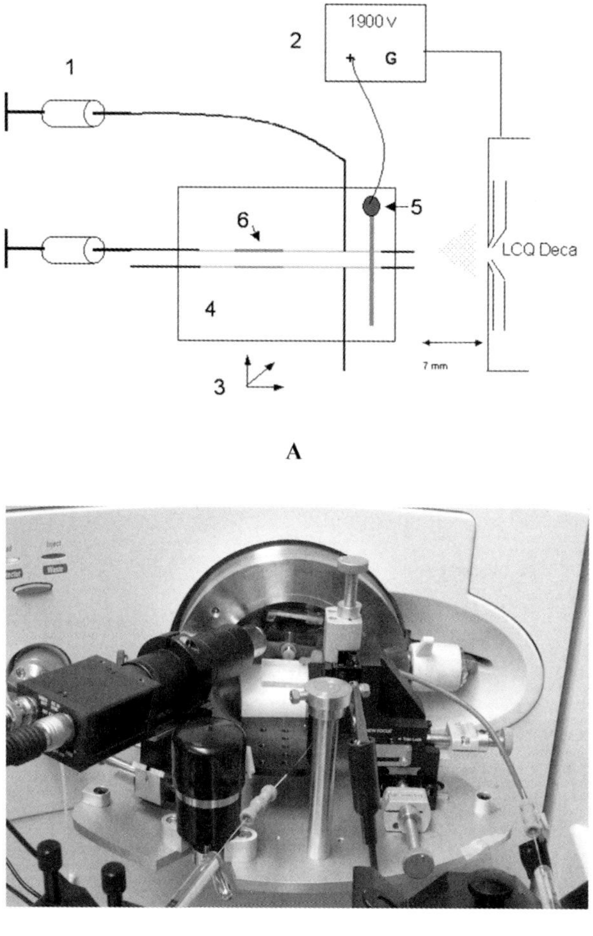

Fig. 2 Experimental setup for coupling microchip to an LCQ Deca IT mass spectrometer. (A) Schematic drawing of the setup. (1) Syringe infusion pump; (2) power supply; (3) *X, Y, Z* stage; (4) microchip; (5) metal screw for voltage application; (6) monolith column. (B) Picture of the experimental setup modified from PicoView system.

5.2.5
SPE-ESI-MS of imipramine

5.2.5.1 Sample loading
After washing with ACN for 10 min at 3 μL/min, the polymer bed was preconditioned with 10 mM ammonium acetate buffer (pH 9.3) for 10 min at the same flow rate. Then a solution of 10 mM ammonium acetate (pH 9.3) containing 10 ng/μL imipramine was continuously pumped through the column inside the main channel at 0.15 μL/min while a makeup solution containing 50% methanol, 50% water, and 0.5% formic acid was infused into the side channel at a flow rate of 0.6 μL/min. At the same time, the liquid from the main channel and the side channel were mixed and electrosprayed at the electrospray tip. MS/MS scan mode with a parent ion m/z 281.3 and scan range of m/z 75–100 was used to monitor the sample loading process.

5.2.5.2 Elution
The monolithic SPE column on the microchip was washed and preconditioned as described in Section 2.5.1. The imipramine stock solution was diluted in 10 mM ammonium acetate (pH 9.3) to a final concentration of 0.5 and 5 ng/μL, respectively, and a volume of 20 μL of the sample was loaded to the SPE column at a flow rate of 0.5 μL/min. After washing the column with 10 mM ammonium acetate (pH 9.3) for 10 min at a flow rate of 3 μL/min, the retained imipramine was eluted with ACN containing 0.1% formic acid, which was pumped into the main channel at a flow rate of 0.15 μL/min. Concurrently, a makeup solution containing 50% methanol, 50% water, and 0.5% formic acid was infused into the side channel. The power supply was turned on at the same time as the syringe pumps. The elution process was monitored using full range MS scan (m/z 250–320) with a scan speed of 1.5 scan/s. The data collection started 3 min after the pumps started to account for the time needed for the elution solution to reach the electrospray tip.

5.2.5.3 Cleanup of human urine sample
The imipramine stock solution was diluted in a freshly collected human urine sample to a concentration of 2 ng/μL. The urine sample was preadjusted to pH 9.3 using ammonium hydroxide before spiking with imipramine. The procedure for washing and precondition steps was the same as described in Section 2.5.2. A volume of 20 μL of the spiked human urine sample instead of ammonium acetate buffer containing imipramine was loaded to monolithic SPE column. The elution process was monitored by a full range MS scan (m/z 250–320).

5.3 Results and discussion

5.3.1 Polymerization of the monolithic column

For glass microchips, the covalent attachment between the monolith and the channel walls was usually achieved by modifying the glass channel surface with silane prime reagent 3-(trimethoxysilyl) propyl methacrylate. But this method cannot be applied to polymer-based microchannels due to the lack of functional groups of plastic surfaces. To form covalent attachment between the monolith and our zeonor channel wall, we used a recently described photografting method [35, 36] to modify the channel wall in the zeonor device with a mixture of 1:1 EDMA and MMA containing 3% benzophenone before polymerization of the monolithic bed. SEM imaging of the polymer beds prepared in microchannels with and without surface modification suggested no gap between the monolith bed and the microchannel in both cases. Moreover, a flow rate as high as 20 μL/min was not able to dislodge the polymer bed without photografting of the channel surface. This is consistent with previous observation by Tan *et al.* [17]. One explanation for this is that the mechanical forces due to the surface roughness of the microchannel in this device were enough to retain the polymer bed in place. Based on these results, we later directly prepared the monolith column in the microchannel without the pre-photografting step.

The pore size of the polymerized monolithic bed is dependent on the composition of monomer solution and polymerization condition [18]. We found that it was particularly sensitive to the ratio of methanol to ethanol used in the mixture. Figures 3A and B shows monoliths polymerized from monomer mixture with different methanol and ethanol ratios. All the other polymerization parameters were the same. As can be seen, a small change in the ratio of methanol to ethanol results in a significant change in the resulting pore size and nodule size. To facilitate flow of solutions through the polymer bed, we chose a ratio of methanol to ethanol 2:1 for the SPE application in this work.

Fig. 3C shows the cross section of a monolithic polymer bed inside a microchannel on the zeonor device. As can be seen, the porous monolithic bed fills the entire embossed channel. The top zeonor substrate and the bottom zeonor substrate are bonded to form one intact piece. Although the dimension of the microchannel after embossing was 150 μm^2, the microchannel decreased to $\sim$ 100 μm after the thermal bonding and annealing steps. The length and location of the polymer bed can be controlled reproducibly with our polymerization procedure. Thus the two sets of monolithic columns with their respective side channels were used interchangeably. The monolith turns white after polymerization. A 5 mm long polymer bed inside a microchannel on the zeonor device is shown in Fig. 3D.

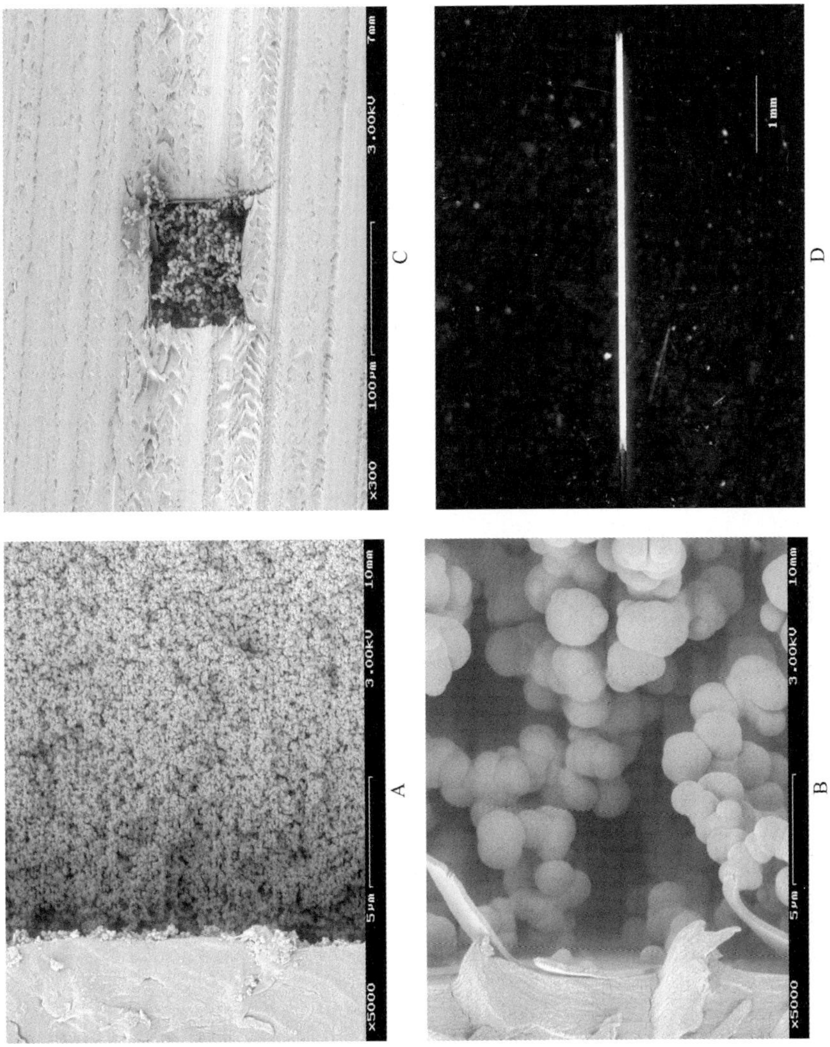

Fig. 3 *In situ* polymerized monolith. (A) Scanning electron microscope (SEM) image of the cross section of a monolithic polymer bed, formed using a mixture of methanol and ethanol with a ratio of 3:2 as porogen. (B) SEM image of the cross section of a monolithic polymer bed, formed using a mixture of methanol and ethanol with a ratio of 2:1 as porogen. (C) SEM image of the monolith column cross section inside a microchannel on a zeonor device. (D) Optical micrograph of a 5 mm long monolithic polymer bed.

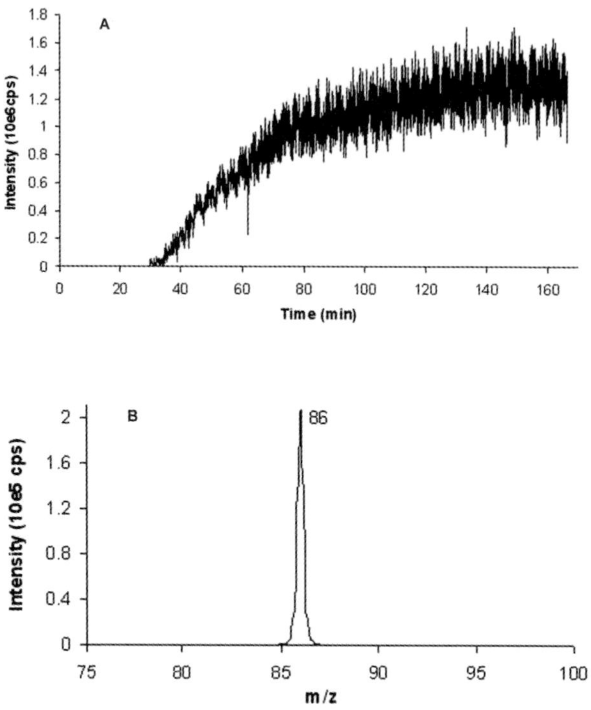

Fig. 4 Adsorption of imipramine on the monolithic column monitored by ESI-MS. Sample containing 10 ng/μL imipramine in 10 mM ammonium acetate buffer (pH 9.3) was infused to pass the monolithic column in the main channel at a flow rate of 0.15 μL/min and a makeup solution containing 50% methanol, 50% water, and 0.5% formic acid was pumped into the side channel at a flow rate of 0.6 μL/min. Electrospray voltage: 1.9 kV. MS/MS scan with parent ion *m/z* 281.3; scan range, *m/z* 75–100; scan speed, 0.85 scan/s. (A) Breakthrough curve from total ion current; (B) mass spectrum averaged from five scans from 145 to 145.1 min in (A).

5.3.2 Imipramine adsorption

The adsorption of a common antidepressant, imipramine, onto the on-chip monolithic column was monitored by MS/MS until the MS signal reached maximum (Fig. 4A). A solution of 10 mM ammonium acetate buffer (pH 9.3) containing 10 ng/μL imipramine was infused into the main channel at a flow rate of 0.15 μL/min while the makeup solution containing 50% water, 50% methanol, and 0.5% formic acid was flowed into the side channel at a flow rate of 0.6 μL/min. A higher flow rate could be used for loading samples, but we used a flow rate of 0.15 μL/min to be compatible with the experimental conditions used for the electrospray tip. Although the sample was in 100% aqueous buffer, it was mixed with the makeup solution from the side channel before it reached the electrospray tip. The resulting solution, containing 60% water and 40% organic solvent, was sprayed at a relatively low voltage (1.9 kV) for the whole period monitored (166 min). Fig. 4B shows the averaged mass spectrum of five MS/MS scans obtained after the column was saturated.

The data in Fig. 4A suggest that all of the imipramine molecules in the buffer bound to the column for the first 30 min. Then some imipramine appears in the effluent. The results suggest that the column was gradually saturated. Based on the dead volume from the fused-silica capillary and the microchannel in the device

($\sim$ 0.6 µL), it took 4 min for the sample to reach the electrospray tip. The actual time to reach 50% of the ion current maximum is 54 min. Based on the flow rate, the imipramine concentration, and the time to reach 50% of the ion current maximum, the estimated capacity of this 5 mm long polymer bed housed in the microchip was 81 ng (2.9×10^{-10} mol).

5.3.3 Elution of imipramine from the on-chip SPE column

Elution of the imipramine from the monolithic column was monitored by ESI-MS through the integrated electrospray tip. The appropriate flow rate for the electrospray tip was 0.2–1 µL/min as suggested by the manufacturer. The flow rate at the tip is the sum of the flow rates from the main channel and the side channel. The connected capillary and the microchannel were filled with 100% aqueous ammonium acetate buffer prior to elution, and the elution was performed with 100% ACN. To monitor the elution process continuously using stable electrospray, makeup solution was introduced from the side channel. The ratio between the flow rate used in the main channel and that used in the side channel alters the final composition of the electrospray solution reaching the electrospray tip. The smaller the ratio of the main channel flow rate to the side channel flow rate, the less the change in the composition of the resulting electrospray solution. However, a smaller ratio will result in more dilution in the analyte concentration coming out of the main channel, which will increase the detection limit. Based on all these considerations and preliminary experiments, we chose a ratio of 1:4 for the flow rate in the main channel to that of the side channel with a total flow rate of 0.75 µL/min for the elution process. The relatively low flow rate used in elution (0.15 µL/min) is advantageous because a large number of MS scans across the elution peak can be acquired.

Fig. 5A shows a representative elution peak from the SPE column loaded with 20 µL of 0.5 ng/µL imipramine monitored by ESI-MS. The mass spectrum at the highest point of the peak is shown in Fig. 5B. There is much less tailing compared to a previous study, which used a much higher through column flow rate (2 µL/min), a larger column diameter (360 µm), and an off-chip metal microspray tip [17].

The concentration factor can be simply calculated [16] by dividing the volume of the imipramine solution used in the adsorption step by the volume of the eluted peak. The volume of the eluted peak can be calculated from the peak width and the flow rate. For the elution peak shown in Fig. 5A, the calculated concentration factor is $\sim$ 267. A much higher concentration enhancement is expected if the column is loaded with an imipramine sample with a lower concentration or a larger volume.

Fig. 6 shows the comparison of the elution peak for the column loaded with different amount of imipramine. A is the same elution peak from Fig. 5A with a sample loading of 10 ng of imipramine. B is the elution peak from the SPE column loaded with 100 ng imipramine, in which the SPE column was saturated. Thus, a much higher and wider peak was resulted in B with much higher sample loadings.

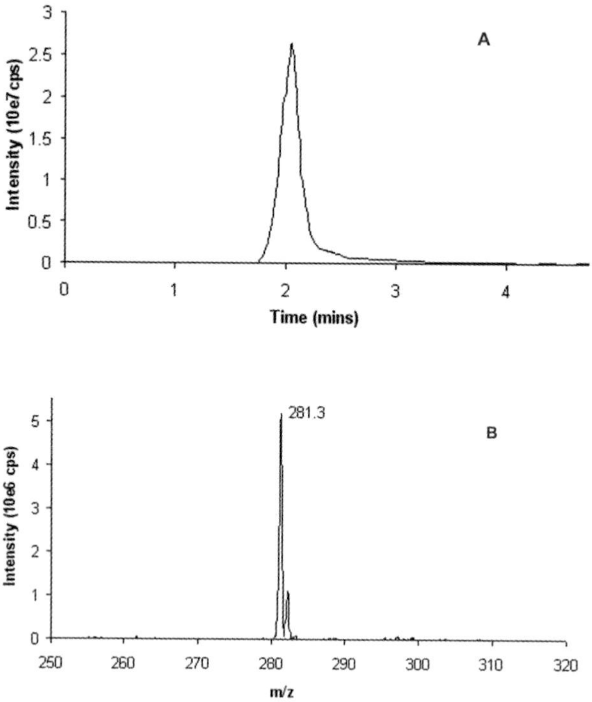

Fig. 5 Elution of imipramine monitored from on-chip monolith column by ESI-MS. SPE conditions: loading, 20 μL of 0.5 ng/μL imipramine in 10 mM ammonium acetate buffer (pH 9.3) at a flow rate of 0.6 μL/min; elution, ACN containing 0.1% formic acid at a flow rate of 0.15 μL/min. MS parameters: full range MS scan (*m*/*z* 250–320) with a scan speed of 1.5 scan/s. (A) Extracted ion current for *m*/*z* 280.8–281.8 from full range scan; (B) mass spectrum at the highest point of the peak in (A).

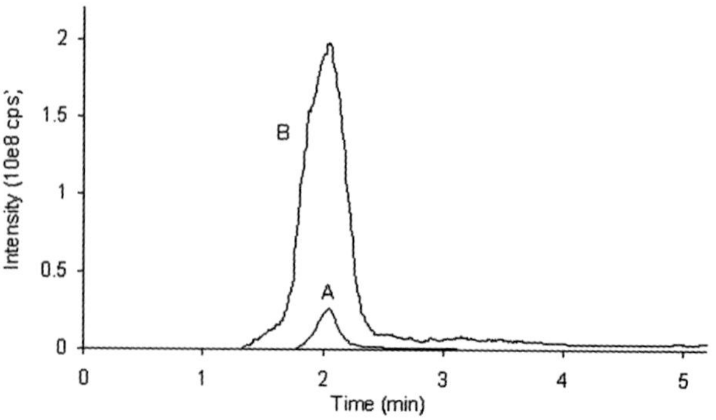

Fig. 6 Comparison of imipramine from on-chip monolith column monitored by ESI-MS. Extracted ion current for *m*/*z* 280.8–281.8 from full range MS scan (*m*/*z* 250–320) with a scan speed of 1.5 scan/s. SPE conditions: (A) loading, 20 μL imipramine in 10 mM ammonium acetate buffer (pH 9.3) at a flow rate of 0.5 μL/min; elution, ACN containing 0.1% formic acid at a flow rate of 0.15 μL/min. (A) Imipramine concentration of 0.5 ng/μL; (B) imipramine concentration of 5 ng/μL.

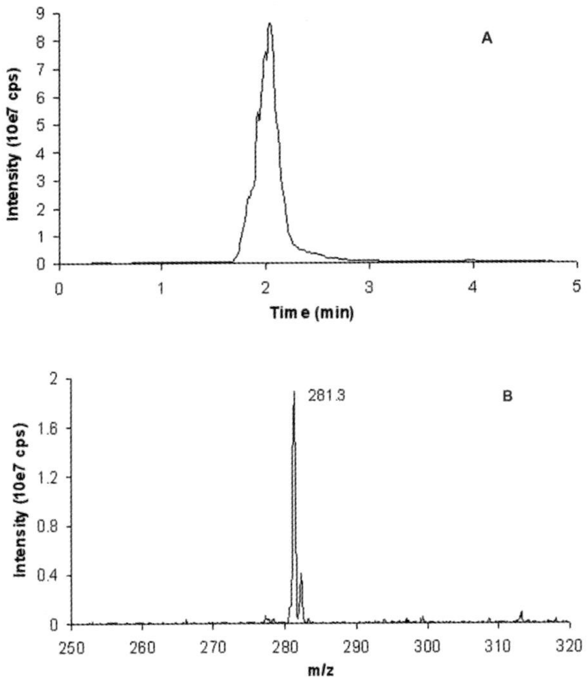

Fig. 7 MS elution profile for imipramine spiked in human urine sample. SPE conditions: loading, 20 μL of 2 ng/μL imipramine spiked in human urine sample (pH 9.3) at a flow rate of 0.5 μL/min; elution, ACN containing 0.1% formic acid at a flow rate of 0.15 μL/min. MS parameters: full-range MS scan (*m*/*z* 250–320) with a scan speed of 1.5 scan/s. (A) Extracted ion current for *m*/*z* 280.8–281.8 from full-range scan; (B) mass spectrum at the highest point of the peak in (A).

5.3.4 Cleanup of urine sample

Although analytes in buffer or solvent are typically used for the evaluation of SPE columns on a microfluidic device, practical applications of such SPE columns are likely to involve cleanup of raw samples. To demonstrate the analytical potential of this device for the pretreatment of real world samples, it was employed for the cleanup of human urine sample spiked with imipramine at a concentration of 2 ng/μL. The spiked urine sample was loaded into the on-chip SPE column with a volume of 20 μL. Fig. 7A shows the elution peak of the adopted imipramine. Fig. 7B shows the mass spectrum at the highest point of the peak. The imipramine spiked in human urine has been concentrated for 208 times by the SPE column.

5.4 Concluding remarks

SPE by *in situ* polymerized porous monolith columns housed in a polymer-based microfluidic device has been successfully coupled to ESI-MS through integrated on-chip electrospray tips. The adsorption of a pharmaceutical test compound imipramine onto the SPE column has been monitored with ESI-MS for a period of 166 min, and the estimated column capacity is 81 ng (2.9×10^{-10} mol) from the breakthrough curve. The elution peak of imipramine from the SPE column, monitored by MS, suggested a 267 times concentration enhancement. In addition, cleanup of human urine samples spiked with this compound has been demonstrated with a 208-fold concentration. Later, this device can be easily integrated with on-chip RPLC by *in situ* polymerization of another monolithic column for separation. Thus, complex samples can be pretreated by SPE, separated by RPLC, and then coupled to ESI-MS for identification and quantitation.

The authors would like to thank Scott Verbridge for cutting the microchip, Erin J. Finehout for help with the mass spectrometer, and Dr. Aimin Tan for helpful discussion about monolith preparation. This work was supported by the Nanobiotechnology Center (NBTC), an STC Program of the National Science Foundation under Agreement No. ECS-9876771. The authors appreciate access and use of the Cornell Nanoscale Science and Technology Facility (a member of the National Nanofabrication Users Network), which is supported by the National Science Foundation under Grant no. ECS-9731293.

5.5 References

[1] Voldman, J., Gray, M. L., Schmidt, M. A., *Annu. Rev. Biomed. Eng.* 1999, *1*, 401–425.
[2] de Mello, A. J., Beard, N., *Lab Chip* 2003, *3*, 11N–19N.
[3] Lichtenberg, J., de Rooij, N. F., Verpoorte, E., *Talanta* 2002, *56*, 233–266.
[4] Fritz, J. S., Macka, M., *J. Chromatogr. A* 2000, *902*, 137–166.
[5] Lion, N., Rohner, T. C., Dayon, L., Arnaud, I. L., Damoc, E., Youhnovski, N., Wu, Z. Y., *et al.*, *Electrophoresis* 2003, *24*, 3533–3562.
[6] Kutter, J. P., Jacobson, S. C., Ramsey, J. M., *J. Microcol. Sep.* 2000, *12*, 93–97.
[7] Christel, L. A., Petersen, K., McMillan, W., Northrup, M. A., *J. Biomech. Eng.-T. ASME* 1999, *121*, 22–27.
[8] Lion, N., Gellon, J. O., Jensen, H., Girault, H. H., *J. Chromatogr. A* 2003, *1003*, 11–19.
[9] Lion, N., Gobry, V., Jensen, H., Rossier, J. S., Girault, H., *Electrophoresis* 2002, *23*, 3583–3588.
[10] Bonneil, E., Li, J. J., Tremblay, T. L., Bergeron, J. J., Thibault, P., *Electrophoresis* 2002, *23*, 3589–3598.
[11] Oleschuk, R. D., Shultz-Lockyear, L. L., Ning, Y. B., Harrison, D. J., *Anal. Chem.* 2000, *72*, 585–590.
[12] Jemere, A. B., Oleschuk, R. D., Ouchen, F., Fajuyigbe, F., Harrison, D. J., *Electrophoresis* 2002, *23*, 3537–3544.
[13] Ekstrom, S., Malmstrom, J., Wallman, L., Lofgren, M., Nilsson, J., Laurell, T., Marko-Varga, G., *Proteomics* 2002, *2*, 413–421.
[14] Bergkvist, J., Ekstrom, S., Wallman, L., Lofgren, M., Marko-Varga, G., Nilsson, J., Laurell, T., *Proteomics* 2002, *2*, 422–429.
[15] Wolfe, K. A., Breadmore, M. C., Ferrance, J. P., Power, M. E., Conroy, J. F., Norris, P. M., Landers, J. P., *Electrophoresis* 2002, *23*, 727–733.
[16] Yu, C., Davey, M. H., Svec, F., Fréchet, J. M. J., *Anal. Chem.* 2001, *73*, 5088–5096.
[17] Tan, A. M., Benetton, S., Henion, J. D., *Anal. Chem.* 2003, *75*, 5504–5511.
[18] Yu, C., Xu, M. C., Svec, F., Fréchet, J. M. J., *J. Polym. Sci. Part A: Polym. Chem.* 2002, *40*, 755–769.
[19] Fintschenko, Y., Choi, W. Y., Ngola, S. M., Shepodd, T. J., *Fresenius' J. Anal. Chem.* 2001, *371*, 174–181.
[20] Throckmorton, D. J., Shepodd, T. J., Singh, A. K., *Anal. Chem.* 2002, *74*, 784–789.
[21] Li, C., Lee, K. H., *Anal. Biochem.* 2004, *333*, 381–388.

[22] Rohr, T., Yu, C., Davey, M. H., Svec, F., Fréchet, J. M. J., *Electrophoresis* 2001, *22*, 3959–3967.

[23] Figeys, D., Ning, Y. B., Aebersold, R., *Anal. Chem.* 1997, *69*, 3153–3160.

[24] Ramsey, R. S., Ramsey, J. M., *Anal. Chem.* 1997, *69*, 1174–1178.

[25] Xue, Q. F., Foret, F., Dunayevskiy, Y. M., Zavracky, P. M., McGruer, N. E., Karger, B. L., *Anal. Chem.* 1997, *69*, 426–430.

[26] Wen, J., Lin, Y. H., Xiang, F., Matson, D. W., Udseth, H. R., Smith, R. D., *Electrophoresis* 2000, *21*, 191–197.

[27] Lazar, I. M., Li, L. J., Yang, Y., Karger, B. L., *Electrophoresis* 2003, *24*, 3655–3662.

[28] Deng, Y. Z., Henion, J., Li, J. J., Thibault, P., Wang, C., Harrison, D. J., *Anal. Chem.* 2001, *73*, 639–646.

[29] Kameoka, J., Craighead, H. G., Zhang, H. W., Henion, J., *Anal. Chem.* 2001, *73*, 1935–1941.

[30] Li, J. J., Thibault, P., Bings, N. H., Skinner, C. D., Wang, C., Colyer, C., Harrison, J., *Anal. Chem.* 1999, *71*, 3036–3045.

[31] Zhang, B., Liu, H., Karger, B. L., Foret, F., *Anal. Chem.* 1999, *71*, 3258–3264.

[32] Zhang, B. L., Foret, F., Karger, B. L., *Anal. Chem.* 2000, *72*, 1015–1022.

[33] Li, J. J., Kelly, J. F., Chemushevich, I., Harrison, D. J., Thibault, P., *Anal. Chem.* 2000, *72*, 599–609.

[34] Yang, Y., Li, C., Kameoka, J., Lee, K., Craighead, H. G., *Lab Chip* 2005, *5*, 869–876.

[35] Rohr, T., Ogletree, D. F., Svec, F., Fréchet, J. M. J., *Adv. Funct. Mater.* 2003, *13*, 264–270.

[36] Chen Li, Y. Y., Craighead, H. G., Lee, K. H., *Electrophoresis* 2005, *26*, 1800–1806.

6
Electrospray interfacing of polymer microfluidics to MALDI-MS*

Ying-Xin Wang, Yi Zhou, Brian M. Balgley, Jon W. Cooper, Cheng S. Lee, Don L. DeVoe

The off-line coupling of polymer microfluidics to MALDI-MS is presented using electrospray deposition. Using polycarbonate microfluidic chips with integrated hydrophobic membrane electrospray tips, peptides and proteins are deposited onto a stainless steel target followed by MALDI-MS analysis. Microchip electrospray deposition is found to yield excellent spatial control and homogeneity of deposited peptide spots, and significantly improved MALDI-MS spectral reproducibility compared to traditional target preparation methods. A detection limit of 3.5 fmol is demonstrated for angiotensin. Furthermore, multiple electrospray tips on a single chip provide the ability to simultaneously elute parallel sample streams onto a MALDI target for high-throughput multiplexed analysis. Using a three-element electrospray tip array with 150 μm spacing, the simultaneous deposition of bradykinin, fibrinopeptide, and angiotensin is achieved with no cross talk between deposited samples. In addition, in-line proteolytic digestion of intact proteins is successfully achieved during the electrospray process by binding trypsin within the electrospray membrane, eliminating the need for on-probe digestion prior to MALDI-MS. The technology offers promise for a range of microfluidic platforms designed for high-throughput multiplexed proteomic analyses in which simultaneous on-chip separations require an effective interface to MS.

6.1
Introduction

Microfluidic systems offer important advantages for biomolecular analysis, including low reagent consumption, small sample volumes with high analyte concentrations, and potential for low-cost mass production. A wide range of microfluidic sys-

* Originally published in Electrophoresis 2005, 26, 3631–3640

Microfluidic Applications in Biology. Edited by Niels Lion, Joël S. Rossier, and Hubert H. Girault

ISBN-10: 3-527-31761-9

tems for bioseparations have been reported, with chip-level optical or electrical sensing typically employed for detection. However, high-resolution molecular analysis ultimately requires the coupling of microchannel separations with MS for obtaining accurate mass measurements. Direct ESI-MS using ESI emitters fabricated in planar microfluidic chips provides a particularly efficient and straightforward interface between microfluidic separations and MS analysis. Extensive studies on direct microfluidic ESI-MS have been reported, including the use of tips fabricated in glass [1, 2] and bulk etched silicon [3]. The integration of silica capillary emitters directly into microchannel exits [4–10] or through a liquid junction [11] has also been reported. Electrospray chips fabricated from polymer materials have recently received significant attention, including the use of shaped tip elements [12–22], integrated wicking structures [23, 24], and the application of porous hydrophobic materials at the emitter [25, 26] to improve electrospray stability and reduce liquid spreading at the tip.

Despite the potential for direct ESI-MS analysis from microfluidic systems, it is often desirable to decouple on-chip separations from MS analysis. For example, there is often an incompatibility between the time scales for biomolecular separations and MS data acquisition. Another important demand for off-line analysis arises from the need for multiple parallel on-chip separations in high-throughput microfluidic systems, in which simultaneous ESI-MS from each separation channel is not feasible. MALDI-MS is a powerful analytical technique commonly employed for off-line MS analysis following capillary separations [27–29]. Although it is a serial process, the high duty cycle of MALDI-MS analysis enables high-throughput for large numbers of samples deposited on a single target plate.

Preparation of MALDI targets is often carried out by the dried-droplet method, in which spotting of an aliquot containing a mixture of sample and matrix solution is followed by air-drying of the deposited spot [30]. While spotting sample from microfluidic chips may be feasible, the quality of MALDI data is highly dependent on the way analyte is prepared on the target plate [31]. Liquid-phase deposition methods including dried-droplet [30], fast solvent evaporation [32], sandwich [33], and two-layer [34] preparation tend to suffer from poor homogeneity of crystallized sample, since matrix and analyte tend to partition during the solvent evaporation process [32, 35], resulting in significant variations in mass resolution, intensity, and selectivity, and preventing meaningful quantitative analysis [30, 36, 37].

As an alternative to mechanical pipetting or spotting, a number of studies have investigated the use of electrospray deposition of analytes onto MALDI targets followed by MALDI-MS analysis. Capillary electrospray has been reported for MALDI target preparation following CE [29, 38], RP-LC [39], and size-exclusion chromatography [28]. As a sample preparation technique, electrospray has been shown to yield uniform films for plasma desorption MS [40–42], and secondary ion MS [43]. A number of studies have shown that electrospray deposition can markedly improve the homogeneity of sample on the MALDI target surface by reducing segregation of matrix/analyte components, leading to greatly enhanced repeatability [35, 44–47], thereby enabling improved quantitative analysis [35, 46]. Furthermore, electrospray deposition has been shown to significantly improve the precision of molecular mass measurements during MALDI-MS [47].

While interfacing microfluidic systems with MALDI-MS has not been explored as extensively as ESI-MS, several examples exist. Ekstrom *et al.* [48, 49] used a microchip-based immobilized enzyme reactor for protein digestion followed by a microfabricated piezoelectric flow-through dispenser to deposit the protein digest onto a high-density nanovial MALDI target plate. Gustafsson *et al.* [50] described the use of centrifugal forces in a disk format to elute protein digest from a microfluidic RP-LC column to an on-chip MALDI target, with sub-fmol mass sensitivity achieved. Murray and coworkers [51] have also described the coupling of microfluidic chips to a continuous-flow MALDI system using a frit-based elution method.

Here we investigate electrospray deposition for MALDI-MS sample preparation as a new approach for efficiently coupling microfluidic systems to MS. Electrospray deposition provides several advantages over other approaches to chip-based MALDI interfaces such as piezoelectric [48, 49] and centrifugal deposition [50]. It does not require the addition of relatively complex external microfabricated elements, as in the case of piezoelectric deposition, and unlike centrifugal deposition it is a generic method which can be applied to a wide range of microfluidic systems. Electrospray deposition also provides excellent uniformity in on-target sample morphology and distribution, while also offering the potential to enable simultaneous deposition from high-density arrays of electrospray tips on a single chip with no cross talk for multiplexed analysis.

6.2 Materials and methods

6.2.1 Electrospray chip fabrication

Electrospray emitters were integrated into polymer microfluidic chips using a recently reported method in which a porous hydrophobic membrane is bonded to the channel exit [25]. Briefly, planar microchannels were fabricated in a polycarbonate (PC) substrate using hot embossing with a silicon template patterned by bulk Si micromachining. The resulting open microfluidic channels with trapezoidal cross section (122 μm wide at the top, 80 μm wide at the bottom, 30 μm deep, 4.5 cm long) were enclosed by thermally bonding with a blank PC substrate with reservoir holes predrilled for electrical and fluidic access at one end of each channel. Capillary connections (Nanoport, Upchurch Scientific, Oak Harbor, WA, USA) were bonded *via* epoxy on top of the reservoir to provide liquid supply from a syringe pump. A platinum electrode contacting the liquid through a capillary T-junction provides the required electrical contact for electrospray deposition. The resulting chip is cut to expose the microchannels at their ends opposite the reservoirs using a CNC milling system. A 50 μm thick PTFE membrane with 70% porosity and 0.22 μm average pore size (GE Osmonics, Minnetonka, MN, USA) is thermally bonded to the exposed microchannel outlets to provide a hydrophobic surface which prevents lateral spreading of liquid exiting the microchannels during electrospray.

In the present study, all tests were performed using virgin electrospray chips. The danger of cross contamination due to residual sample molecules bound to the microchannel walls or hydrophobic membrane surface dictates the use of disposable single-use chips for practical applications.

6.2.2 Instrumentation

A schematic of the experimental apparatus is shown in Fig. 1. The microfluidic chip was fixed on an insulated plate mounted on a three-axis translational stage for alignment of the emitter tips to a stainless steel MALDI target (MTP 384 target plate, Bruker Daltonics, Bremen, Germany). The *x-y-z* stage allows precise adjustment, accurate to 10 μm, between the spray tip and grounded MALDI target. A 500 μL gastight syringe driven by a syringe pump (PHD2000; Harvard Apparatus, Holliston, MA, USA) delivered analyte solution through a capillary T-junction to a Nanoport fitting (Upchurch Scientific) interfaced to an on-chip reservoir. The reservoir was located 4.5 cm from the channel exit at the electrospray tip. A high voltage power supply (CZE 1000R; Spellman High Voltage Electronics, Plainview, NY, USA) was used to apply an electrical bias to the analyte solution through a platinum wire inserted into the T-junction located approximately 3 cm from the reservoir. A fused-silica capillary with 100 μm ID (Polymicro Technologies, Phoenix, AZ, USA) was used throughout the setup. Total ion current generated from the spray between the tip and MALDI target was determined using a multimeter (34401A, Agilent, Palo Alto, CA, USA) by measuring the voltage across a 100 kΩ resistor connecting the MALDI target to a common ground. A potential ranging from 2 to 3 kV was applied to the analyte solution to achieve electrospray, with a stable total ion current between 42 and 55 nA maintained for all tests.

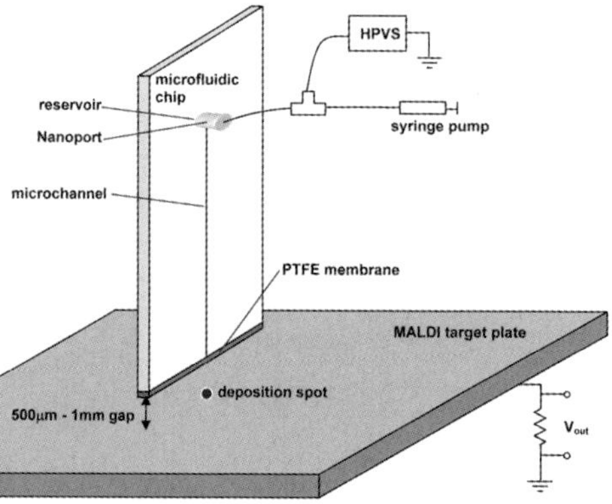

Fig. 1 Schematic of experimental apparatus for electrospray deposition interface between a polymer microfluidic chip and MALDI target. A three-axis micrometer stage (not shown) was employed for precise control over the electrospray gap.

6.2.3
Reagents

Peptides (angiotensin I, MW 1.29 kDa; bradykinin, MW 1.24 kDa; and fibrinopeptide, MW 1.72 kDa), proteins (BSA, MW 66.0 kDa; bovine heart cytochrome *c*, MW 12.3 kDa; and ovalbumin, MW 45 kDa), α-cyano-4-hydroxycinnamic acid (CHCA) matrix solution, and all other chemical reagents (methanol, ACN, acetic acid, and TFA) were purchased from Sigma-Aldrich (St. Louis, MO, USA).

6.2.4
Sample preparation and MALDI-MS analysis

Matrix solution was added after sample deposition using a custom capillary spotting tool following the underlayer target preparation method, in which analyte is pipetted onto the target before addition of the matrix solution [52]. This approach is well suited for microfluidic applications, since it eliminates the need to premix matrix and analyte solutions prior to deposition which would lead to increased complexity of the microfluidic chip, and would contribute to band-broadening for applications involving sample separation prior to electrospray deposition. A 52 μm ID, 360 μm OD fused-silica capillary (Polymicro Technologies) was secured on a union (Upchurch Scientific) mounted on a three-axis stage. The exit end of the capillary tubing was positioned at an angle of approximately 60° to the target surface, and the exit end was aligned to the electrospray deposition spot using a stereomicroscope. Matrix solution was delivered to the deposited spots at a flow rate of 70 nL/min. The matrix deposition time was selected in order to achieve an analyte:matrix ratio of 1:1 by volume. The deposited spots were allowed to dry at room temperature. No obvious spreading of the matrix solution beyond the initial droplet diameter was observed during this process.

MALDI-MS measurements were performed with a Bruker Autoflex MALDI-TOF mass spectrometer (Bruker Daltonics) equipped with a 337 nm nitrogen laser and operated in the linear mode. Each mass spectrum was produced by averaging 30 laser shots.

6.3
Results and discussion

During microfluidic electrospray deposition, the analyte is sprayed from the surface of the porous hydrophobic PTFE membrane along the exit face of the polymer microfluidic chip, resulting in a circular deposition pattern on the stainless steel MALDI target plate. Stability of the electrospray current is dependent on a number of factors, including bias voltage, counterelectrode spacing, flow rate, and solvent conditions, as described previously for the microfluidic membrane-based electrospray emitters [25]. The goal of the present study is not to detail the optimal electrospray conditions, but to evaluate the performance of the deposition process, deposited sample morphology, and MALDI-MS analysis under a set of nominal electrospray conditions. All analyses presented in the following were performed using a buffer solution of 50% deionized (DI) water, 49% methanol, and 1% acetic acid.

6.3.1 Deposited film morphology

Sample preparation for MALDI-MS plays a critical role in determining the sensitivity and quality of the mass spectral data [31]. Using angiotensin as a sample peptide, six different matrix solutions were evaluated to determine a suitable matrix for further characterization. Solutions were prepared with CHCA at 10 μg/μL in an organic solvent solution with an acidic component. Variations in the primary organic solvent component (ACN or methanol), solvent content (70 or 50%), and acidic component (acetic acid or TFA) were considered.

Observation of the deposited film morphology was performed using scanning electron microscopy. Overall, matrix solutions with higher organic solvent content resulted in greater inhomogeneity in the resulting peptide/matrix film, presumably due to enhanced resolvation and resulting segregation of deposited peptides. Compared to methanol as the solvent, ACN was generally found to result in more homogeneous films. In addition, matrix solutions with higher solvent content exhibited significant interference peaks which were not present with the lower solvent content compositions. The best matrix solution evaluated was 50% ACN, 40% water, 10% acetic acid, which provided consistently uniform films with nearly constant film morphology across the deposited spot, and produced the smallest deposited spot diameters with no observed spreading of analyte from the initial as-deposited diameter, suggesting that this matrix produces the highest on-target concentration of analyte. For these reasons, this matrix was used exclusively for further testing.

Electron micrographs reveal a significant difference in crystallization between the electrosprayed sample and mechanically spotted sample, as shown in Fig. 2. In both cases, 10.5 pmol of angiotensin was deposited onto the MALDI target, followed by spotting 3.5 μL of matrix solution. Compared to the electrospray-deposited peptides, which exhibit excellent uniformity and an average crystal size of several microns, the spotted sample shows significantly larger individual crystals, about 25 μm, and significant variation in crystal distribution across the deposition spot. While it is well-known that components of the matrix/analyte solution tend to segregate as the solvent evaporates, no significant segregation was observed during the tests upon addition of matrix solution to the electrospray-deposited analyte. A possible explanation for this observation is that crystallization of the evenly deposited analyte films occurs without substantial resolvation in the matrix solution, thereby preventing any diffusion and segregation of the mixture components as observed in traditional dried-droplet sample preparation. In contrast, previous reports of electrospray deposition for MALDI sample preparation employ codeposition of sample and matrix solution to achieve desolvation prior to the matrix/analyte reaching the MALDI target [35, 47, 53].

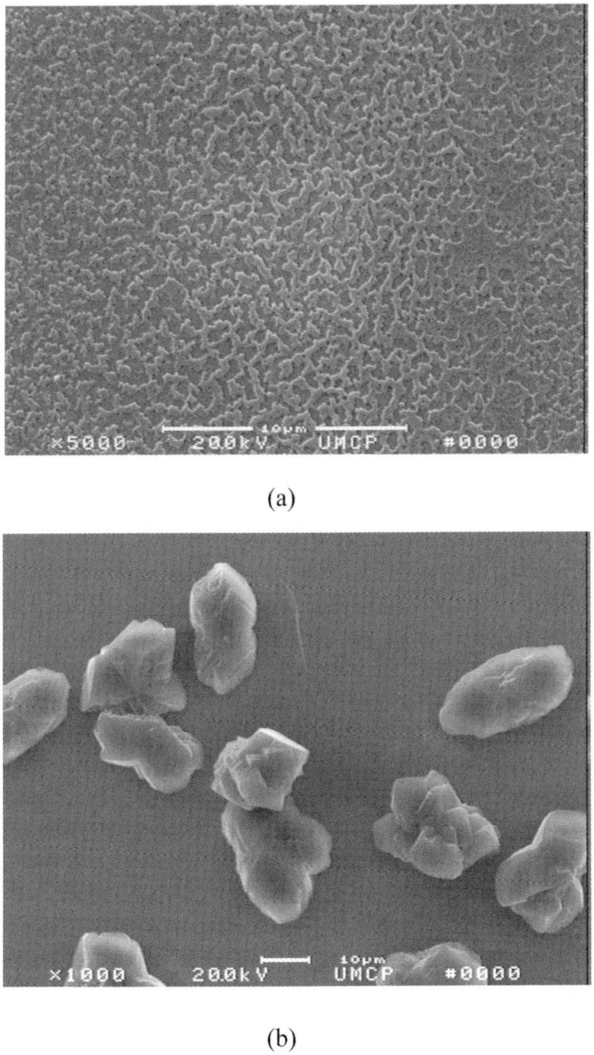

Fig. 2 Scanning electron micrographs of 10.5 pmol angiotensin deposited onto an uncoated MALDI target by (a) electrospray deposition and (b) mechanical spotting, followed by addition of CHCA matrix solution.

A comparison of signal reproducibility within the sample was examined by plotting analyte ion signal intensity against laser position for both the electrospray-deposited and spotted sample, with the results shown in Fig. 3. Each spot was prepared with a total of 5 pmol angiotensin followed by the addition of 2 μL matrix solution. At this high loading level, the electrosprayed sample exhibits nearly constant analyte ion signal intensity across the deposition spot, while the spotted sample shows up to a tenfold variation in signal intensity. Sample-to-sample reproducibility was monitored by measuring the variations in the analyte ion peak intensity on five different deposition spots prepared by electrospray deposition and the mechanical spotting method. The RSD for the electrosprayed sample was found to be 1.4%,

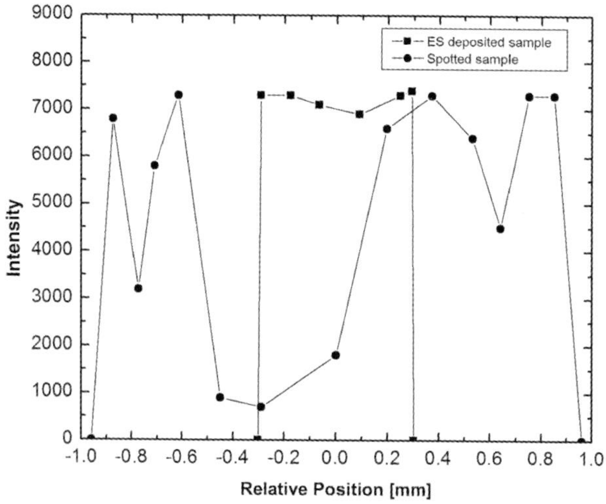

Fig. 3 MALDI peak intensity profiles across 5 pmol angiotensin spots deposited by both electrospray and mechanical spotting.

significantly lower than that for the spotted sample at 49%. The results demonstrate that microchip electrospray deposition gives marked improvements in sample-to-sample reproducibility over mechanical spotting. This observation is consistent with previous studies which have reported greatly enhanced reproducibility from capillary-based electrospray preparation of MALDI targets [35, 44–47].

6.3.2
Chip-to-target spacing

In the above comparison between electrospray deposition and mechanical spotting, the spotted sample was found to wet the surface and spread far beyond the local deposition region (approximately defined by the elution capillary outer diameter of 360 μm) with a measured spot diameter of nearly 2 mm. In contrast, the electrospray-deposited sample was constrained to a diameter of 500–600 μm for the 1 mm chip-to-target spacing. Since a smaller spot size implies increased analyte concentration, and thus improved signal intensity for low-abundance analytes, this represents a significant benefit for the electrospray deposition approach.

Chip-to-target spacing is expected to affect the deposited spot size on the MALDI target. Spot size measurements were performed by evaluating angiotensin ion signal intensity along the diameter of each spot, and calculating the diameter limits when the MALDI-MS signal dropped to four times the noise level. As expected, smaller chip-to-target spacing was found to result in a smaller deposition spot size. The relationship between spacing and spot diameter is nearly linear over the tested range as shown in Fig. 4, from a diameter of 545 μm at 1.1 mm spacing to 170 μm at 0.45 mm spacing. For the experimental microfluidic apparatus used for this test, electrical discharge was often observed for spacings below 0.45 mm.

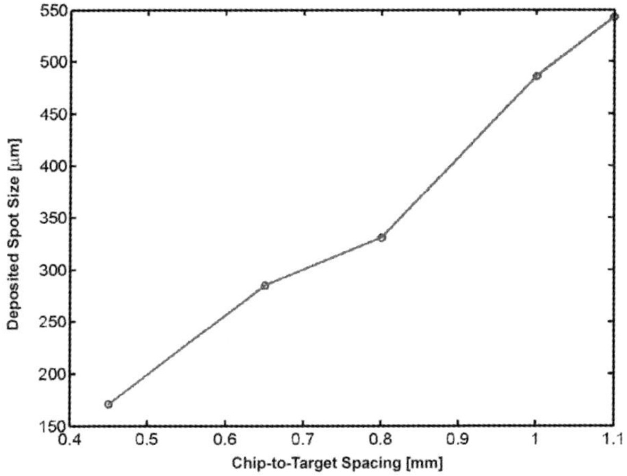

Fig. 4 Relationship between chip-to-target spacing and resulting angiotensin spot size on the MALDI target. Spot diameters were measured by determining the limits when the MALDI-MS signal dropped to four times the noise level.

6.3.3 Peptide concentration

To evaluate the effects of local peptide concentration on ion signal intensity, the total amount of angiotensin deposited by electrospray deposition from the microfluidic chip was varied, followed by MALDI-MS analysis. Agiotensin was diluted to concentrations from 50 down to 0.05 pmol/µL, and deposited at a flow rate of 70 nL/min for 1 min. The corresponding amount of angiotensin deposited on the target ranges from 3.5 pmol down to 3.5 fmol. As shown in Fig. 5a, ion intensity remains relatively constant down to a concentration of 2.5 pmol/µL (total peptide loading of 175 fmol), beyond which the average signal begins to drop. The relatively constant intensity at higher loadings is due to the formation of multilayer peptide films on the MALDI target, resulting in saturation of the MALDI-MS signal. For the 3.5 fmol spectrum, the average S/N was between 4 and 5, putting this measurement near the LOD.

As expected, increasing variability in signal intensity was observed as sample loading was reduced, due to the formation of discontinuous sample films on the target surface. To validate this assumption, consider that 175 fmol was the lowest loading for which constant intensity was measured. Given an effective molecular diameter around 1.2 nm for angiotensin, and assuming this diameter expands to twice this value upon impacting the target plate, the target area covered by angiotensin may be calculated to be 0.24 mm^2, which is nearly equal to the total area of the deposited spot (0.28 mm^2). This result suggests that 175 fmol loading produces a nearly continuous monolayer on the target surface, while lower loading levels will produce regions with no sample, and MALDI-MS spectra with increasing variations in measured intensity.

A dilution study was also performed using the mechanical spotting method, with experimental conditions identical to the previous morphology study but with angiotensin concentration ranging from 0.05 to 50 pmol/µL. In this case, no ion

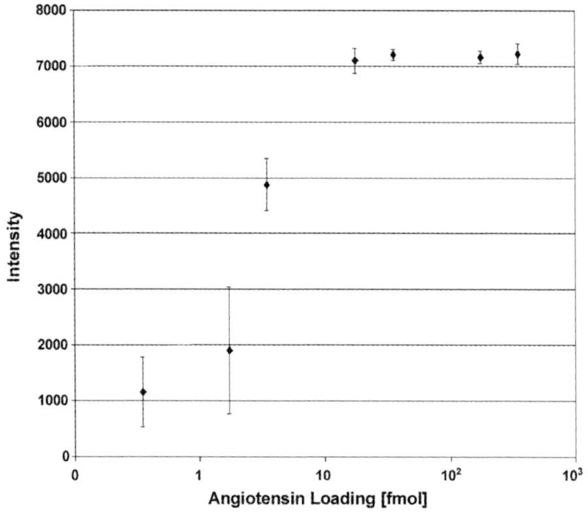

(a)

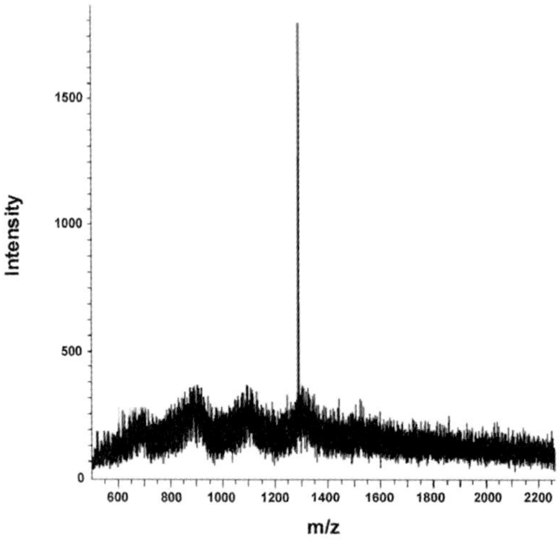

(b)

Fig. 5 (a) MALDI-MS peak intensities measured for varying amounts of deposited angiotensin and (b) measured MALDI-MS spectrum for 3.5 fmol deposition.

signal for angiotensin could be measured following repeated laser probing of the sample for depositions of 35 fmol and below. Thus, the electrospray deposition method enabled the measurement of angiotensin for an initial sample concentration at least ten times lower than the mechanical spotting approach. Despite the demonstrated improvement over mechanical spotting, the measured detection limit is substantially larger than recently reported MALDI-MS detection limits between 0.7 and 25 amol for several neuropeptides deposited by electrospray from silica capillaries [53]. Lower detection limits from the microchip format should be feasible by reducing the chip-to-target spacing in order to increase on-target concentration of deposited analyte.

6.3.4
Protein analysis

The analysis of intact proteins deposited from the electrospray chip was also investigated. Stable operation of the microfluidic electrospray emitters requires the use of a PTFE membrane as a hydrophobic layer at the channel exit [25] in order to prevent lateral spreading of the emitted fluid. To explore whether proteins may bind to the hydrophobic membrane surface, BSA was loaded on an electrospray chip at a concentration of 1 mg/mL and deposited onto the MALDI target at a flow rate of 70 nL/min for 120 min. The chip was moved to a new deposition location in 5 min increments, for a total of 350 ng BSA deposited *per* spot. For the first 5 min following elution of BSA from the electrospray tip, no BSA signal was observed. At 10 min the BSA peak was clearly visible, and from 20 to 120 min the peak intensity saturated with an average value approximately 20% higher than the initial value at 10 min, indicating that the majority of available binding sites became saturated between the first 5 and 10 min of deposition. A MALDI target was also prepared by deposition of 350 ng BSA using the mechanical spotting method. Typical spectra from the spotted target and the electrospray target exhibited comparable maximum signal intensity at the stated loading level.

While the hydrophobic nature of the PTFE membrane remains a potential source of sample loss due to protein adsorption (between 350 and 700 ng for BSA based on the measured binding capacity), it also makes the electrospray membrane an ideal support for the adsorption of proteolytic enzymes such as trypsin, enabling in-line protein digestion during the electrospray deposition process. The concept is similar to previous work in which a hydrophobic PVDF membrane sandwiched between two PDMS-based microfluidic channels was successfully used as a trypsin support to realize a miniature membrane reactor for in-line protein digestion [54]. To prepare the PTFE membranes in the present study, 2 mg/mL trypsin in 10 mM ammonium acetate buffer was delivered through the microfluidic chip at a flow rate of 70 nL/min for 1 h, followed by flushing with 10 mM ammonium acetate buffer for 20 min to remove any unbound trypsin. Denatured and reduced cytochrome c and ovalbumin at 1 mg/mL were pumped through individually prepared electrospray chips at 70 nL/min, and deposited onto a MALDI target for 1 min. The electrospray deposition with in-line trypsin digestion was performed at room temperature. The resulting MALDI-MS spectra are shown in Fig. 6. The total protein digest deposited in each target spot is 70 ng, or about 5.7 pmol cytochrome *c* and 1.6 pmol ovalbumin. Peptide coverage for both proteins over three runs was between 22 and 41%, with between 6 and 12 peptide matches. In both cases, no signal was observed for the intact proteins (data not shown), indicating that complete digestion was achieved during the electrospray process. Furthermore, good digestion efficiency appears to be maintained in the presence of the high electric fields used during the electrospray process. No significant changes in the spectra were observed over multiple tests during a 3 h period of continuous electrospray deposition.

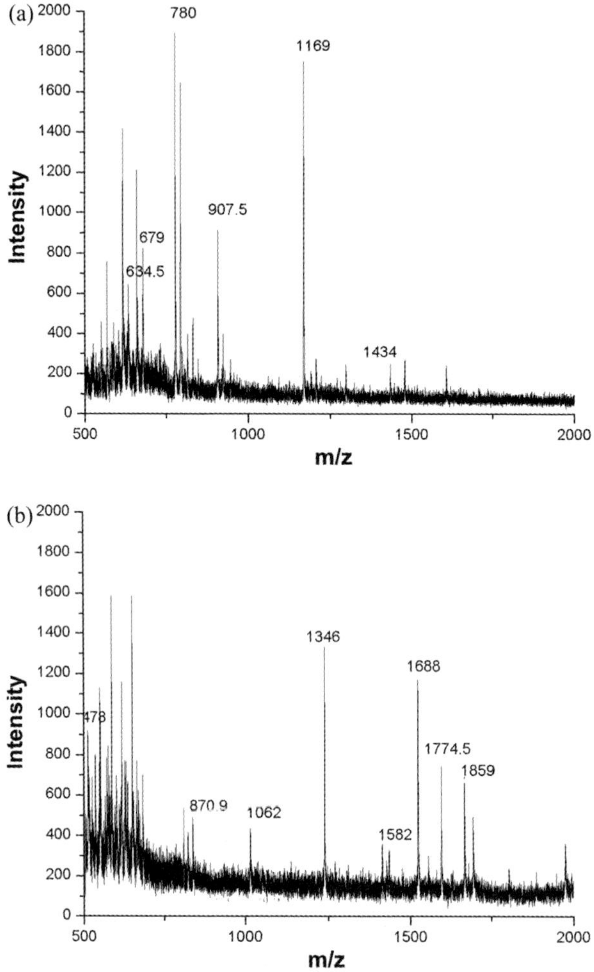

Fig. 6 MALDI-MS spectra for in-line tryptic digests of (a) cytochrome *c* and (b) ovalbumin. Approximately 5.7 pmol cytochrome *c* and 1.6 pmol ovalbumin were deposited in each spot, with peptide coverage for both proteins between 22 and 41% over three independent runs.

To ensure that the observed digestion was not the result of residual trypsin bound to the PC microchannel, a 100 μm wide, 50 μm deep, 5 cm long channel was filled with 2 mg/mL trypsin. The solution was allowed to stand in the channel for 30 min. The channel was flushed with water and filled with 0.2 mg/mL denatured cytochrome *c* in electrospray buffer. The solution was allowed to stand in the channel for 10 min, and was then collected and analyzed by LC-MS/MS using an Ultimate HPLC (LC Packings, Sunnyvale, CA, USA) connected to a 50 μm diameter, 10 cm long pulled tip C18 column. A linear gradient was delivered over 45 min from 5 to 65% ACN containing 0.02% heptafluorobutyric acid and 0.02% formic acid at a flow rate of 200 nL/min. The eluants were monitored using a qTOF micro (Waters, Milford, MA, USA) mass spectrometer and mass spectra were acquired from 500 to 1900 *m/z* for 1 s, followed by three data-dependent MS/MS scans from 50 to 1900 *m/z* for 3 s each. The resulting spectra showed only intact proteins, with no cytochrome *c* peptides observed, indicating that either the amount of channel-bound trypsin is negligible, or diffusion kinetics within the channel severely limits the degree of digestion.

6.3.5 Multiplexed electrospray deposition

As revealed in Fig. 4, a chip-to-target spacing of 0.45 mm resulted in a deposition spot diameter of 170 μm. This deposition resolution suggests the potential for simultaneously depositing analyte from multiple closely spaced electrospray tips without contamination between adjacent spots. To explore this concept, an electrospray chip containing three parallel microchannels was fabricated. Each channel possessed cross-sectional dimensions 100 μm wide at the top, 50 μm wide at the bottom, and 35 μm deep. Spacing between the edges of each channel was 150 μm. Peptide solutions of 50 pmol/μL bradykinin, angiotensin, or fibrinopeptide were introduced into specific channels as depicted in Fig. 7. Individual flow rates for each channel were adjusted to initiate stable electrospray. Up to 20% variation in flow rate was required to maintain stable electrospray from all channels simultaneously. An average flow rate of 70 nL/min was applied for 3 min with a tip-to-target spacing of 1 mm. Following deposition, a 200 nL droplet of matrix solution was added to each spot individually using the capillary spotter to prevent diffusion of peptides between spots.

The observed variation in required flow rate is believed to be due to differences in the effective electrical and flow resistance imposed by the bonded PTFE films, which can exhibit different amounts of pore collapse during the bonding process. Improvements in uniformity are anticipated through the use of engineered membranes with higher compressive strength, such as hydrophobic glass, silicon, or silicon carbide.

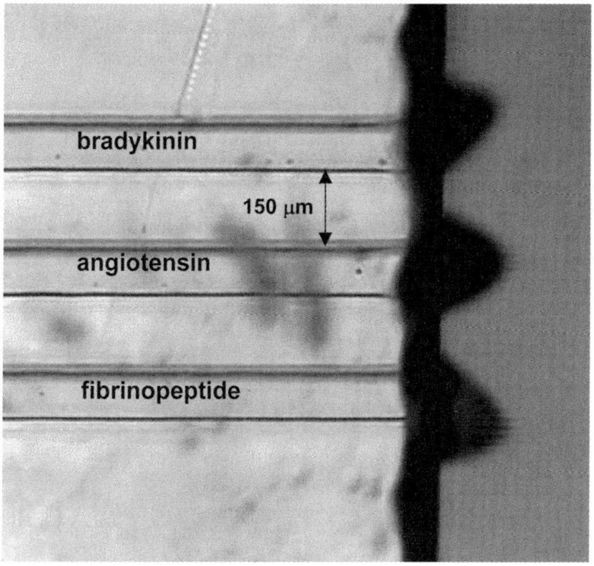

Fig. 7 Optical micrograph showing simultaneous ESI from three parallel 100 μm wide microchannels spaced 150 μm apart. Due to the small channel spacing, electrostatic interactions are sufficiently strong to result in distortion of the Taylor cones.

As shown in Fig. 7, charge interactions between fluid eluted from each electrospray tip result in a slight distortion of the two outer Taylor cones. Because the lateral electrostatic forces are balanced, the center cone remains unaffected. Although not directly observed, it is assumed that the ionic streams emitted from the cone tips interact in a similar manner during the deposition process. This is supported by the observation that the centers of the deposited peptide spots are spaced approximately 1 mm apart, substantially larger than the initial channel spacing. Despite the observed distortions in the electrospray deposition process, the resulting peptide spots remain circular with a diameter of around 500 μm. Further experimental work will be needed to determine if the observed elution stream deflections may be reduced or eliminated by increasing the number of electrospray tips or by reducing the chip-to-target spacing.

MALDI-MS analysis was performed by sampling along a line through the centers of the spots. The resulting measurements of ion intensity are shown in Fig. 8. The mass spectra across each spot (inset) reveal no observable contamination between adjacent peptides, indicating that no significant cross talk occurs during deposition or matrix crystallization. In the regions between adjacent spots, no signal above noise is detected for any of the deposited peptides.

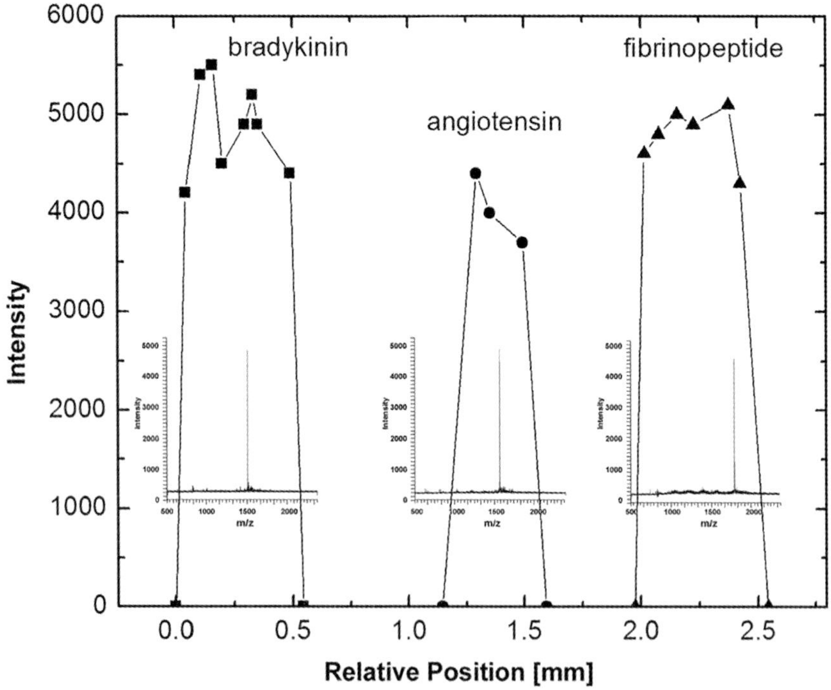

Fig. 8 Ion signal intensities measured across peptide spots deposited from the multichannel electrospray emitter array. Sample spectra for each peptide are shown inset.

6.4 Concluding remarks

The combination of electrospray target preparation with off-line MALDI-MS is a new approach for efficiently integrating multiplexed microfluidic systems with MS. Unlike direct microfluidic ESI-MS, which requires serial coupling of individual separation channels with MS, electrospray deposition followed by MALDI-MS offers a path toward true multiplexing of on-chip peptide and protein separations for high-throughput analysis. While several other approaches to interfacing microfluidics with MALDI-MS have been investigated, one advantage of the electrospray deposition technique is that the morphology of deposited sample exhibits significantly improved homogeneity compared to mechanical sample spotting. Although this observation has been reported for capillary-based electrospray deposition of premixed matrix and sample solutions, the present work demonstrates excellent MALDI-MS signal with matrix solution added only after electrospray deposition. This feature is significant for microfluidic analytical platforms, since it can eliminate the broadening of narrow sample bands which would otherwise result from the on-chip mixing of matrix and sample. Similarly, the demonstrated ability to perform real-time tryptic digestion, using the electrospray membrane as an in-line proteolytic reactor, eliminates the need for on-target digestion, which can lead to undesirable inhomogeneities and broadening of the deposited sample spots. The ability to maintain small spot diameters further enables high on-target sample concentrations for improved MALDI-MS detection of low-abundance analytes. Additionally, since the deposition is not discretized, an arbitrarily small volume of analyte solution can be eluted from the microfluidic chip onto each target spot, providing improved compatibility with high-resolution separations where mixing of separated species is undesirable prior to MALDI-MS analysis.

The authors acknowledge support from the National Institutes of Health through NCI Grant No. CA094400, the Defense Advanced Research Projects Agency through Contract No. DAAH01-03-C-R182, and the US Army through Contract No. W911SR-04-C-0007.

6.5 References

[1] Xue, Q., Foret, F., Dunayevskiy, Y. M., Zavracky, P. M., McGruer, N. E., Karger, B. L., *Anal. Chem.* 1997, *69*, 426–430.

[2] Ramsey, R. S., Ramsey, J. M., *Anal. Chem.* 1997, *69*, 1174–1178.

[3] Shultz, G. A., Corso, T. N., Prosser, S. J., Zhang, S., *Anal. Chem.* 2000, *72*, 4058–4063.

[4] Lazar, I. M., Ramsey, R. S., Sundberg, S., Ramsey, J. M., *Anal. Chem.* 1999, *71*, 3627–3631.

[5] Chan, J. H., Timperman, A. T., Aebersold, R., *Anal. Chem.* 1999, *71*, 4437–4444.

[6] Li, J., Kelly, J. F., Chernushevich, I., Harrison, D. J., Thibault, P., *Anal. Chem.* 2000, *72*, 599–609.

[7] Jiang, Y., Wang, P., Locascio, L. E., Lee, C. S., *Anal. Chem.* 2001, *73*, 2048–2053.

[8] Figeys, D., Aebersold, R., *Anal. Chem.* 1998, *70*, 3721–3727.

[9] Xu, N., Lin, Y. H., Hofstadler, S. A., Matson, D., Call, C. J., Smith, R. D., *Anal. Chem.* 1998, *70*, 3553–3556.

[10] Meng, Z., Qi, S., Soper, S. A., Limbach, P. A., *Anal. Chem.* 2001, *73*, 1286–1291.

[11] Kameoka, J., Craighead, H. G., Zhang, H., Henion, J., *Anal. Chem.* 2001, *73*, 1935–1941.

[12] Bings, N. H., Wang, C., Skinner, C. D., Colyer, C. L., Thibault, P., Harrison, D. J., *Anal. Chem.* 1999, *71*, 3292–3296.

[13] Wen, J., Lin, Y., Xiang, F., Matson, D. W., Udseth, H. R., Smith, R. D., *Electrophoresis* 2000, *21*, 191–197.

[14] Licklider, L., Wang, X., Desai, A., Tai, Y., Lee, T. D., *Anal. Chem.* 2000, *72*, 3627–3631.

[15] Rohner, T. C., Rossier, J. S., Girault, H. H., *Anal. Chem.* 2001, *73*, 5353–5357.

[16] Yuan, C., Shiea, J., *Anal. Chem.* 2001, *73*, 1080–1083.

[17] Tang, K., Lin, Y., Matson, D. W., Kim, T., Smith, R. D., *Anal. Chem.* 2001, *73*, 1658–1663.

[18] Kim, J., Knapp, D. R., *J. Am. Soc. Mass Spectrom.* 2001, *12*, 463–469.

[19] Gobry, V., Van Oostrum, J., Martinelli, M., Rohner, T. C., Reymond, F., Rossier, J. S., Girault, H. H., *Proteomics* 2002, *2*, 405–412.

[20] Arscott, S., LeGac, S., Druon, C., Tabourier, P., Rolando, C., *Sens. Actuators B* 2004, *98*, 140–147.

[21] Muck, A., Svatos, A., *Rapid Commun. Mass Spectrom.* 2004, *18*, 1459–1464.

[22] Fortier, M.-H., Bonneil, E., Goodley, P., Thibault, P., *Anal. Chem.* 2005, *77*, 1631–1640.

[23] Kameoka, J., Orth, R., Ilic, B., Czaplewski, D., Wachs, T., Craighead, H. G., *Anal. Chem.* 2002, *74*, 5897–5901.

[24] Hsu, F.-L., Chen, C.-H., Yuan, C.-H., Shiea, J., *Anal. Chem.* 2003, *75*, 2493–2498.

[25] Wang, Y.-X., Cooper, J. W., Lee, C. S., DeVoe, D. L., *Lab Chip* 2004, *4*, 263–267.

[26] Koerner, T., Turck, K., Brown, L., Oleschuk, R. D., *Anal. Chem.* 2004, *76*, 6456–6460.

[27] Rejtar, T., Hu, P., Juhasz, P., Campbell, J. M., Vestal, M. L., Preisler, J., Karger, B. L., *J. Proteome Res.* 2002, *1*, 171–179.

[28] Lou, X. W., van Dongen, J. L., *J. Mass Spectrom.* 2000, *35*, 1308–1312.

[29] McLeod, G. S., Axelsson, J., Self, R., Derrick, P., *J. Rapid Commun. Mass Spectrom.* 1997, *11*, 1785–1793.

[30] Karas, M., Hillenkamp, F., *Anal. Chem.* 1988, *60*, 2299–2301.

[31] Cohen, S. L., Chait, B., *Anal. Chem.* 1996, *68*, 31–37.

[32] Vorm, O., Roepstorff, P., Mann, M., *Anal. Chem.* 1994, *66*, 3281–3287.

[33] Li, L., Golding, R. E., Whittal, R. M., *J. Am. Chem. Soc.* 1996, *118*, 11662–11663.

[34] Dai, Y. Q., Whittal, R. M., Li, L., *Anal. Chem.* 1996, *68*, 2494–2500.

[35] Hensel, R. R., King, R. C., Owens, K. G., *Rapid Commun. Mass Spectrom.* 1997, *11*, 1785–1793.

[36] Strupat, K., Karas, M., Hillenkamp, F., *Int. J. Mass Spectrom. Ion Processes* 1991, *111*, 89–102.

[37] Amado, F. M. L., Domingues, P., Santana-Marques, M. G., Ferrer-Correla, A. J., Tomer, K. B., *Rapid Commun. Mass Spectrom.* 1997, *11*, 1347–1352.

[38] DeVault, G. L., Sepaniak, M. J., *Electrophoresis* 2000, *21*, 1320–1328.

[39] Wall, D. B., Berger, S. J., Finch, J. W., Cohen, S. A., Richardson, K., Chapman, R., Drabble, D., Brown, J., Gostick, D., *Electrophoresis* 2002, *23*, 3193–3204.

[40] Roepstorff, P., Nielsen, P. F., Sundqvist, B. U. R., Hakansson, P., Jonsson, G., *Int. J. Mass Spectrom. Ion Processes.* 1978, *78*, 229–236.

[41] McNeal, C. J., Macfarlane, R. D., Thurston, E. L., *Anal. Chem.* 1979, *51*, 2036–2039.

[42] Bondarenko, P. V., Zubarev, R. A., Knysh, A. N., Rozynov, B. V., *Biol. Mass Spectrom* 1992, *21*, 323–330.

[43] Standing, K. G., Beavis, R., Ens, W., Schueler, B., *Int. J. Mass Spectrom. Ion Phys.* 1983, *53*, 125–134.

[44] Axelsson, J., Hoberg, A. M., Waterson, C., Myatt, P., Shield, G. L., Varney, J., Haddleton, D. M., Derrick, P., *J. Rapid Commun. Mass Spectrom.* 1997, *11*, 209–213.

[45] Hanton, S. D., Clark, P. A. C., Owens, K. G., *J. Am. Soc. Mass Spectrom.* 1999, *10*, 104–111.

[46] Go, E. P., Shen, Z., Harris, K., Siuzdak, G., *Anal. Chem.* 2003, *75*, 5475–5479.

[47] Hanton, S. D., Hyder, I. Z., Stets, J. R., Owens, K. G., Blair, W. R., Guttman, C. M., Giuseppetti, A. A., *J. Am. Soc. Mass Spectrom.* 2004, *15*, 168–179.

[48] Ekström, S., Önnerfjord, P., Nilsson, J., Bengtsson, M., Laurell, T., Marko-Varga, G., *Anal. Chem.* 2000, *72*, 286–293.

[49] Ekström, S., Ericsson, D., Önnerfjord, P., Bengtsson, M., Nilsson, J., Marko-Varga, G., Laurell, T., *Anal. Chem.* 2001, *73*, 214–219.

[50] Gustafsson, M., Hirschberg, D., Palmberg, C., Jörnvall, H., Bergman, T., *Anal. Chem.* 2004, *76*, 245–250.

[51] Musyimi, H., Narcisse, D. A., Xhang, X., Soper, S. A., Murray, K. K., *Anal. Chem.* 2004, *76*, 5968–5973.

[52] Moseley, M. A. III, Sheeley, D. M., Blackburn, R. K., Johnson, R. A., Merrill, B. M., in: Larsen, B., McEwen, C. N. (Eds.), *Mass Spectrometry of Biological Materials*, 2nd Edn., Marcel Dekker, NY 1998, p. 162.

[53] Wei, H., Nolkrantz, K., Powell, D. H., Woods, J. H., Ko, M.-C., Kennedy, R. T., *Rapid Commun. Mass Spectrom.* 2004, *18*, 1193–1200.

[54] Gao, J., Xu, J., Locascio, L., Lee, C. S., *Anal. Chem.* 2001, *73*, 2648–2655.

7
Nanoliquid chromatography-mass spectrometry of oligosaccharides employing graphitized carbon chromatography on microchip with a high-accuracy mass analyzer*

Milady Niñonuevo, Hyunjoo An, Hongfeng Yin, Kevin Killeen, Rudi Grimm, Robert Ward, Bruce German, Carlito Lebrilla

The nanoLC separations of oligosaccharides using microchip-based columns are described. Mixtures of alditols from mucins and human milk are separated on graphitized carbon. The nanoLC-MS device showed high mass accuracy for the oligosaccharides ranging between 1 and 6 ppm on routine analyses. The high mass accuracy readily allowed identification of oligosaccharide peaks and the determination of their compositions. High retention time reproducibility was exhibited by the microchip LC. Little variation was observed for standard sample either alone or in a complex heterogeneous mixture. The nanoLC-MS exhibits excellent capabilities in profiling mixtures of oligosaccharides.

7.1
Introduction

Separation of the oligosaccharides are complicated by the number of isomers as well as the minor structural variations (*e.g.*, linkage and anomeric character) that produce little physical changes in structure. The traditional separation of oligosaccharides involves the use of amide columns [1]. They are generally useful for both neutral and anionic oligosaccharides. However, amide columns have severely limited lifetimes and their separating efficiencies degrade significantly over time. A more robust medium for separation is the RP column such as C18. These columns have been used to separate complex mixtures but have generally difficulty in separating complex oligosaccharide mixtures. A separating medium that is both effective for separating complex oligosaccharide mixtures and has great stability and long life is graphitized carbon chromatography (GCC) [2]. Based on our experience, GCC columns generally last significantly longer than amide columns,

* Originally published in Electrophoresis 2005, 26, 3641–3649

Microfluidic Applications in Biology. Edited by Niels Lion, Joël S. Rossier, and Hubert H. Girault

ISBN-10: 3-527-31761-9

while producing effective separation of both neutral and anionic oligosaccharides. There have been a number of recent reports showing the effectiveness of GCC separation of oligosaccharides including isomers of neutral and anionic species [3].

Nanoflow LC (nanoLC) has received considerable attention for the efficient separation of peptides. When coupled with MS and employing an RP it provides a method for the rapid identification of proteins based on peptide sequences.

NanoLC-MS of oligosaccharide has seen limited applications. Wuhrer *et al.* [4] used nanoLC with an amide column to separate *N*-glycans. A 300-nL/min flow rate was employed on 75 μm inner diameter columns. Each peak was accompanied by a shoulder that they attributed to possibly the α- and β-anomeric forms at the reducing end. The mass spectra yielded the sodium-coordinated species. Glycans in the low femtomole quantities were analyzed without derivatization. To account for the shifts in retention times, a dextran ladder was used as an internal standard. GCC columns have also been used in the microLC and nanoLC applications for the separation of oligosaccharides. Itoh *et al.* [5] used a microbore GCC column with 1 mm inner diameter column and a flow rate of 50 μL/min to separate *N*-linked oligosaccharides. The use of GCC allowed effective separation of mixtures of *N*-linked oligosaccharide with a triple-stage quadrupole mass spectrometer for mass analyses.

The nanoLC separation of oligosaccharides has recently been reported by Karlsson *et al.* [6]. With a column inner diameter of 150 mm and a flow rate of 600 nL/min they were able to separate a number of *N*- and *O*-linked oligosaccharides as well as neutral and anionic species of both. A 3-D quadrupole IT was used for online MS analyses of the oligosaccharides in the negative mode. By extrapolation, they determined that sensitivity could be as low as 1 fmol.

In this report, we describe the separation of oligosaccharides with an on-chip column, which is 50 mm (*L*) × 75 μm (*W*) × 50 μm (*D*) and packed with graphite material. The description of the microchip is provided in an earlier publication [8]. With a flow rate of 300 nL/min it provides an effective separation of oligosaccharides. The mass analyzer is an orthogonal TOF, which provides mass accuracy to less than 5 ppm under wide dynamic range. The combination of effective separation and accurate mass analyses provides a new tool for profiling complicated mixtures of oligosaccharides.

7.2 Materials and methods

7.2.1 Oligosaccharides

The model oligosaccharide (OS) maltohexaose was obtained from commercial sources (Sigma, St. Louis, MO, USA) and used without further purification. To produce the maltohexaose alditol, the oligosaccharides were reduced under NaOH/$NaBH_4$ using standard procedures. *O*-linked oligosaccharides were obtained from mucins using the procedures employed in earlier publications [3, 7]. This involved dialysis of the glycoproteins and release of the oligosaccharides using NaOH/

$NaBH_4$. For analyses, the samples were diluted to an approximate concentration of 10^{-6} M. The milk samples were obtained from the Mother's Milk Bank of San Jose, CA, USA and Austin, TX, USA through our collaborators in the Department of Food Science and Nutrition. Oligosaccharides were extracted with chloroform:methanol (3:1 v/v). The emulsion was centrifuged and the organic layer was discarded. The aqueous layer was collected and freeze-dried. The OS fraction was then reduced by alkaline sodium borohydride, desalted, and fractionated into neutral and anionic OS using graphitized carbon black cartridge-SPE.

7.2.2 NanoLC/MS analyses

All solvents were of the highest HPLC grade and used without further purification. For the sample loading, an Agilent 1100 pump (Agilent Technologies, Palo Alto, CA, USA) operating at 5 μL/min with water + 0.1% formic acid was connected to an Agilent 1100 μ-well plate autosampler. Injection volume was typically 1–5 μL. NanoLC separation was performed on a μ-fluidic HPLC chip made by a laser-ablated and laminated biocompatible polyimide film. The chip consists of an integrated sample loading structure, a packed LC separation column, and nanoelectrospray tip. It is hydraulically interfaced with LC pumps and an autosampler through a face-seal rotary valve [8]. An Agilent 1100 nL pump was used to deliver the LC gradient at 300 nL/min. The chip was interfaced to an Agilent TOF MS for online nanoESI. Both on-chip enrichment column and on-chip LC separation column were packed with graphitized carbon media obtained uncharacterized from Alltech (Deerfield, IL, USA). Separation was performed by a binary gradient consisting of two solutions (A) water + 0.1% formic acid and (B) 90% ACN/water + 0.1% formic acid. A 40 min gradient was ran from 2 to 42% B, and to 90% B at the completion of the run.

7.3 Results

7.3.1 Analyses of model oligosaccharides

The nanoLC chromatogram of maltohexaose, a glucose hexamer, on the GCC column is shown in Fig. 1. To effect this separation, approximately 100 fmol was injected into the column to obtain the chromatogram. The peak shape was ideal showing symmetry and a width at half-height of 20 s. The GCC chromatography showed significant improvements over the nanoLC employing amide columns. The satellite peaks that were reported by Wuhrer *et al.* [4] and by Karlsson *et al.* [6] that were attributed either to α- or β-anomers were not observed in the chromatogram, however this may be sample-dependent. The corresponding mass spectrum (Fig. 2) shows the quasimolecular ions as the proton-coordinated (*m*/*z* 991.330), the ammonium-coordinated species (*m*/*z* 1008.356), and the sodium-coordinated species (*m*/*z* 1013.312) (Tab. 1).

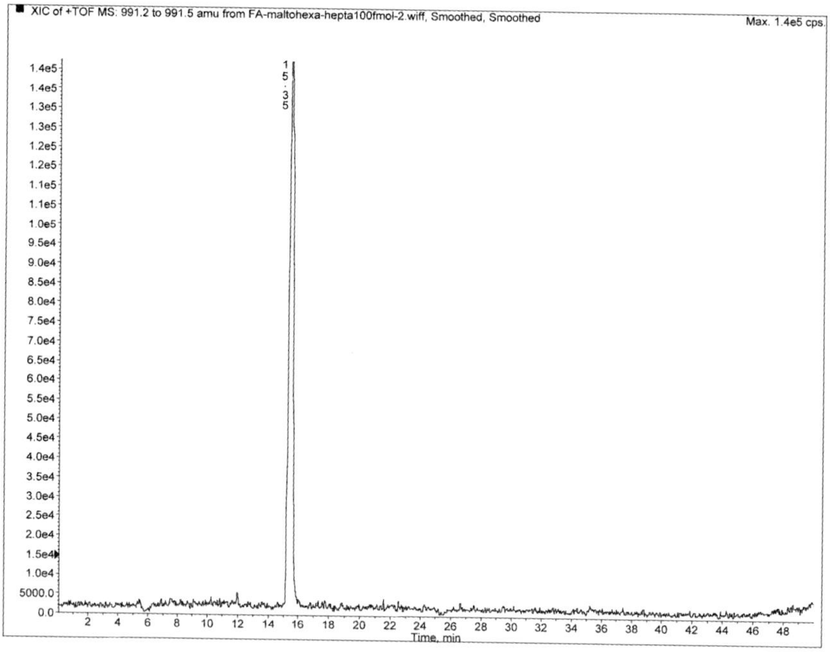

Fig. 1 NanoLC trace of maltohexaose standard using GCC on microchip. Approximately 100 fmol was used in the injection.

Often oligosaccharides are analyzed as reduced species due to the release procedure. *O*-Linked oligosaccharides are released primarily as the reduced alditol. Additionally, alditols have two mass units higher than the nonreduced residues, making it easier to distinguish the presence of the reducing end in MS/MS spectra. While MS/MS was not possible on this mass analyzer, chromatographic separation followed by in-source fragmentation can produce somewhat similar effects. This application and others including the implementation of chip LC with mass analyzers capable of online MS/MS analyses will be the subject of future studies. Reduction of maltohexaose yields the alditol, which was analyzed using the same conditions as that for the unreduced species. The separation showed the same quality peaks as that of the parent compound (RT = 15.31 min). The mass spectrum yielded the same quasimolecular ions corresponding to the protonated, ammonium-coordinated, and sodium-coordinated species. The summary of results for both maltohexaose standards including the measured masses and theoretical masses for the protonated and ammonium-coordinated species is shown in Tab. 1. The ammonium adduct ions were observed even when only formic acid was used as an LC mobile phase additive. The ammonium ions, however, became dominant when 5 mM ammonium acetate was used as LC mobile phase additive. The mass accuracy of all measured masses in these standards was found to be less than 6 ppm. The background phthalate ions (391.2842) were used as an internal reference mass. The drawback of such an approach is that the background ion is not consistently observed throughout the LC gradient and often varies between LC/MS runs.

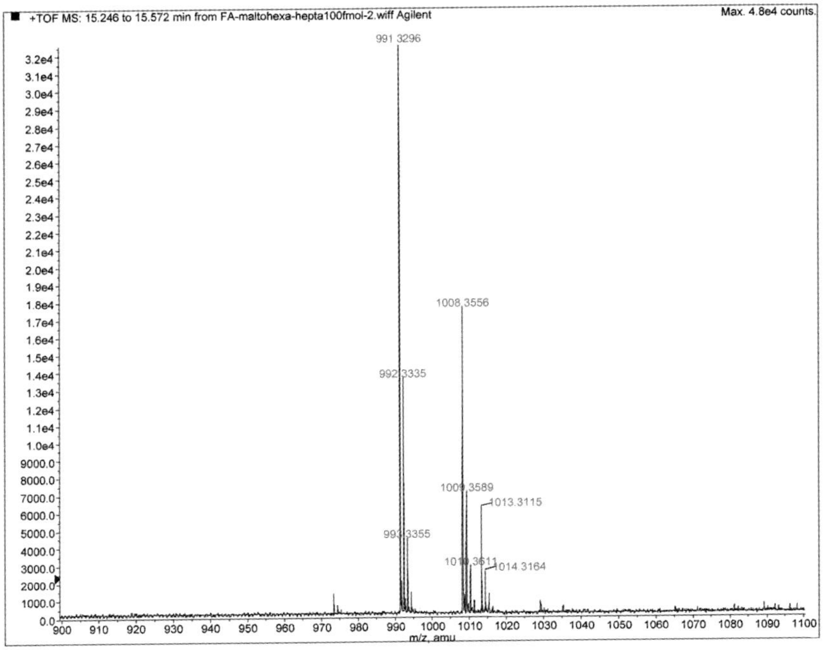

Fig. 2 Mass spectrum of maltohexaose obtained from nanoLC in Fig. 1. Three quasimolecular ion peaks corresponded to the protonated, ammoniated, and sodiated species.

Tab. 1 Quasimolecular ions obtained from maltohexaose and maltohexaose alditol

Maltohexaose	Theoretical mass	Observed mass	Mass error, ppm
Protonated	991.3347	991.3296	−5.2
Ammonium adduct	1008.3613	1008.3556	−5.7
Alditol form, protonated	993.3504	993.3454	−5.1
Alditol form, ammonium adduct	1010.3769	1010.3709	−6.0

Precise retention times along with accurate masses can be used to rapidly identify known oligosaccharides. One advantage of GCC over the amide column for separating oligosaccharide is the robustness of the GCC support. GCC columns under normal HPLC conditions can last several years without degrading based on our experience. Experiments involving multiple injections were performed to determine the variation in retention times. Five separate samples of maltohexaose alditol were injected into the column in the same day. For the maltohexaose alditol an average retention time of 16.05 min was obtained with an SD of 0.04, illustrating high reproducibility with a variation of less than 0.5%. Fig. 3 shows the overlay of the five

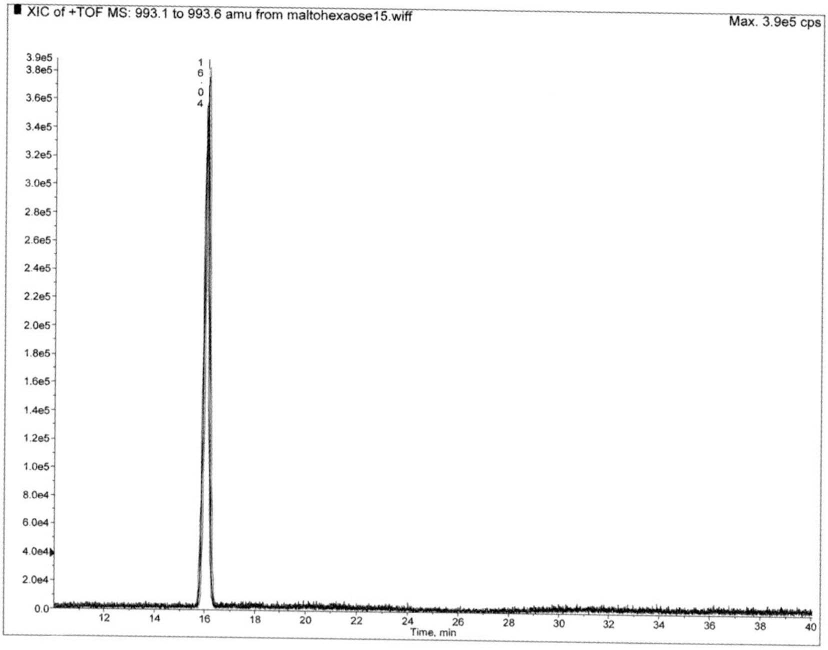

Fig. 3 Extracted ion chromatogram (EIC) consisting of five replicate LC runs (overlayed) of 100 fmol of maltohexaose alditol.

chromatograms representing an injection of 100 fmol of material. The peak area shows similarly small variations of $5.29 \pm 00.13 \times 10^6$ U with variations less than 5%.

The high detectability of oligosaccharides in this method is illustrated in Fig. 4a. With 10 fmol of material injected into the column, an S/N of 10 was obtained in the chromatogram. The mass spectrum showed the same S/N sign for the singly charged protonated species (Fig. 4b). The retention time of the 10 fmol sample (15.91 min) varied only slightly from the average of 16.05 min for the five runs with 100 fmol of samples. The detection limit was consistent with those reported by Wuhrer on a nanoLC amide column and by Karlsson on a larger bore GCC column. This further illustrated that high detectability can be obtained in LC separation without derivatization when a mass analyzer is used for detection.

7.3.2 Analyses of complex mixtures

7.3.2.1 O-linked mucin oligosaccharides

The advantages of this technique were readily apparent in complex mixtures containing scores of components. Since the identification of oligosaccharide isomers is based on LC retention time, it is of crucial importance to have the highest reproducibility. The LC retention time reproducibility of single oligosaccharide components is demonstrated above in Fig. 3. Such reproducibility can only be achieved

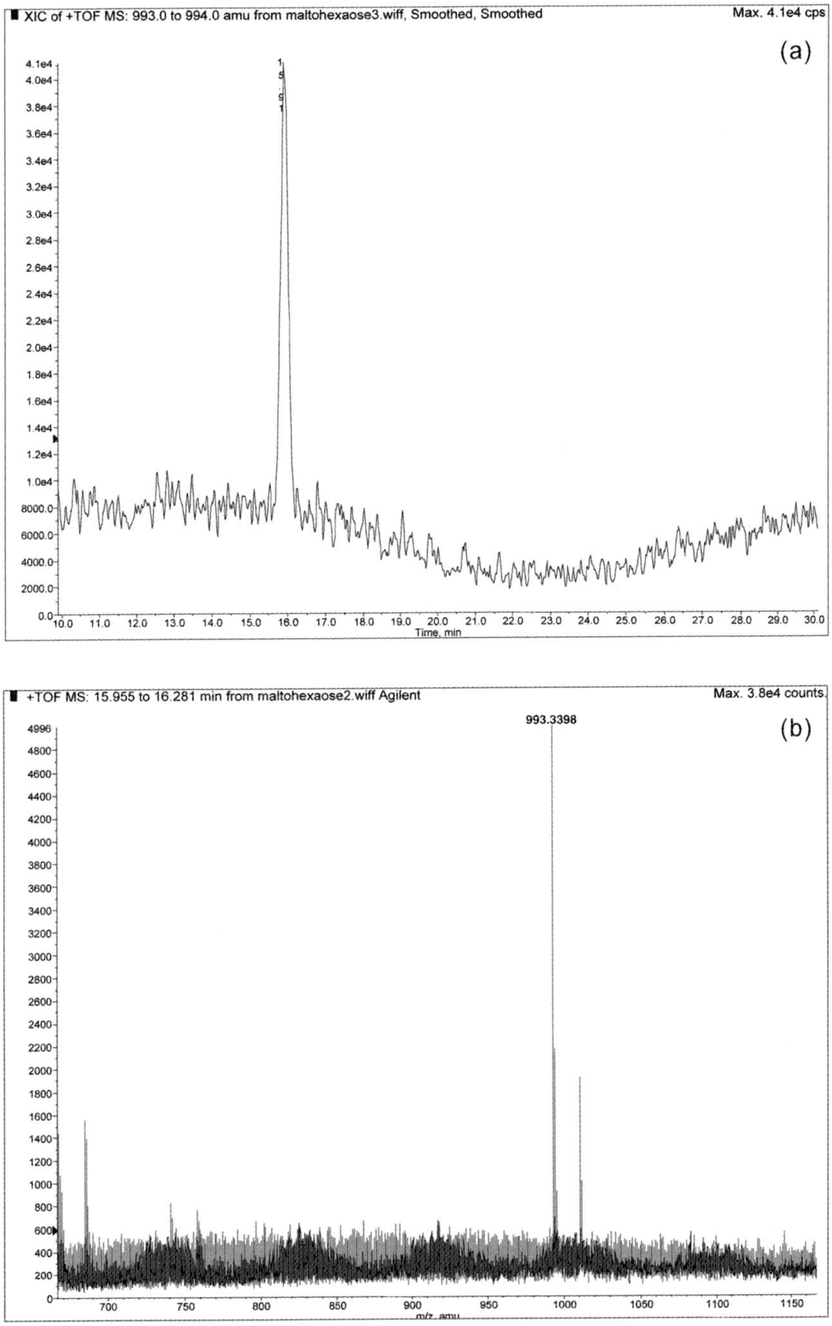

Fig. 4 (a) EIC of maltohexaose consisting of 10 fmol injection. (b) Corresponding mass spectrum of maltohexaose on the TOF mass analyzer.

with accurate and reproducible delivery of the LC mobile phase. The delivery system implemented with this device is highly stable up to 300 nL/min. It is also worth mentioning that the ambient temperature should be kept constant to achieve the most reproducible retention times.

An important aspect of any separation system is its ability to handle complex mixtures. A system where the component maintains its retention times between pure and in a complex and heterogeneous mixture is highly preferable. Maltohexaose was doped in a mixture containing several dozens of *O*-linked oligosaccharides released from mucin glycoproteins. Fig. 5 shows a plot of *m*/*z* *versus* retention time with 100 fmol sample of maltohexaose spiked into the mixture and the mixture separated by the microchip LC. The arrow points to the maltohexaose compound. The LC retention times were relatively unchanged from 15.34 min for the pure and 15.37 min for the complex mixture. Spiking the sample in this way can also provide direct quantitation of each component. This is possible as we have previously shown that the neutral oligosaccharides of similar sizes have essentially identical ionization efficiencies.

The *O*-linked oligosaccharides in these experiments were released by a standard $NaOH/NaBH_4$ method to yield the reduced alditols. The mixture has previously been examined using HPLC (GCC) and offline MALDI coupled with Fourier transform-ion cyclotron resonance-MS (FT-ICR-MS). The mixture contained nearly 100 different components consisting of neutral and anionic species. Because of the complexity of the mixture, many chromatographic peaks were found to be overlapping. However, an inspection of a single component was found to be more informative in evaluating the performance of the technique with regard to this complicated mixture. Shown in Fig. 6 is the extracted ion chromatogram (EIC) of *m*/*z* 1246.5 (MW 1245.5) with several baseline-resolved isomers. Seven isomers were identified with all being baseline-resolved. The chromatogram corresponded to the overlay of five LC runs and exhibited very little variation between runs.

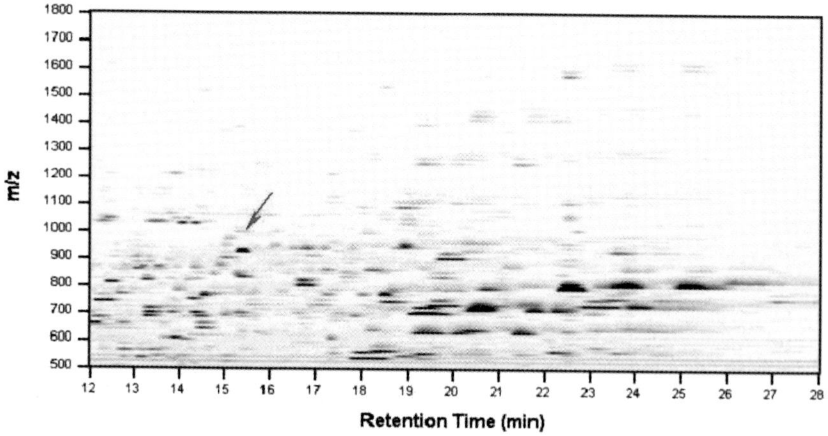

Fig. 5 *m*/*z* *versus* retention time plot of a mixture of mucin oligosaccharides spiked with 100 fmol maltohexaose. The arrow points to the maltohexaose sample.

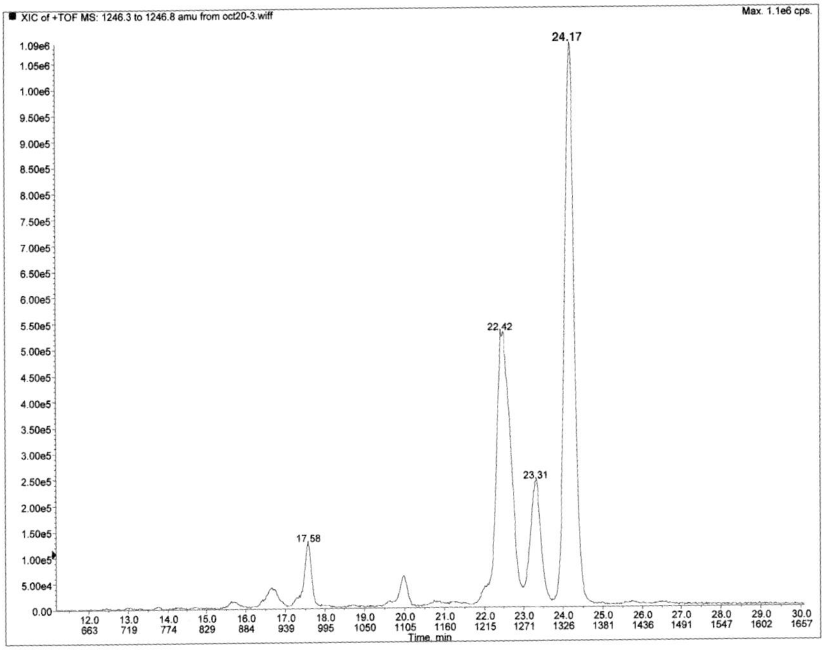

Fig. 6 EIC of m/z 1246.5 showing several isomers with specific retention times in mixture of *O*-linked oligosaccharides from mucin glycoproteins.

The number of isomers for 1245.5 found in the nanoLC/MS analysis exceeded the FT-ICR-MS study where only three isomers were found. The offline MALDI experiment was performed with a standard-size GCC column and the same gradient. In general, online MS analyses yielded an additional 30% more oligosaccharides, distinct masses with distinct retention times, than the offline HPLC-MALDI-MS analysis.

7.3.2.2 Human milk oligosaccharides (HMOs)

Another highly complex mixture is that of oligosaccharides from human milk. HMOs are the third most abundant group of compounds in human milk. The ability to profile HMOs will allow studies of oligosaccharide variability in individuals as well as to determine the biological role of HMOs. A similar analysis of milk oligosaccharides containing over 150 different oligosaccharide components was performed. HMOs pooled from several individuals were reduced to produce the alditol species. The oligosaccharides were reduced to facilitate the interpretation of tandem mass spectra in the FT-ICR and obtain structural information. The goal is to annotate the oligosaccharides in human milk for rapid identification. The base peak chroma-

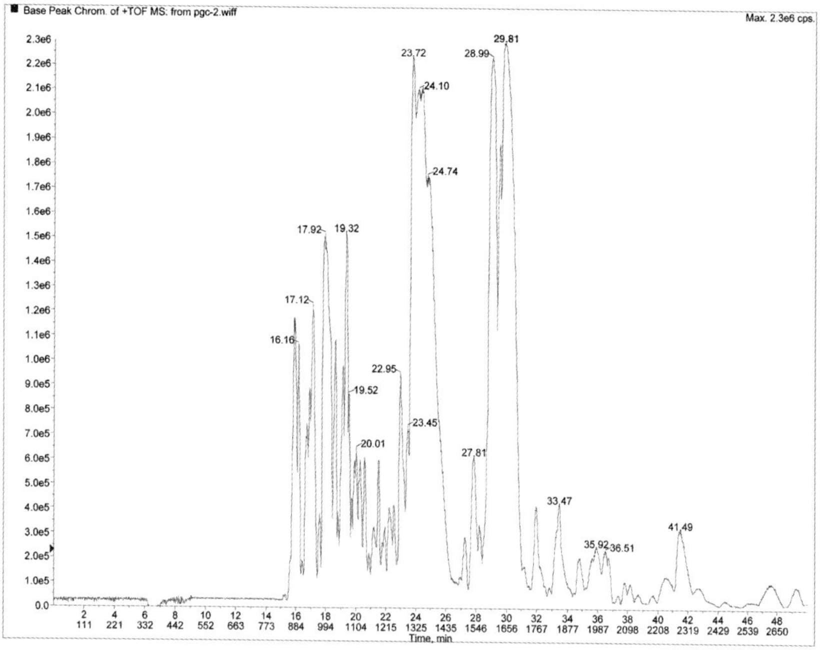

Fig. 7 Base peak chromatogram of oligosaccharides isolated from human milk.

togram of this sample is shown in Fig. 7. A corresponding study is currently being performed with the offline HPLC and FT-ICR-MS. It appears that online analysis is superior as we observe nearly 30% more compounds with the nanoLC.

Tab. 2 lists 21 masses that correspond to 97 different structures. For the time being, only masses below m/z 2000 were considered. There are more masses above 2000 that have not yet been examined. There are at least 86 reported structures [8–10]. Clearly, this type of profiling finds significantly more species that can be found with existing methods. For these samples the average mass error was found to be typically less than 2 ppm. The high mass accuracy allows the composition of each mass to be determined. An EIC of the mass the singly charged ion m/z 1221.5 is shown in Fig. 8. The saccharide composition, based on the mass, corresponded to 4Hex, 2HexNAc, and 1Fuc. This varies from the theoretical composition by less than 1 ppm. The chromatogram (Fig. 8) shows at least six distinct isomers (Tab. 3). Three structures corresponding to this mass have been reported in literature so that three of the structures are new. The structures of three of the isomers are shown to illustrate the complexity and the variability in oligosaccharide structures (structures 1–3).

Tab. 2 Masses observed in nanoLC of milk oligosaccharides that correspond to 97 reported structure

	Mass (observed)	Mass (theoretical)	Error	Error, ppm	Intensity	Oligosaccharide composition			
						Hex	Hex-NAc	Fuc	Neu-Ac
1	709.262	709.2640	0.00201	2.8404	134.29	3	1	0	0
2	855.320	855.3219	0.00191	2.2385	41.91	3	1	1	0
3	1000.358	1000.3594	0.00141	1.4141	285.35	3	1	0	1
4	1001.378	1001.3798	0.00181	1.8121	6.42	3	1	2	0
5	1058.401	1058.4013	0.00031	0.2972	96.85	3	2	1	0
6	1074.396	1074.3962	0.00021	0.1997	509.41	4	2	0	0
7	1146.417	1146.4173	0.00031	0.2744	28.87	3	1	1	1
8	1204.458	1204.4592	0.00121	1.0084	9.96	3	2	2	0
9	1220.455	1220.4541	−0.00089	−0.7255	1455.17	4	2	1	0
10	1365.490	1365.4916	0.00161	1.1824	144.10	4	2	0	1
11	1366.511	1366.5120	0.00101	0.7425	636.50	4	2	2	0
12	1382.517	1382.5069	−0.01009	−7.2950	34.87	4	2	2	0
13	1439.526	1439.5284	0.00241	1.6774	163.02	5	3	0	0
14	1511.547	1511.5495	0.00251	1.6636	288.85	4	2	1	1
15	1512.567	1512.5699	0.00291	1.9269	85.47	4	2	3	0
16	1585.583	1585.5863	0.00331	2.0905	248.58	5	3	1	0
17	1731.641	1731.6442	0.00321	1.8564	269.78	5	3	2	0
18	1876.678	1876.6817	0.00371	1.9793	25.62	5	3	1	1
19	1877.698	1877.7021	0.00411	2.1913	130.86	5	3	3	0
20	2023.756	2023.7600	0.00401	1.9837	7.03	5	3	4	0
21	2169.814	2169.8179	0.00391	1.8041	7.50	5	3	5	0

Tab. 3 LC retention time, measured mass, and mass error for oligosaccharide isomers at 1220.5 Da

RT, min	Observed mass	Theoretical mass	Error, ppm	Intensity
15.90	1220.453	1220.4541	−0.9	44.1
16.78	1220.452	1220.4541	−1.7	283.2
16.87	1220.454	1220.4541	−0.1	310.7
17.96	1220.455	1220.4541	0.7	1455.2
19.05	1220.452	1220.4541	−1.7	241.8
20.20	1220.451	1220.4541	−2.6	21.5
20.72	1220.453	1220.4541	−0.9	31.3
21.49	1220.452	1220.4541	−1.7	174.9

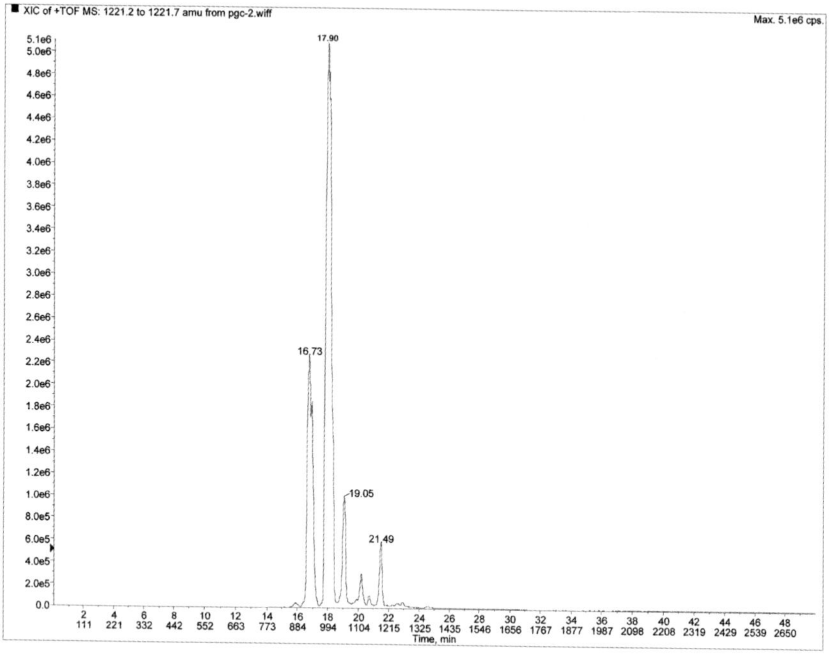

Fig. 8 EIC of m/z 1221.5 from human milk showing several isomers.

7.4 Discussion

GCC in nanoLC separation provides an effective method for the separation of oligosaccharides. Isomers are resolved and retention times are highly reproducible. The high sensitivity of nanoLC-MS even for underivatized oligosaccharides provides a new tool for oligosaccharide profiling. The combination of accurate mass and reproducible retention times provides an effective tool for the profiling of oligosaccharides, even when the number of components is large. However, there are a few concerns that need to be addressed in future publications. While the retention time reproducibility is high, it is not necessarily reproducible from chip to chip as would be expected in any chromatographic separation device. It will be necessary to employ a similar dextran ladder to make retention times portable from one chip to another. The other aspect is the robustness of the graphitized carbon media. We have been able to use essentially the same chip throughout the study. The chip and the separating media can readily take numerous injections. These and other issues regarding chip lifetimes with regard to changes in pH, sample saturation, and the injection of heterogeneous mixtures of oligosaccharides (neutral, sialylated, and sulfated) will be addressed in future publications.

Funding provided by NIH is gratefully acknowledged.

7.5 References

[1] Mechref, Y., Novotny, M. V., *Chem. Rev.* 2002, *102*, 321–369.

[2] Davies, M. J., Smith, K. D., Carruthers, R. A., Lawson, A. M., Hounsell, E. F., *J. Chromatogr.* 1993, *646*, 317–326.

[3] Zhang, J. H., Xie, Y., Hedrick, J. L., Lebrilla, C. B., *Anal. Biochem.* 2004, *334*, 20–35.

[4] Wuhrer, M., Koeleman, C. A. M., Deelder, A. M., Hokke, C. N., *Anal. Chem.* 2004, *76*, 833–838.

[5] Itoh, S., Kawasaki, N., Ohta, M., Hyuga, M., Hyuga, S., Hayakawa, T., *J. Chromatogr. A* 2002, *968*, 89–100.

[6] Karlsson, N. G., Wilson, N. L., Wirth, H. J., Dawes, P., Joshi, H., Packer, N. H., *Rapid Commun. Mass Spectrom.* 2004, *18*, 2282–2292.

[7] Zhang, J., Hedrick, J., Lebrilla, C. B., *Anal. Chem.* 2004, *76*, 5990–6001.

[8] Jensen, R. G., *Handbook of Milk Composition*, Academic Press, New York 1995.

[9] Guerardel, Y., Morelle, W., Plancke, Y., Lemoine, J., Strecker, G., *Carbohydr. Res.* 1999, *320*, 230–238.

[10] Pfenninger, A., Karas, M., *J. Am. Soc. Mass Spetrom.* 2002, *13*, 1341–1348.

[11] Yin, H., Killeen, K., Brennen, R., Sobak, D., Werlich, M., van de Goor, T., *Anal. Chem.* 2005, *77*, 527–533.

8
Chip electrospray mass spectrometry for carbohydrate analysis*

Alina D. Zamfir, Laura Bindila, Niels Lion, Mark Allen, Hubert H. Girault, Jasna Peter-Katalinić

Currently two types of chip systems are used in conjunction with MS: out-of-plane devices, where hundreds of nozzles, nanospray emitters are integrated onto a single silicon substrate from which electrospray is established perpendicular to the substrate, and planar microchips, embedding a microchannel at the end of which electrospray is generated in-plane, on the edge of the microchip. In the last two years, carbohydrate research greatly benefited from the introduction and implementation of the chip-based MS. In two laboratories the advantages of the chip electrospray in terms of ionization efficiency, sensitivity, reproducibility, quality of data in combination with high mass accuracy, and resolution of detection were systematically explored for several carbohydrate classes: *O*- and *N*-glycopeptides, oligosaccharides, gangliosides and glycoprotein-derived *O*- and *N*-glycans, and glycopeptides. The current state-of-the-art in interfacing the chip electrospray devices to high-performance MS for carbohydrate analysis, and the particular requirements for method optimization in both positive and negative ion modes are reviewed here. The recent applications of these miniaturized devices and their general potential for glycomic-based surveys are highlighted.

8.1
Introduction

In the development of miniaturized microfluidic systems a significant progress in the past decade took place primarily due to numerous advantages of microchip analysis, including the ability to analyze minute samples, the increased throughput and performance, as well as reduced costs. In addition, several laboratory procedures such as sample preparation, purification, separation, and detection have

* Originally published in Electrophoresis 2005, 26, 3650–3673

Microfluidic Applications in Biology. Edited by Niels Lion, Joël S. Rossier, and Hubert H. Girault

ISBN-10: 3-527-31761-9

been integrated onto a single microchip unit forming the so-called micro-total analysis systems (μTAS) to allow enhancement in sensitivity, speed, and accuracy [1–4]. In the last years, by introducing MS as a detector, the potential and the applicability of μTAS have been significantly extended [5, 6]. On the other hand, the option for miniaturized, integrated devices for sample infusion into MS was driven by several technical, analytical, and economical advantages such as simplification of the laborious chemical and biochemical strategies required currently for MS research, minimization of sample handling and sample loss, elimination of possible cross-contamination, increased ionization efficiency, high quality of spectra, possibility for unattended high-throughput experiments, and the flexibility provided by the microchips for coupling to different MS configurations.

In the field of ESI–MS, two types of chip-based devices are currently being investigated. The first category is represented by the out-of-plane devices, where 100 or 400 nanospray emitters are integrated onto a single silicon substrate, from which electrospray is established perpendicular to the substrate [7]. These devices are amenable to high-throughput sample delivery to ESI-MS by automated infusion [8, 9] and have the potential to completely replace flow-injection analysis assays. Moreover, the technical quality of the nanosprayers obtained by advanced silicon microtechnology is so high, and the experiments so reproducible, that such devices have been found in some instances to give more robust and quantitative analyses than LC- or CE-MS and to be able to suppress the need for separation prior to MS analysis [10, 11].

Due to efficient ionization properties, silicon-based nanoESIchip/MS preferentially forms multiply charged ions, and the in-source fragmentation of labile groups attached to the main structural backbone is minimized, enhancing the sensitivity of the analysis.

The second category consists of planar or thin microchips, made from glass [12, 13] or polymer [14, 15] material, embedding a microchannel at the end of which electrospray is generated in-plane, on the edge of the microchip. During the last years, the progress in polymer-based microsprayer systems was promoted by development of simpler methods for accurate plastic replications and ease to create lower-cost disposable chips.

Though compatible to automation for high-throughput analysis, these designs are best suited to act like μTAS by integration of other analytical functions prior to sample delivery to the MS system, such as sample cleanup [16], analyte separation by CE [17], and chemical tagging [18]. Moreover, in comparison with glass nanospray capillaries, these thin microsprayers were found to provide superior stability of the spray with time, improved S/N at various flow rates.

Several approaches for interfacing MS to silicon, glass, or polymer chips, providing flow rates which include them in categories from nano- to microsprayers, were reported [19]. Different configurations such as single and triple quadrupole, TOF, IT, and ultrahigh-resolution Fourier-transform ion cyclotron resonance (FTICR) mass spectrometers were adapted to chip-based ESI and contributed significant benefits for various studies. However, as nicely illustrated already by a number of reviews [8, 11, 20–22] the general areas of implementation and applications of chip technologies either silicon-based, glass, or polymer microchips have by far been primarily genomics, drug discovery, and proteomics.

Carbohydrates are the most abundant class of organic components found in living organisms, characterized by a high degree of structural complexity. They are present in nature as either oligosaccharides or glycoconjugates in which the oligosaccharide portion is covalently linked to a protein and/or a lipid and occur ubiquitously displayed on macromolecules and the surface of cells. This biopolymer category is involved in basic biological functions, such as antigen recognition machinery, cellular adhesion of bacteria and viruses, and protein folding, stability, and trafficking [23]. Particular structures were found to represent biomarkers of severe diseases and others to play an essential role in fertilization and embryogenesis [24].

Despite the high biological importance of carbohydrates and the performance exhibited by chip-based MS methodologies, the glycomics field has benefited from the chip technology only to a limited extent. Optimization of ESIchip/MS for operation in the negative ion mode to detect carbohydrate ions was considered a challenging task mostly due to the relative low ionization efficiency that ultimately leads to decreased sensitivity. Besides, each class of carbohydrates was shown to require particular and defined conditions for chip ionization followed by the MS detection. These conditions are strongly dependent on the type of the labile peripheral attachments on the sugar chain such as fucosylation, sialylation, sulfation, phosphorylation, the ionizability of the functional groups, the hydrophilic and/or hydrophobic nature, branching of the sugar chains *etc.* [25, 26].

However, during the past years, efforts have been invested toward implementation of chipESI-MS/MS in glycomics. Initial experiments for testing of general techniques and methodology provided clear-cut evidence for high potential in applications of this technology to research projects requiring compositional and structural analysis of *N*- and *O*-glycans, *O*-glycopeptides, and gangliosides.

In this context, the purpose of a review highlighting the first achievements and trends in chip-based ESI-MS of carbohydrates is primarily to offer a general background for its better understanding and the motivations for such an analytical option. We believe that from this review the reader will learn the basic requirements for successful chip MS analysis of different carbohydrate categories and disclose the advantages of chip-based ESI-MS devices for their compositional and structural assays.

8.2 Thin chip polymer-based electrospray MS

8.2.1 Coupling of the thin chip microsprayer system to high-performance MS for carbohydrate analysis

For screening and sequencing of carbohydrates, recently, a disposable polymer microchip with integrated microchannels and electrodes was coupled to both a quadrupole TOF (QTOF) and an FTICR mass spectrometer [27, 28].

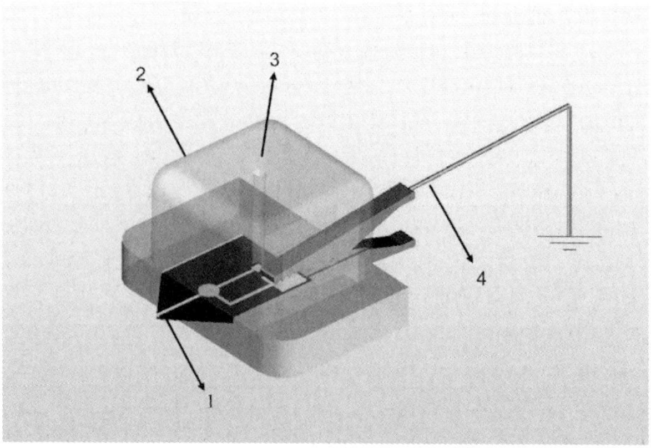

Fig. 1 Schematic of the polymer chip holder with integrated reservoir. (1) Polymer chip; (2) sandwich chip holder; (3) sample reservoir; (4) conductive wire. Reproduced from [28]; with permission.

The chip presented in Fig. 1 was microfabricated by semiconductor techniques including photolithography [29]. For the microchip fabrication, the starting material was a polyimide foil of 75 µm thickness, which is coated on both sides with 5 µm copper. A photoresist was patterned on the copper-coated polyimide foil through a printed slide acting as a mask. Photoresist was then developed and chemical etching was used to remove the deprotected copper where microchannels are to be patterned. Polyimide was plasma-etched to the desired depth. As both sides of the substrate were exposed to the plasma, through-holes were fabricated to act as sample reservoirs and/or provide access to the microchannel.

The final microchannels were 120 µm wide, 45 µm deep (nearly "half moon" cross section), with 100 µm gold-coated microelectrodes placed at the bottom of the microchannel. A 35 µm polyethylene/polyethylene terephthalate was laminated to close the channels. As described by Gobry *et al.* [15], one end of each channel was manually cut in a tip shape, so that the outlet of the microchannel was located on the edge of the chip. For sample dispensing, either a reservoir was pasted over the inlet of the microchannel or the chip was sandwiched in a home-made chip holder with an integrated reservoir. For MS coupling one end of each channel was manually cut in a tip shape, which was visually inspected with a stereomicroscope. This way, the outlet of the microchannel is located on the edge of the chip, providing an in-plane electrospray.

The thin polymer chip was coupled to a QTOF-MS (Micromass, UK) with Z-spray ion source geometry [27]. For sample loading, a microvial reservoir was pasted over the inlet-hole of the microchannel and the whole chip/reservoir assembly was mounted onto the mass spectrometer. An optimal coupling able to promote efficient ionization and sensitive detection of carbohydrates was found to require (i) stable electrical contact, (ii) careful positioning of the chip toward the MS sampling cone, (iii) adjustment of the ESI voltage and sampling cone voltages within 2–3 kV and 80–100 V, respectively, and (iv) systematic control of the solvent/analyte desolvation process. In order to realize the electrical contact to the ESI power supply, the ESI-QTOF sampler was removed and the chip system was directly connected to the ESI high-voltage plate, which is a fixed part of QTOF conventional ESI source. Exchange between the original source and chip system interface did not claim for any definitive dismantling or special mechanical modifications to either of the original assemblies and no further modifications on the QTOF-MS were reported as necessary. The position of chip emitter was adjusted in the vicinity of the entrance hole at a distance of about 5 mm from the QTOF-MS sampling cone. The ESI field was established by applying a voltage (2–3 kV) on the microchip electrode and an attractive potential (100–150 V) on the MS counterelectrode (cone). The voltage on the microelectrode was applied by using a conductive wire with one terminal connected to the chip electrode and the other fixed on the ESI high-voltage plate of the QTOF-MS instrument. Under the formed high electric field, the fluid was electrokinetically driven through the chip microchannel from its inlet-hole toward the outlet-edge where the electrospray process occurred [27]. In the case of QTOF-MS, the ESI source is configured in *Z* geometry by the orthogonal alignment of the sampling capillary and MS counterelectrode.

In Fig. 2, a photography of the QTOF source assembly with mounted polymer chip is presented. The photography has been taken during the thin chipESI-QTOF-MS running, immediately after application of the negative voltages on both chip and counterelectrode. The ESI plume is clearly visible demonstrating the instant initiation of the electrospray. Under the same ESI-QTOF-MS conditions, the reported [27] in-run reproducibility in terms of sensitivity, spray stability, number of detected components/fragments, ion intensity, and charge state was almost 100%, while the day-to-day reproducibility was 95–100%.

The first interfacing of chip electrospray to FTICR-MS was reported in 2003 by Rossier *et al.* [30]. The system described by the authors included the thin polymer microchip placed into a specially designed holder and a 7 T FTICR MS (Bruker Daltonics) operating in the positive ion mode. In this study, the first application of the chip-FTICR system to proteome analysis and in particular to complex tryptic peptide mixture of the human microtubule-associated tau protein, considered as a target in neurofibrillary tangles characteristic of Alzheimer disease, is described. In the (+)chipESI-FTICR

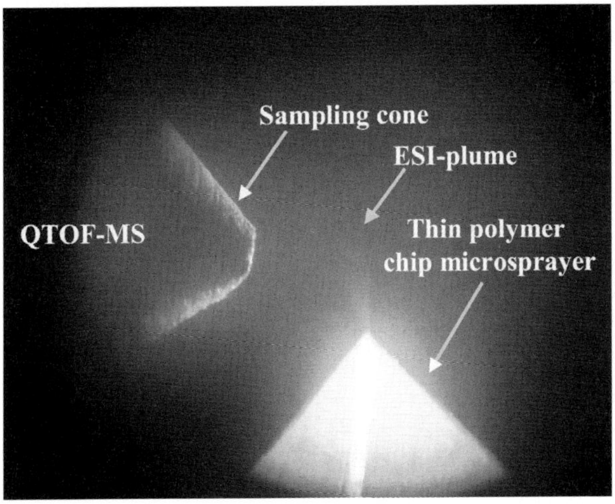

Fig. 2 Photography of the thin polymer chip/ESI-QTOF-MS coupling, taken immediately after the application of the ESI voltages. Reproduced by permission of The Royal Society of Chemistry [27].

mass spectrum the primary structure including 18 serine (Ser) and threonine (Thr) phosphorylations could be directly identified, which clearly demonstrated the high analytical performance of the chipESI-FTICR-MS in proteomics.

Last year, an Apex II FTICR mass spectrometer equipped with a 9.4 T superconducting actively shielded magnet and an Apex II Apollo ESI ion source was interfaced to the disposable thin polymer microchip, and the system was for the first time optimized in the negative ion mode and introduced in glycomics [28]. For sample loading and FTICR interfacing, the chip was sandwiched in a home-made chip holder with an integrated reservoir (Fig. 1) and positioned into the holder with the microchannel in contact with the reservoir and the front part extruding a few millimeters. Unlike the case of ESI-QTOF source, in the FTICR Apex II Apollo system the sampling capillary is grounded and the electric field is created by applying the ESI voltage on the MS metal-coated glass capillary. Therefore, for polymer chip interfacing to FTICR-MS, the chip was grounded *via* a conductive wire connected to the terminal gold-coated electrode. Another peculiarity of the FTICR ESI source is its direct spray geometry as a result of axial alignment of the sampling capillary and MS entrance. The microchip system was coupled to this ion source by an in-laboratory constructed mounting system consisting of a metal plate connected to the source by two 90° brackets. The in-house made mounting system was described as a robust and viable interfacing of the polymer chip to the FTICR-MS instrument.

8.2.2 Applications

8.2.2.1 Complex biological mixtures of glycopeptides

In human urine, carbohydrates are catabolic products excreted as either oligosaccharides or glycopeptides and their concentration, amount, and structure are known to vary under different physiological and/or pathological conditions. For this reason, monitoring the glycopeptide expression in human urine by MS could potentially become a method of diagnostic importance [31]. Various glycoconjugate pools extracted and purified from normal human urine and urine of patients suffering from diseases caused by aberrant glycosylation were already submitted to a number of MS-based analytical experiments to assess their potential for glycoscreening [32–35].

The thin chip microsprayer/QTOF-MS coupling described above was optimized in the negative ion mode for determination of *O*-GalNAc-Ser/Thr expression in the normal human urine [27]. A mixture of *O*-GalNAc glycosylated amino acids and peptides was dissolved in pure methanol to a concentration of 5 pmol/μL and an aliquot of 10 μL was dispensed into the reservoir pasted over the inlet of the chip microchannel. The negative ion mode electrospray process was initiated at 2.8 kV ESI voltage and 100 V potential of the sampling cone. At these values of source parameters a constant and stable spray accompanied by a high intensity of the total ion current was generated. The signal was acquired over 20 scans (40 s) which, at the flow rate of about 200 nL/min provided by the microchip under these conditions, was equivalent to a sample consumption of 0.66 pmol. The spectrum combined over 20 scans exhibited a high S/N and a number of 26 different saccharide components. Fifteen species detected as singly and/or doubly charged ions expressed *O*-GalNAc-Ser/Thr core-motif extended by either sialylation or fucosylation. The mixture was found to be dominated by Ser- and Thr-linked disialo saccharides with chain lengths ranging from tetra- to octasaccharide. Interestingly, two new structures which could originate from glycopeptides with extended and further modified chains, not detectable before by any MS-related methods, were detected as doubly charged ions.

In order to test the limit of the microchip sensitivity for glycoconjugate detection, the solution was further diluted in pure methanol yielding an aliquot at 1.25 pmol/μL concentration. The signal was acquired over 20 scans, which was equivalent to 0.16 pmol sample consumption for this experiment. Even under these restrictive concentration conditions, a fair S/N was reported and 13 different components in the mixture could be still identified. All microchip spectra exhibited an interesting peculiarity related to ionic charge-state distribution. Thus, it was found that at 100 V cone potential, the charge distribution was not shifted toward lower values as observed in the capillary-based ESI-QTOF-MS experiments and the intensity of the signals corresponding to the doubly charged ions was still the highest ones. Moreover, the advantage of the microchip ESI-MS regarding the minimization of the in-source decay of the labile attachments such as *N*-acetyl neuraminic acid (Neu5Ac) was observed.

To investigate the possibility of rapid and accurate glycopeptide sequencing by MS/MS using the polymer chip for sample infusion, a doubly charged ion at *m/z* 532.08, assigned according to the *m/z* value to the already known structure of $NeuAc_2$Hex-HexNAc-Thr (Hex = hexose, HexNAc = *N*-acetyl hexosamine), was isolated and submitted to low-energy CID (−)microchip ESI-QTOF-MS/MS. Ion acceleration energy, collision gas pressure, and precursor ion isolation parameters were carefully adjusted to provide the full set of structural information upon the molecule. The product ion spectrum (Fig. 3) was obtained after 30 scans (1 min) of signal acquisition under the sample consumption of 1.23 pmol. A high S/N and significant number of diagnostic elements valuable for assessing the sialylation pattern and linkages specific to the core 1 type of *O*-glycosylated molecules were obtained.

In another study, the combination of the thin chip microsprayer performances with high accuracy and resolution of the FTICR-MS was designed and implemented for the screening in negative ion mode of several urine glycopeptide mixtures [28]. Special considerations of the thin chipESI/FTICR-MS parameters with respect to the direct spray configuration of the FTICR instrument were necessary here. The initiation of the electrospray was achieved at 1500–2500 V applied to the transfer capillary, while the fine positioning of the microsprayer toward the MS inlet turned out to be crucial for efficient transfer of the ionic species into the MS and long-term stability of the electrospray. The system was tested for compositional mapping of a mixture of *O*-glycosylated amino acids and peptides extracted and purified from urine of a patient suffering from Schindler disease and an age-matched healthy individual for comparative assay [28]. Schindler disease is a rare inherited metabolic disorder characterized by a deficiency of the lysosomal enzyme

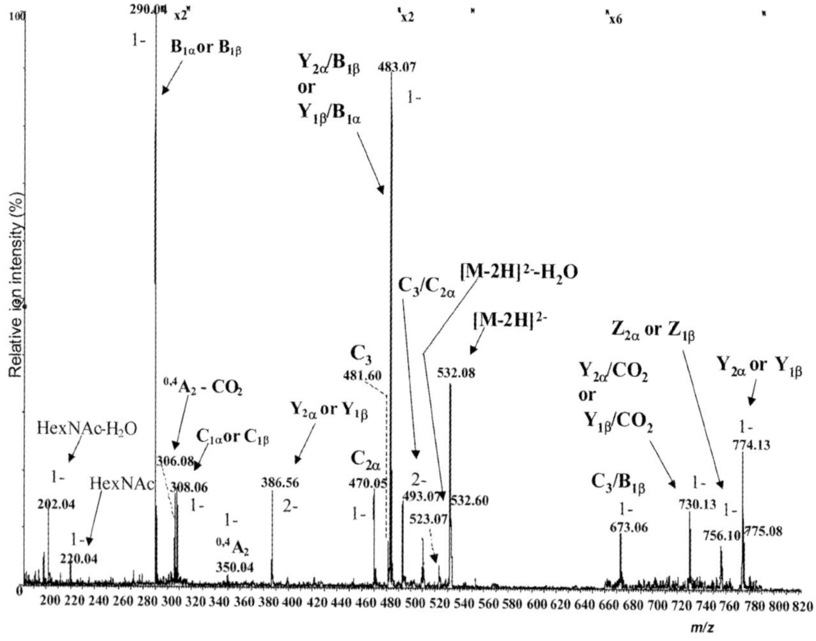

Fig. 3 Microchip ESI/QTOF MS/MS of the Neu5Ac$_2$HexHexNAc-Ser doubly charged ion at m/z 532.08. ESI voltage: 2.8 kV; sampling cone potential: 100 V; solvent: methanol; average sample concentration: 5 pmol/μL; collision energy: 40 eV; signal acquisition: 30 scans; average sample consumption: 1.23 pmol. Nomenclature for assignment of fragment ions is according to [61, 62]. Reproduced by permission of The Royal Society of Chemistry [27].

α-*N*-acetylgalactosaminidase (NAGA), which leads to an abnormal accumulation of sialylated and asialo-glycopeptides and oligosaccharides with α-*N*-acetyl-galactosaminyl residues. The deficient NAGA is causing not only a 100 times higher concentration of *O*-glycans in patient urine than in healthy controls, but it was also found to give rise to unspecifically cleaved products consisting of longer saccharide chains [31, 36–38].

In Fig. 4, the (−)thin chipESI/FTICR mass spectrum of the *O*-glycosylated amino acids and peptides disease is depicted. After 50 scans of signal recording under a flow rate of approximately 200 nL/min, 27 ionic species could be already identified and straightforwardly assigned with an average mass accuracy of 5.41 ppm and a maximum sensitivity of 3 pmol/μL (Fig. 4). In comparison with the chipESI/FTICR spectrum of the normal urine glycopeptides, a higher content of pentasaccharides was observed, as documented by the doubly charged ion at m/z 562.193 and the corresponding singly charged ion at m/z 1125.386, assigned to Neu5AcHex$_2$HexNAc$_2$-Ser. The doubly charged ion at m/z 569.199 along with its singly charged counterpart at m/z 1139.401 was assigned to the homologous structure, Neu5AcHex$_2$HexNAc$_2$-Thr. Additionally, the hexasaccharides linked to

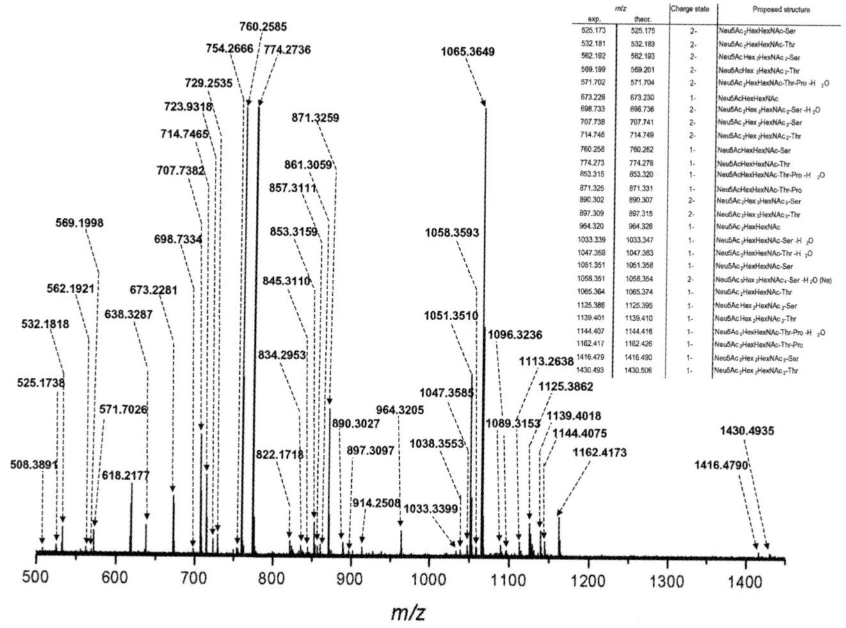

m/z exp.	m/z theor.	Charge state	Proposed structure
525.173	525.175	2-	Neu5Ac$_2$HexHexNAc-Ser
532.181	532.183	2-	Neu5Ac$_2$HexHexNAc-Thr
562.192	562.193	2-	Neu5AcHex$_2$HexNAc$_2$-Ser
569.199	569.201	2-	Neu5AcHex$_2$HexNAc$_2$-Thr
571.702	571.704	2-	Neu5Ac$_2$HexHexNAc-Thr-Pro -H$_2$O
673.228	673.230	1-	Neu5AcHexHexNAc
698.733	698.736	2-	Neu5Ac$_2$Hex$_2$HexNAc$_2$-Ser -H$_2$O
707.738	707.741	2-	Neu5Ac$_2$Hex$_2$HexNAc$_2$-Ser
714.746	714.749	2-	Neu5Ac$_2$Hex$_2$HexNAc$_2$-Thr
760.258	760.262	1-	Neu5AcHexHexNAc-Ser
774.273	774.278	1-	Neu5AcHexHexNAc-Thr
853.315	853.320	1-	Neu5AcHexHexNAc-Thr-Pro -H$_2$O
871.325	871.331	1-	Neu5AcHexHexNAc-Thr-Pro
890.302	890.307	2-	Neu5Ac$_2$Hex$_3$HexNAc$_3$-Ser
897.309	897.315	2-	Neu5Ac$_2$Hex$_3$HexNAc$_3$-Thr
964.320	964.326	1-	Neu5Ac$_2$HexHexNAc
1033.339	1033.347	1-	Neu5Ac$_2$HexHexNAc-Ser -H$_2$O
1047.358	1047.363	1-	Neu5Ac$_2$HexHexNAc-Thr -H$_2$O
1051.351	1051.358	1-	Neu5Ac$_2$HexHexNAc-Ser
1058.351	1058.354	2-	Neu5Ac$_3$Hex$_2$HexNAc$_4$-Ser -H$_2$O (Na)
1065.364	1065.374	1-	Neu5Ac$_2$HexHexNAc-Thr
1125.386	1125.395	1-	Neu5AcHex$_2$HexNAc$_2$-Ser
1139.401	1139.410	1-	Neu5AcHex$_2$HexNAc$_2$-Thr
1144.407	1144.416	1-	Neu5Ac$_2$HexHexNAc-Thr-Pro -H$_2$O
1162.417	1162.426	1-	Neu5Ac$_2$HexHexNAc-Thr-Pro
1416.479	1416.490	1-	Neu5Ac$_2$Hex$_2$HexNAc$_2$-Ser
1430.493	1430.506	1-	Neu5Ac$_2$Hex$_2$HexNAc$_2$-Thr

Fig. 4 (−)Microchip ESI/FTICR MS of a fraction of *O*-glycosylated peptides extracted from urine of a patient diagnosed with Schindler disease. ESI voltage: 1.5–2.5 kV. Capillary exit: −79 V. Average sample concentration: 20 pmol/μL in methanol. Number of scans: 50. Average sample consumption: 2 pmol. Inset: table with assignment of the ions detected in the mixture by microchip ESI/FTICR-MS in the negative ion mode. Reproduced and adapted with permission from [28].

Ser and Thr bearing two sialic acid moieties were detected at higher abundance than in the glycopeptide mixture from normal urine, and a nonasaccharide bearing three sialic acid moieties was for the first time detected at appreciable abundance as a doubly charged ion at m/z 1058.354 and assigned to the sodiated dehydrated Neu5Ac$_3$Hex$_2$HexNAc$_4$Ser with a mass accuracy of 4.6 ppm.

8.2.2.2 **Gangliosides**

Gangliosides are glycosphingolipids (GSLs) consisting of a mono- to polysialylated oligosaccharide chain of different lengths attached to a ceramide portion of variable composition with respect to the type of sphingoid base and fatty acid residues. They are abundant components of the central nervous system and their compositions undergo changes during brain development, maturation, aging, and neurodegeneration processes [39–41]. For these reasons, gangliosides are considered as tissue stage or diagnostic markers and potentially therapeutic agents. In human brain, region-specific differences in ganglioside composition and quantity, as well as in their distribution and expression have been demonstrated primarily by TLC, immunochemical, and immunohistochemical methods besides FAB-MS [42, 43] and capillary-based electrospray [44]. In all ESI-MS-related studies using either QTOF-MS [44], off-line CE/QTOF-MS, or FTICR-MS [45, 46], requirements for elevated values of ionization parameters and an extended time for CID-MS/MS signal acquisition, implying more sample consumption for identification and structural elucidation, were defined.

To explore the performance of the microchip ESI/QTOF-MS approach and the detection limits for structural characterization of gangliosides, a fraction of polysialylated gangliosides (GT1), isolated from the total native ganglioside mixture of normal adult human brain, was considered. For (−)microchip ESI/QTOF-MS analysis, GT1 sample was dissolved in pure methanol to a concentration of 5 pmol/μL and an aliquot of 10 μL was loaded into the microchip reservoir. To optimize the ionic current value, the (−)ESI voltage and cone potential were increased following a ramping procedure. The total ion chromatogram (TIC) profile indicated a maximum of ionic current, a sustained spray, and an efficient ionization at 3 kV ESI and 100 V sampling cone voltages. After 4 min of signal acquisition at a flow rate of 200 nL/min a spectrum of high S/N was obtained (Fig. 5), although a fair S/N was observed already after 60 scans (30 s). A reproducible compositional mapping of eight GT1 molecular components, reflecting different lipid variants in the fraction, was obtained from both triply and doubly charged molecular ions. Interestingly, three minor GT1 molecular species not detectable by capillary infusion could be documented by the low-abundant molecular ions at *m/z* 1094.20, 1108.20, and 1109.21, assigned by calculation to GT1 (d18:1/21:1), GT1 (d18:1/23:1), and GT1 (d18:1/23:0), respectively.

In the same study [27], the thin chipESI/QTOF-MS was explored to address the fundamental issue of efficient fragmentation analysis by CID-MS/MS of ganglioside species, known to require an extended signal acquisition time under variable collision energies to generate enough fragment ions for high sequence coverage. By this novel approach, the instability of the electrospray and/or signal interruption, frequently reported in the case of capillary-based ESI [44], could be overcome, and a stable TIC MS/MS was shown to be crucial for different sequencing events. A triply charged ion at *m/z* 717.50, corresponding to the GT1 (d18:1/20:0), could be successfully sequenced (Fig. 6). The fragmentation process gave rise to sialylated fragment ions indicative for the carbohydrate sequence of the GT1, as well as for characterization of its ceramide portion. The complete fragmentation pattern of the molecule (Fig. 7) could be deduced from the abundant Y- and Z-type ions formed from the nonreducing end, while the X-type ring cleavages appeared at lower abundance. Informative C- and B-type ions observed to be similarly favored under given conditions were obtained. Moreover, a number of fragment ions deriving from different sugar portion GT1 isomers, in particular *m/z* 364.09 ($B_{2\alpha}$), 966.39 $^{2,4}X_{2\alpha}$ specific for GT1c, and *m/z* 493.10 ($B_{2\alpha}/B_{1\alpha}$) specific for GT_{α}, respectively, could be clearly discriminated here, supporting the presence of structural isomers or isobars as distinct species.

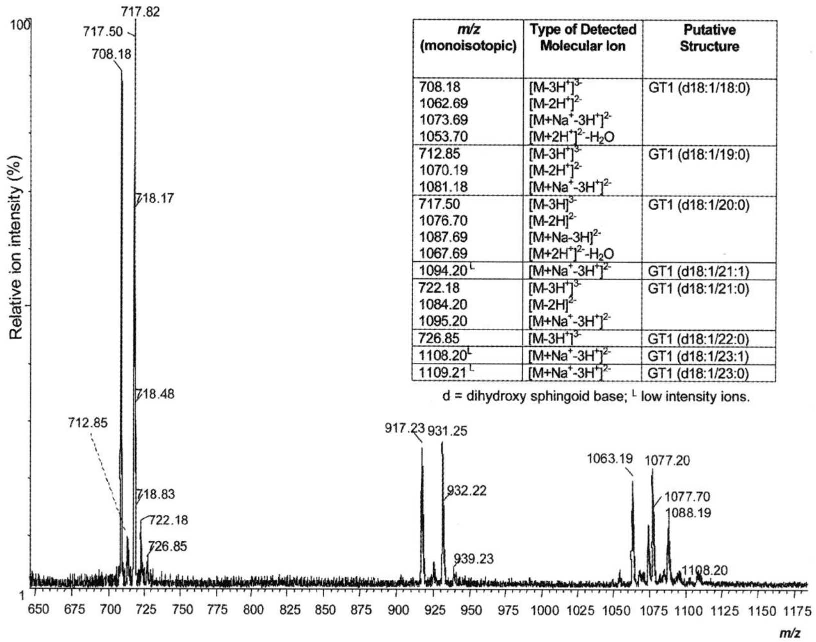

m/z (monoisotopic)	Type of Detected Molecular Ion	Putative Structure
708.18	$[M\text{-}3H^+]^{3-}$	GT1 (d18:1/18:0)
1062.69	$[M\text{-}2H^+]^{2-}$	
1073.69	$[M\text{+}Na^+\text{-}3H^+]^{2-}$	
1053.70	$[M\text{+}2H^+]^{2-}\text{-}H_2O$	
712.85	$[M\text{-}3H^+]^{3-}$	GT1 (d18:1/19:0)
1070.19	$[M\text{-}2H^+]^{2-}$	
1081.18	$[M\text{+}Na^+\text{-}3H^+]^{2-}$	
717.50	$[M\text{-}3H]^{3-}$	GT1 (d18:1/20:0)
1076.70	$[M\text{-}2H]^{2-}$	
1087.69	$[M\text{+}Na\text{-}3H]^{2-}$	
1067.69	$[M\text{+}2H^+]^{2-}\text{-}H_2O$	
1094.20 L	$[M\text{+}Na^+\text{-}3H^+]^{2-}$	GT1 (d18:1/21:1)
722.18	$[M\text{-}3H^+]^{3-}$	GT1 (d18:1/21:0)
1084.20	$[M\text{-}2H]^{2-}$	
1095.20	$[M\text{+}Na^+\text{-}3H^+]^{2-}$	
726.85	$[M\text{-}3H^+]^{3-}$	GT1 (d18:1/22:0)
1108.20 L	$[M\text{+}Na^+\text{-}3H^+]^{2-}$	GT1 (d18:1/23:1)
1109.21 L	$[M\text{+}Na^+\text{-}3H^+]^{2-}$	GT1 (d18:1/23:0)

d = dihydroxy sphingoid base; L low intensity ions.

Fig. 5 Microchip ESI/QTOF MS1 of a ganglioside fraction derived by combining the TIC MS scans at 3 kV ESI voltage and 100 V sampling cone potential. (a) *m/z* range: (650–1175). Inset: the compositional mapping of the purified native GT1 ganglioside fraction (exhibiting HPTLC migration properties of the GT1b species) separated from the total ganglioside mixture isolated from adult human brain tissue as detected by (−)microchip ESI-QTOF-MS1. Reproduced and adapted by permission of The Royal Society of Chemistry [27].

8.3 Fully automated chip-based electrospray MS

8.3.1 Coupling of the fully automated chip-based nanoESI system to high-performance MS for carbohydrate analysis

NanoMate™ 100 (Advion BioSciences), the world's first fully automated nanoelectrospray system, is a robotic device that provides an automated nanoelectrospray ion source for mass spectrometers [8]. In this system, by an automatic infusion, samples at low flow rates (50–100 nL/min) are admitted directly into MS [9, 10]. In addition to the robot itself, the key component of the system is the ESI chip. The robot holds a 96-well sample plate and a 96-pipette tip tray. Automated sample analysis is achieved by loading a disposable, conductive pipette tip on a movable sampling probe, aspirating sample *via* a syringe pump, and moving the sampling probe to engage against the back of the ESI chip. ESI process is initiated by applying a head pressure and voltage to the sample in the pipette tip. Each

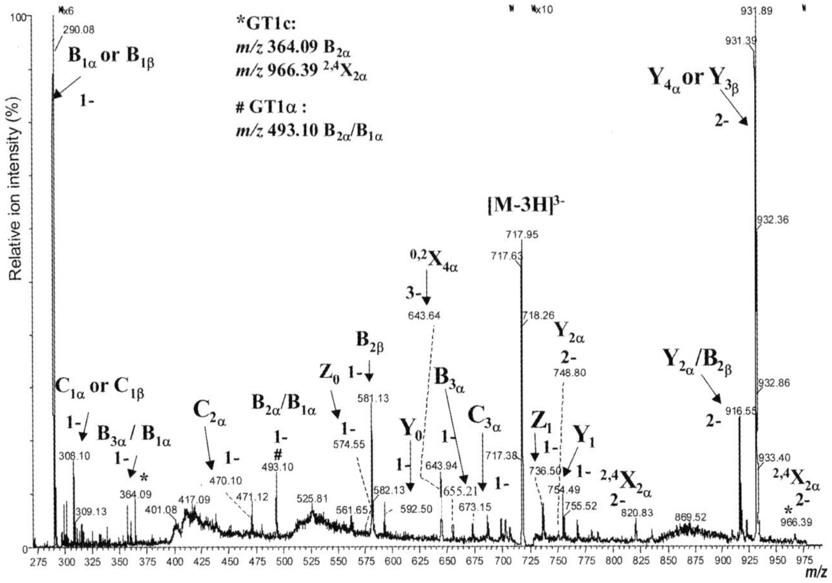

Fig. 6 Microchip ESI/QTOF-MS/MS of the triply charged ion at *m/z* 717.50 corresponding to GT1 (d18:1/20:0). ESI voltage: 3 kV. Sampling cone potential: 100 V. Spectrum is derived by combining the scans corresponding to 40 and 70 eV in the TIC MS/MS. Nomenclature for assignment of fragment ions is according to [63–65]. Reproduced by permission of The Royal Society of Chemistry [27].

nozzle and tip is used only once inorder to eliminate carryover and contamination typical to conventional autosamplers. The ESI chip is an array of nanoESI nozzles of 10 μm internal diameter etched in a planar silicon chip. The chip is fabricated from a monolithic silicon substrate using deep reactive ion etching (DRIE) and other standard microfabrication techniques [7]. The inert coating on the surface allows a variety of acidic and organic compositions and concentrations to be used to promote ionization without any degradation of the nozzle [7]. As visible in Fig. 8, a channel extends from the nozzle through the microchip.

In conventional electrospray devices the electric field is defined by the potential difference between the ESI tip (fluid potential) and the mass spectrometer inlet. A unique feature of the ESI chip is the incorporation of the ESI ground potential into the spray nozzle so that the electric field around the nozzle tip is formed from the potential difference between the conductive silicon substrate and the voltage applied to the fluid *via* the conductive pipette tip [7–11]. In such a configuration, the distance that defines the electric field is about 1000 times shorter than the one between the nozzle and mass spectrometer. Therefore, the mass spectrometer position and voltage, though crucial for efficient ion transfer into analyzer, do not play any role in formation of the chip electrospray, thus essentially decoupling the ESI process from the inlet of the mass spectrometer.

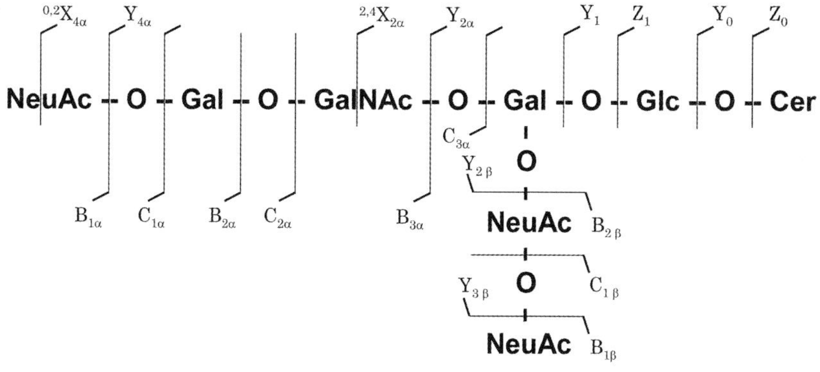

Fig. 7 Structure and the corresponding fragmentation scheme of the GT1b species. Product ions detected from the spectrum depicted in Fig. 6 are assigned. Nomenclature for assignment of fragment ions is according to [61–65]. Reproduced by permission of The Royal Society of Chemistry [27].

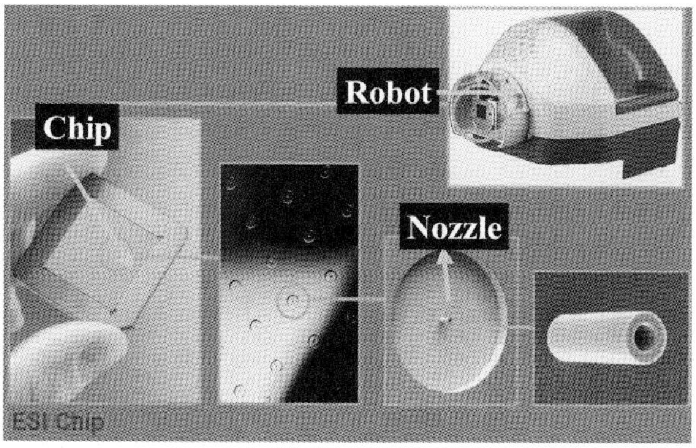

Fig. 8 NanoMate™ 100 incorporating ESI chip technology. Courtesy of Advion BioSciences.

For carbohydrate analysis, the NanoMate robot was first mounted onto the QTOF and IT mass spectrometers [47–49] *via* special brackets, allowing adjustment of the robot position with respect to the sampling cone. Recently, the first coupling of the NanoMate robot to the FTICR-MS at 9.4 T and sustained off-resonance CID (SORI CID) MS/MS was reported [50]. The new assembly was optimized in the negative ion mode to combine automated sample delivery on the chip along with maximal sensitivity, ultrahigh-resolution, accurate mass determination provided by the FTICR instrument, and efficient MS/MS in one system. In this study a specially designed interface was constructed by Bruker Daltonik (Bremen, Germany) in order to obtain a viable coupling of the NanoMate system to the Bruker Apex II Apollo ion source. An interesting coupling feature derived from the NanoMate ionization principle was the possibility to initiate and maintain a constant electrospray under a very low, if any, voltage on the metal-coated glass capillary of the FTICR instrument.

8.3.2
Applications

8.3.2.1 Complex biological mixtures of glycopeptides

The feasibility and analytical potential of the fully automated chip-based ESI/QTOF-MS for high-performance glycoscreening, sequencing, and identification was for the first time explored on a complex glycopeptide mixture of *O*-glycosylated sialylated amino acids and peptides obtained from urine of a patient suffering from Schindler disease [48]. Using the negative ion mode detection and methanol as a solvent, the NanoMate provided a sample infusion at 100 nL/min flow rate. For a sample concentration of approximately 2–3 pmol/μL, spectra of high S/N were obtained within less than 1 min of acquisition, indicating a two to three times higher sensitivity than the one obtained in a similar study by using the conventional capillary-based nanoESI in Z-spray geometry (unpublished results).

The reduced S/N of low-abundant molecular precursor ions in complex mixtures obtained by classical nanoESI-MS/MS can generally be overcome by extended accumulation in an off-line approach [44], but it is frequently hindered by spray and/or signal instability. The performance of the NanoMate system to provide long-lasting electrospray signal, rendered reliable conditions for high sensitivity, particularly useful for detection and sequencing of minor glycopeptide components, previously accessible for fragmentation from such complex mixtures only by the on-line CE/ESI-QTOF-MS [33].

Singly and doubly charged ions derived from tri- to octasaccharide peptides or free oligosaccharides were detected, originating from tri- (singly charged), tetra- (singly and doubly charged), penta- (doubly charged), and hexasaccharide (doubly charged) *O*-linked either to Ser or Thr, whereas species from hepta- and octasaccharide peptides were observed as less abundant doubly charged pseudomolecular ions. A large number of minor sialylated oligosaccharide components were identified as well. A low-abundant doubly charged ion detected at *m/z* 890.32 in MS1 was subjected to the chip nanoESI-MS/MS by CID at low collision energies (Fig. 9) and, accordingly, the fragmentation pattern (Fig. 10) could be identified and assigned to the octasaccharide $Neu5Ac_2Gal_{32}GalNAc$-Ser. All these data demonstrated for the first time the ability of the NanoMate chipESI/QTOF-MS methodology to identify minor glycoconjugate species, or those having labile attachments like sialic acid previously undetectable by ESI–MS only [33].

The off-line CE-MS method is considered in some cases a convenient approach due to its higher flexibility toward system optimization than the one provided by the on-line methods. For this reason, the off-line coupling of CE to (−)nanoESI-QTOF-MS has been lately intensively developed and implemented for glycoconjugate analysis. Its potential for separation, screening, and structural elucidation of different types of glycoconjugates, such as glycosaminoglycans [51, 52], *O*-glycosylated peptides [53], and gangliosides [54], was demonstrated in detail. However, using the CE instrument as a fraction collector, the sample concentration is lower-

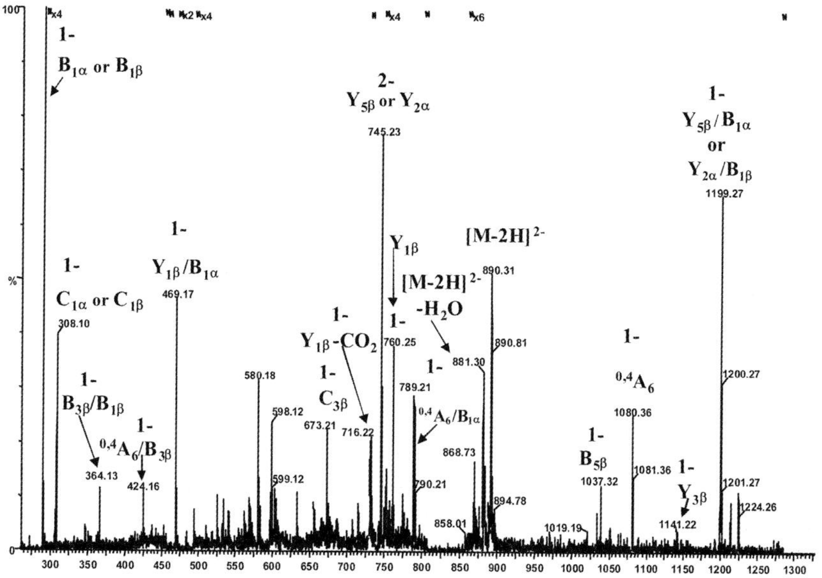

Fig. 9 Automated chip-based (−)nanoESI/QTOF-MS/MS of Neu5Ac$_2$Gal$_3$GlcNAc$_2$GalNAc-Ser detected as a doubly charged ion at m/z 890.32. Spectrum is the sum of scans at collision energy 40 and 70 eV. Number of scans: 200. Average sample concentration: 3 pmol/μL. Average sample consumption for the MS/MS experiment: 2 pmol. Nomenclature for assignment of fragment ions is according to [61, 62]. Reprinted from [47], with permission.

ed because sample reservoirs contain electrolyte, so that volumes of a few nanoliters are diluted into tens-of-microliters. Lack of sensitivity is therefore a major drawback of this approach.

The second limitation of the off-line approach is the low throughput, as a consequence of both the time-consuming collection of fraction and the additional treatments of the sample sometimes necessary to be applied prior to MS. Among the postseparation treatments, increasing the sample concentration by solvent evaporation is the most frequent and time-consuming one. These two inherent problems associated to the off-line CE-MS coupling could be to a significant extent overcome by automated chip-based negative ion ESI-QTOF-MS in off-line conjunction with CE [55]. In this study, for testing the compatibility of the off-line CE coupling to the (−)nano ESIchip NanoMate/QTOF-MS for ionization and detection in a given CE buffer system, a multicomponent mixture of *O*-glycosylated amino acids and peptides from urine of patients suffering from Schindler disease was dissolved in 0.1 M formic acid/ammonia pH 2.8 to a concentration of 5 pmol/μL and infused automatically into MS. A constant and stable spray over 10 min analysis time and high ionic current indicated the compatibility of this buffer system with automatic chip-based ESI and that the analyte molecules could be efficiently ionized. In addition, the flow rate under these buffer conditions was

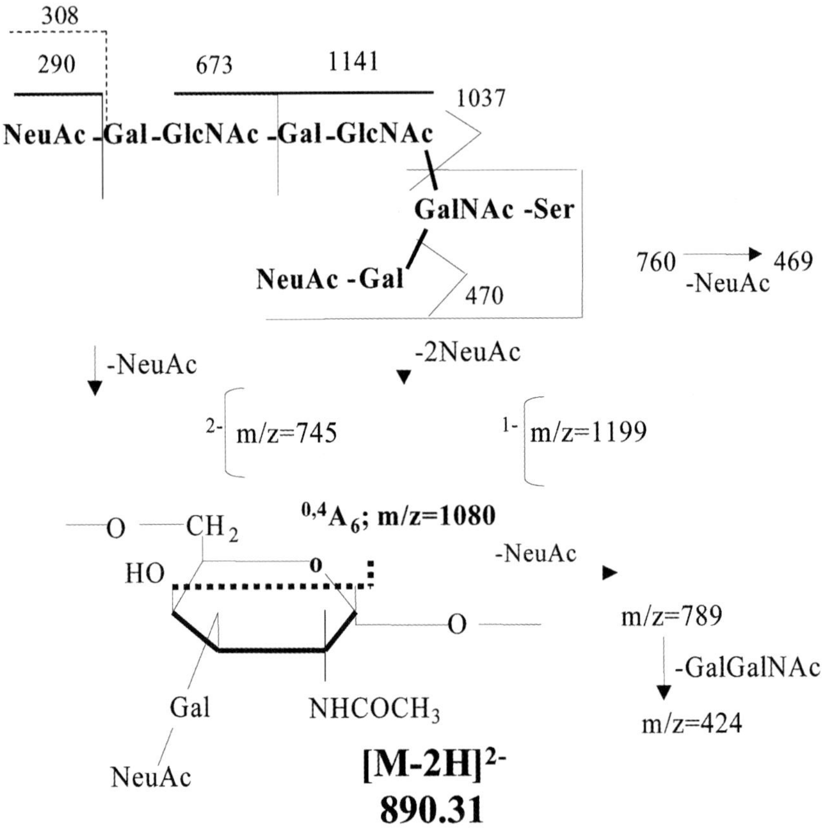

Fig. 10. Fragmentation scheme of $Neu5Ac_2Gal_3GlcNAc_2GalNAc$-Ser detected as a doubly charged ion at *m/z* 890.31. Reprinted with permission from [47].

found reduced to 75 nL/min. Following the CE-UV analysis in the reverse polarity, CE fractions were collected, loaded onto the 96-well plate of the robot, and automatically subjected to the (−)nanoESIchip NanoMate/QTOF-MS and MS/MS (Fig. 11) without the need of preconcentration or any other postseparation treatment which would significantly increase the experiment time. Two major doubly charged ions at *m/z* 525.28 and 532.29 accompanied by their respective singly charged counterparts at *m/z* 1051.23 and 1065.23 corresponding to Ser- and Thr-linked $Neu5Ac_2HexHexNAc$ (HexNAc = *N*-acetyl hexosamine) have been separated and detected in the second CE fraction. Ser- and Thr-linked Neu5AcHexHexNAc were also observed as singly charged ions at *m/z* 760.24 and 774.28, respectively. In lower abundance, the Ser-linked $Neu5Ac_2Hex_2HexNAc_2$ was detected as a doubly charged ion at *m/z* 707.75. Although previously [53] the off-line CE/MS was reported to require rather high amounts of glycopeptide sample for detection, the final concentration of the fraction was here approximately 3 pmol/μL. In the case of

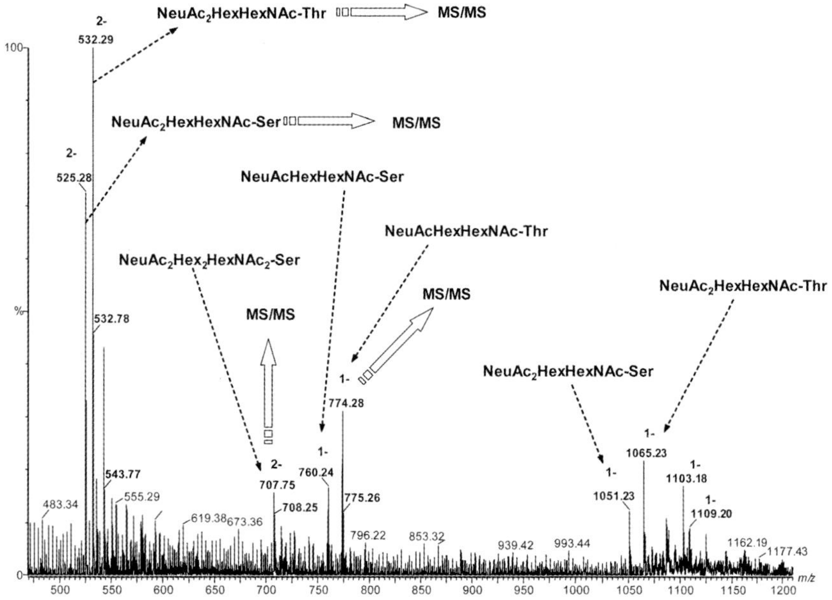

Fig. 11. (−)NanoESIchip/QTOF mass spectrum of the second CE-collected fraction at 80 min after injection. Solvent: 0.1 M formic acid buffered with ammonia to pH 2.8; fraction/buffer concentration: ~3 pmol/µL. Number of scans: 100. Average sample consumption for MS experiment: 0.4 pmol. Copyright © 2004 John Wiley & Sons. Reproduced with permission from [55].

ammonium formate, the 75 nL/min flow rate exhibited by the NanoMate chip system yielded a fair S/N by only 30 scans of signal accumulation (Fig. 11) under the sample consumption of approximately 0.4 pmol, which is a superior sensitivity in comparison with the capillary-based experiments [53, 54]. Due to the high spray stability, the MS/MS investigation of a low abundant doubly charged ion at m/z 707.75 attributed to $Neu5Ac_2Hex_2HexNAc_2$-Ser was greatly enhanced (Fig. 12). The low signal intensity corresponding to this ion could be compensated by the longer acquisition time (15 min) of the MS/MS signal. By this procedure, a sufficient number of sequence ions for clear-cut structural elucidation of the molecule could be generated and detected (Fig. 13). It is noteworthy to mention that in this study, the total amount of sample consumed for generating one MS and three MS/MS of high S/N and sufficiently informative was only 12 pmol.

In another study, the newly conceived NanoMate/FTICR coupling was tested for screening and sequencing by sustained off-resonance (SORI)-CID of *O*-glycosylated peptides and amino acids extracted from urine of a patient suffering from Schindler disease and from urine of a healthy age-matched individual for comparative analyses [50]. The mixtures were dissolved in methanol to different concentrations and transferred onto the glass microtiter plate of the NanoMate robot. The highest sensitivity obtained in this set of experiments was 0.5 pmol/µL, which

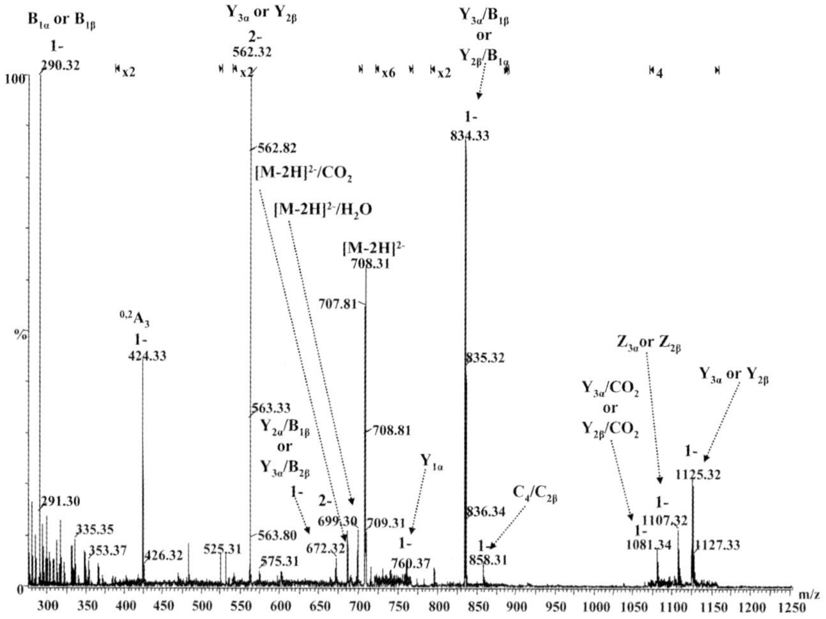

Fig. 12. (−)NanoESIchip/QTOF-MS/MS of the doubly charged ion at m/z 707.75 assigned to Neu5Ac$_2$Hex$_2$HexNAc$_2$-Ser detected in the second CE-collected fraction. Collision energy: 20–40 eV; collision gas pressure: 12 psi. For precursor ion isolation, LM (low-mass resolution) and HM (high-mas resolution) parameters were set at 10 and 10, respectively. Number of scans: 400. Average sample consumption for MS/MS experiment: 3 pmol. Nomenclature for assignment of fragment ions is according to [61, 62]. Reproduced from [55], with permission.

represents a value with one order of magnitude higher than the one achieved in a similar study carried out by using capillary ESI infusion in the same Apollo Apex II configuration [32]. For the initiation of the chip spray, the fine tuning on the x-, y-, and z-axes of the chip nozzle toward the FTICR-MS inlet has been reported to be a critical step [50]. In addition, it was shown that a significant increase of the spray stability and intensity of the MS signal could be achieved by applying a low potential of 50 V to the ion transfer capillary of the FTICR-MS instrument.

In Fig. 14, the (−)chipESI/FTICR-MS of an *O*-glycopeptide mixture from patient's urine is depicted, where a significantly higher number (50) of components than previously detected [32] was found. These glycoforms were in general Ser- and Thr-linked disialo saccharides expressing various saccharide chain lengths ranging from tetra- to octasaccharide. MS experiments carried out under the same conditions of sample solution and ionization/detection indicated a run-to-run and nozzle-to-nozzle reproducibility of nearly 100%. The particular advantage of the NanoMate system to exhibit a high ionization yield and a preferential formation of multiply charged ions is clearly illustrated by the spectrum in Fig. 14. In comparison with previously reported data obtained by regular capillary-based (−)nanoESI-FTICR-MS

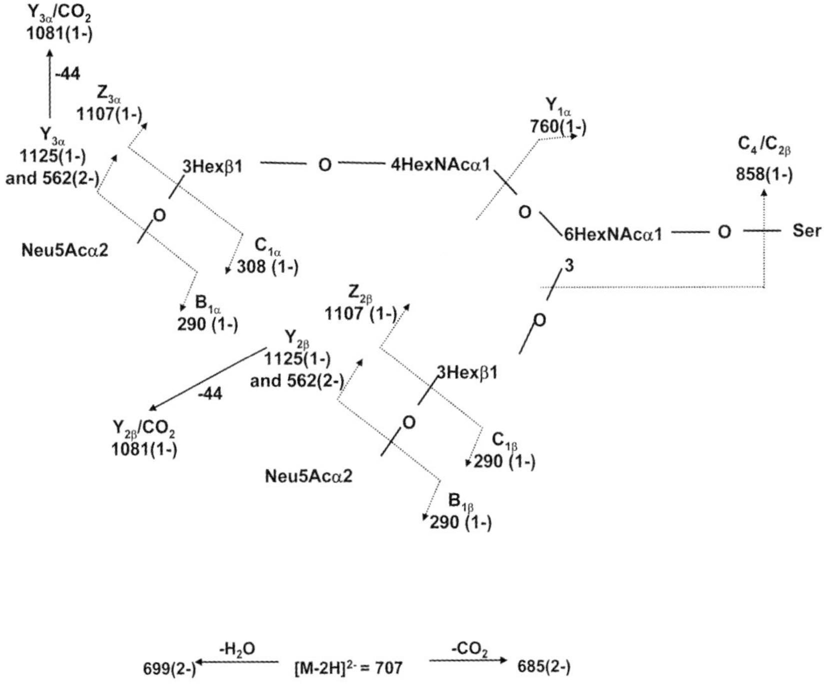

Fig. 13. Fragmentation scheme of the doubly charged ion at *m/z* 707.75 assigned to $Neu5Ac_2Hex_2HexNAc_2$-Ser detected in the second CE-collected fraction. Nomenclature for assignment of fragment ions is according to [61, 62]. Reproduced from [55], with permission.

on a similar glycopeptide mixture [32], the (−)nanoESIchip/FTICR-MS was found to provide a higher ionization efficiency of even less abundant glycopeptide species. In addition, by (−)nanoESIchip/FTICR-MS a considerable number of triply charged ions could be formed and detected under a high S/N, whereas by capillary-based (−)nano-ESI-FTICR-MS no triply charged glycopeptide ions were formed/detected [32].

Moreover, four new components, not identified by any other method before [32–35], were detected by this novel approach and identified under a high mass accuracy: to sodiated Ser- and Thr-linked (disialo) octasaccharides and (trisialo) undecasaccharides, respectively (Fig. 14).

For identification of potential diagnostic marker components, a mixture of *O*-glycopeptides extracted from urine of an age-matched healthy control person, identically prepared as described before, was subjected to (−)ESIchipNanoMate/FTICR-MS and SORI-CID MS/MS under the same experimental conditions as those from patient urine [50]. The MS pattern was different: in this mixture a reduced number of glycoconjugate species has been found, where those with shorter chains and lower degree of sialylation were dominating. By the NanoMate/FTICR SORI-CID MS/MS experiment, the structure of the Ser-linked (disialo) hexasaccharide components could be straightforwardly determined.

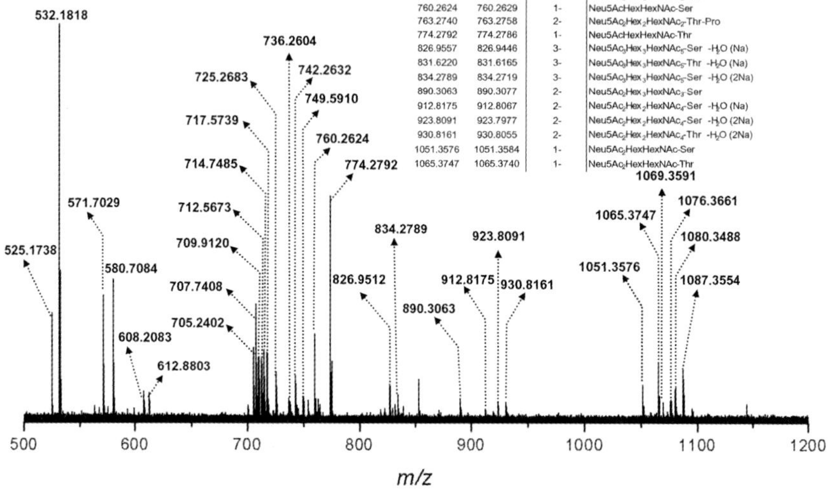

m/z exp.	m/z theor.	Charge state	Proposed structure
525.1738	525.1755	2-	Neu5Ac$_2$HexHexNAc-Ser
532.1818	532.1834	2-	Neu5Ac$_2$HexHexNAc-Thr
571.7029	571.7045	2-	Neu5Ac$_2$HexHexNAc-Thr-Pro -H$_2$O
580.7084	580.7097	2-	Neu5Ac$_2$HexHexNAc-Thr-Pro
608.2083	608.2020	3-	Neu5Ac$_2$Hex$_2$HexNAc$_4$-Ser -H$_2$O (Na)
612.8803	612.8739	3-	Neu5Ac$_2$Hex$_2$HexNAc$_4$-Thr -H$_2$O (Na)
707.7408	707.7416	2-	Neu5Ac$_2$Hex$_2$HexNAc$_2$-Ser
714.7485	714.7495	2-	Neu5Ac$_2$Hex$_2$HexNAc$_2$-Thr
760.2624	760.2629	1-	Neu5AcHexHexNAc-Ser
763.2740	763.2758	2-	Neu5Ac$_2$Hex$_2$HexNAc$_2$-Thr-Pro
774.2792	774.2786	1-	Neu5AcHexHexNAc-Thr
826.9557	826.9446	3-	Neu5Ac$_3$Hex$_3$HexNAc$_5$-Ser -H$_2$O (Na)
831.6220	831.6165	3-	Neu5Ac$_3$Hex$_3$HexNAc$_5$-Thr -H$_2$O (Na)
834.2789	834.2719	3-	Neu5Ac$_3$Hex$_3$HexNAc$_5$-Ser -H$_2$O (2Na)
890.3063	890.3077	2-	Neu5Ac$_2$Hex$_3$HexNAc$_3$-Ser
912.8175	912.8067	2-	Neu5Ac$_2$Hex$_2$HexNAc$_4$-Ser -H$_2$O (Na)
923.8091	923.7977	2-	Neu5Ac$_2$Hex$_2$HexNAc$_4$-Ser -H$_2$O (2Na)
930.8161	930.8055	2-	Neu5Ac$_2$Hex$_2$HexNAc$_4$-Thr -H$_2$O (2Na)
1051.3576	1051.3584	1-	Neu5Ac$_2$HexHexNAc-Ser
1065.3747	1065.3740	1-	Neu5Ac$_2$HexHexNAc-Thr

Fig. 14. Automated (−)chipESI/FTICR mass spectrum of a fraction from urine of patients suffering from Schindler disease. Apex II metal-coated glass capillary voltage: 50 V. Capillary exit: −79 V. Sample concentration: 5 pmol/μL in methanol. Number of scans: 150. Average sample consumption for MS experiment: 2.5 pmol. Inset: Table: Ions detected and identified with a mass accuracy below 12 ppm in the mixture of O-glycosylated peptides at 5 pmol/μL. Nomenclature for assignment of fragment ions is according to [61, 62]. Reproduced and adapted with permission from [50].

8.3.2.2 Combination of fully automated chip ESI-MS and automated software assignment for identification of heterogeneous mixtures of *N*-, *O*-glycopeptides, and oligosaccharides

A urine fraction containing sialylated carbohydrates from a patient presenting a clinical picture associated congenital disorders of glycosylation (CDG) was submitted to glycoscreening by fully automated chip-based (−)nanoESI/QTOF-MS [47]. The sample with average concentration of 5 pmol/μL in methanol was loaded onto the 96-well plate of the NanoMate robot. Following automatic infusion, the mass spectrum acquired for 3 min displayed a high level of heterogeneity concerning the type of glycans and glycopeptides, as well as the degree of sialylation (Fig. 15). For preliminary analysis and assignment of glycoforms, a computer algorithm for calculation of glycoconjugate composition was developed [47, 56] thus providing the first two-stages platform for completely automatic and high-throughput analysis: (1) robotized sample delivery and infusion into MS followed by (2) automatic identification of components using software-assisted assignment. The experimental data for the input were the *m*/*z* values and

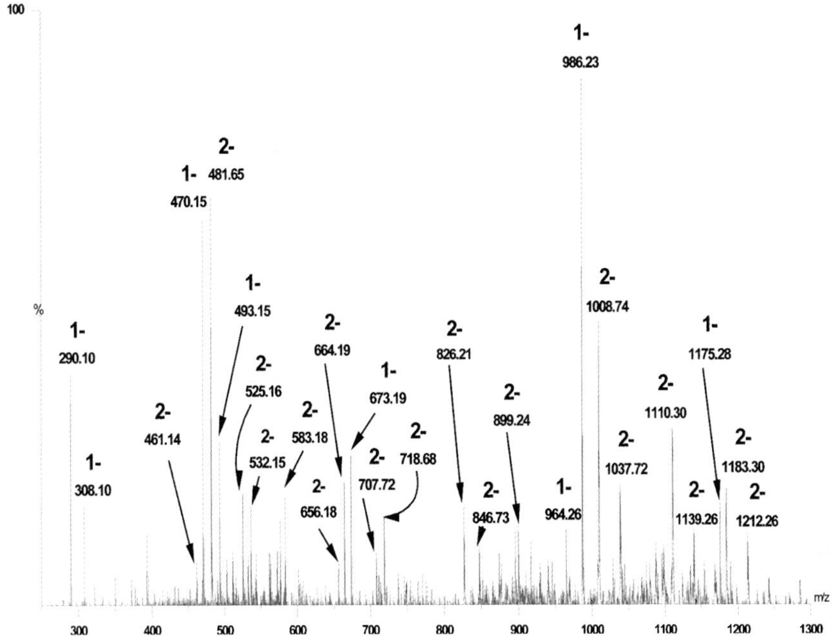

Fig. 15. Fully automated chip-based (−)nanoESI/QTOF MS of the mixture containing glycans and glycopeptides from urine of a patient diagnosed with CDG. Average sample concentration: 5 pmol/μL; number of scans: 100. Average sample consumption for MS experiment: 2 pmol. Sampling cone potential: 50 V. Reprinted from [47], with permission.

the respective charge states provided by chip MS and MS/MS experiments. The capacity of the computer program was to calculate and propose all possible structures containing hexose (Hex), HexNAc, deoxyhexose (dHex), and Neu5Ac as monosaccharide building-block units. In the second step, the modeling of the theoretical fragmentation pattern from different structures with the same *m/z* was processed. The data were presented as a combination of ions B, C, Y, and Z, related to the cleavage of glycosidic bonds, as well as the A- and X-ring cleavage ions, aligned to the proposed structure of the chosen molecular ion. In Tab. 1 the software first-step MS analysis along with the proposed composition of single components from the chip-based (−)nanoESI/QTOF spectrum presented in Fig. 15 is given. The most prominent structures in the mixture were directly identified by this approach.

As an example, the doubly charged molecular ion at *m/z* 1008.74 (Fig. 15) showing the composition of either $Neu5Ac_2Hex_5HexNAc_3$ or $Neu5Ac_2Hex_3HexNAc_1dHex_5$ was submitted to the automated chip-based (−)nanoESI/QTOF-MS/MS by CID at low energies. The fragment ions of Neu5AcHex, Neu5AcHexHexNAc, $Neu5AcHex_2HexNAc$, and $Neu5Ac_2Hex_5HexNAc_2$ as C_2, C_3, C_4, and C_5 were assigned by this software to a truncated *N*-linked biantennary glycan missing the terminal HexNAc at the reducing terminus.

Tab. 1 Computer-assisted assignment of the major ions detected by automated chip-based (−)nanoESI/QTOF-MS in the mixture of glycans and glycopeptides from urine of a patient diagnosed with CDG.

$[M-2H]^{2-}$ *m/z*	$[M-H]^{-}$ *m/z*	Composition according to calculation	Theoretical *m/z*
	290.10	NeuAc ($-H_2O$)	290.09
	308.10	NeuAc	308.10
461.14		$NeuAc_2Hex_2$	461.15
	470.15	NeuAcHex	470.15
481.65		$NeuAc_2HexHexNAc$	481.66
	493.15	NeuAcHexNAc ($-H_2O$)	493.17
525.16		$NeuAc_2HexHexNAc$-Ser	525.18
532.15		$NeuAc_2HexHexNAc$-Thr	532.18
583.18		$NeuAc_2HexHexNAc_2$	583.20
656.18		$NeuAc_2dHexHexHexNAc_2$	656.23
664.19		$NeuAc_2Hex_2HexNAc_2$	664.23
	673.19	NeuAcHexHexNAc	673.23
707.72		$NeuAc_2Hex_2HexNAc_2$-Ser	707.74
718.68		$NeuAc_2Hex_2HexNAc_2$-Ser (Na)	718.74
826.21		$NeuAc_2Hex_4HexNAc_2$	826.28
846.73		$NeuAc_2Hex_3HexNAc_3$	846.80
899.24		$NeuAc_2dHexHex_4HexNAc_2$	899.31
	964.26	$NeuAc_2HexHexNAc$	964.33
	986.23	$NeuAc_2HexHexNAc$ (Na)	986.33
1008.74		$NeuAc_2Hex_5HexNAc_3$	1008.84
1037.72		$NeuAcdHexHex_5HexNAc_4$	1037.87
1110.30		$NeuAc_2Hex_5HexNAc_4$	1110.38
1139.26		$NeuAcdHexHex_5HexNAc_5$	1139.41
	1175.28	$NeuAc_2HexHexNAc$-Ser (2Na + P)	1175.32
1183.30		$NeuAc_2dHexHex_5HexNAc_4$	1183.41
1212.26		$NeuAcdHex_2Hex_5HexNAc_5$	1212.43

Reprinted from [47], with permission.

8.3.2.3 Characterization of *N*-glycosylation microheterogeneity and sites in intact glycoproteins

Glycosylation of proteins is one of the most ubiquitous post-translational modification. It is highly regulated and changes during differentiation, development, under different physiological and cell culture conditions, and in disease.

Diversity of glycoprotein structure resides on number and position of glycosylation sites within the polypeptide backbone and the type of sugar attached. Besides the microheterogeneity on the same glycosylation site, a number and type of glycan chain phosphorylation, sulfation, methylation, and acetylation may be found on glycan residues [57–59]. Therefore, a demand for a complete analysis of a single glycoprotein would have to provide information regarding microheterogeneity at single glycosylation sites as well as a structural characterization of the appropriate carbohydrate moiety.

Due to recent advances in automated chipESI-MS and the development of small-scale sample preparation procedures, the characterization of glycosylations became amenable to high-throughput experiments, in which the necessary sample amount was decreased to the low picomolar level. In the first study carried out for the determination of the glycosylation site by fully automated chip nanoESI-MS, the NanoMate 100 robot was coupled to the front of a Finnigan LTQ IT *via* a specially designed mounting bracket [49]. The tryptic digest of bovine pancreatic ribonuclease B was dissolved in 50% methanol with 0.1% formic acid for positive ion mode and in 50% methanol with 0.1% ammonium hydroxide for negative ion mode to the concentration of 1 pmol/μL. Five microliters of sample was loaded onto the 96-well plate of the robot and infused into the LTQ IT MS at a flow rate of 100 nL/min. From the full-scan MS the complex peptide mass fingerprint of ribonuclease B was derived. Several peptides were sequenced by MS/MS using collision energies of 20–25% and a maximum scan time of 50 ms in combination with the automated search program BioWorks 3.1 (Finnigan LCQ). Nine out of 14 possible tryptic peptides were identified, resulting in a protein sequence coverage of 87%. The unidentified ions, showing a pattern typical for high-mannose type of glycosylation, were subjected to multiple stage MS, up to MS^5. As shown in Fig. 16a by fully automated chip-based infusion in combination with multiple stage CID MS, spectra of high quality and rich in information related to protein glycosylation could be obtained. Since the *N*-glycopeptides tend to fragment mostly at their oligosaccharide portion, it is often difficult to gather the data necessary for the characterization of their amino acid sequence. In the case described here the amino acid sequence was obtained in the MS^4 experiment as illustrated in Fig. 16b.

Fully automated chip-based nanoESI coupled to Q-FTICR at 9.4 T was recently employed for the first direct identification of multiple *O*-glycosylation sites in hinge region (HR) of immunoglobulin A1 (IgA1) isolated from myeloma as reported by Renfrow *et al.* [60]. The method was designed to provide heterogeneity assay of the tryptic glycopeptide expressing various *O*-glycoforms and particularly detection and localization of the Gal-deficient IgA1 extracted, separated, and purified from myeloma IgA1 HR and the structural elucidation of the glycopeptides by means of electron capture dissociation (ECD)-, infrared multiphoton dissociation (IRMPD)-, and activated ion ECD (AI-ECD)-MS/MS. The optimization of the methodology in terms of sequencing analysis has been carried out on a mixture of synthetically obtained peptides and glycopeptides of the IgA1 HR type. The samples were prepared at a concentration of 5 mM in 1:1 water/methanol, 2% acetic acid and microelectrosprayed from an in-house made fused silica capillary at a flow rate of 300 nL/min. By employing IRMPD-MS/MS the assay of the peptide sequence could be achieved, while by ECD-MS/MS the glycosylation sites of single glycopeptide and/or mixture of glycosylated peptides was unambiguously determined. For the tryptic-isolated myeloma IgA1 HR glycopeptides, the samples were desalted by C_{18} ZipTip into 30 μL of a 4:1 ACN/water solution containing 0.1% formic acid, and submitted to microelectrospray and automated chip-based ESI-FTICR-MS and MS/MS. The (+)ESI-FTICR-MS of isolated IgA1 HR peptide features a series of triply charged ions separated by one-third of the masses of GalNAc and

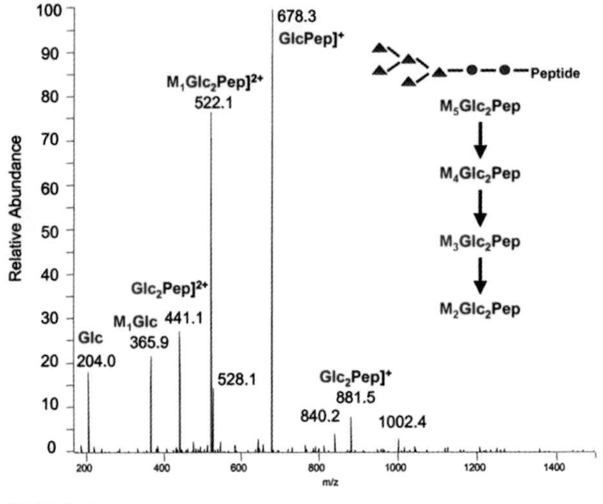

TABLE I

Amino Acid Sequence and the Predicted Peptides from a Trypsin Digestion of Bovine Ribonuclease B

ETAAAK	FER	**QHMDSSTSAASSSNYCNQMMK**	SR	**NLTK**	DR	CK
PVNTFVHESLADVQAVCSQK	**NVACK**	**NGQTNCYQSYSTMSITDCR**	**ETGSSK**			
YPNCAYK	**TTQANK**	**HIIVACEGNPYVPVHFDASV**				

Identified peptides are shown in bold; *N*-glycosylation site is shown in red.

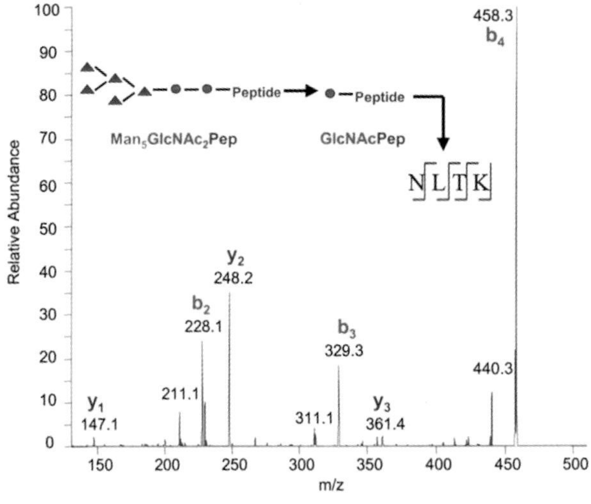

Fig. 16. (a) Fully automated nanoESINanoMate/IT MS^5 for the stepwise removal of the mannose residue from the precursor ribonuclease B *N*-glycopeptide. (b) Fully automated nanoESINanoMate/IT MS^4 derived from ribonuclease B related *N*-glycopeptide for the identification of amino acid sequence. Inset: amino acid sequence and predicted peptides from tryptic digestion of ribonuclease B. Nomenclature for assignment of fragment ions is according to [66]. Reprinted and adapted with permission from [49].

Gal units. This reveals on one hand the high heterogeneity of the differentially glycosylated species, and on the other hand it can be deduced that the in-source fragmentation is reduced almost to zero, since no ion corresponding to non-glycosylated species could be detected and identified. Based on the amino acid sequence of the isolated IgA1 HR determined by IRMPD-MS/MS, building-block analysis of the monosaccharide units, and high mass accuracy and resolution of the detection, the number of GalNAc and Gal residues attached to the peptide back bone could be predicted. This assessment indicated that two of the predicted glycopeptides could bear a Gal-deficient glycan part.

When applying the same sequencing protocol as for synthetic peptides and glycopeptides for analysis of the isolated IgA1 glycopeptides by IRMPD- and ECD-MS/MS, only a limited number of fragment ions could be generated. Therefore, a combined sequencing technique based on tandem IRMPD and ECD events, called AI-ECD, was optimized to provide the full set of diagnostic ions for the localization of glycosylation sites. Thus, the quadrupole and stored waveform inverse Fourier transform (SWIFT) isolated precursor ions were subjected for 300 ms to photon irradiation at 6.4 or 8 W laser power, followed immediately by electron irradiation event for 10–20 ms at cathode heating power of 11 W, which corresponds to an electron emission current of 300 nA. The AI-ECD FTICR MS/MS of an isolated [V^{216}–L^{246}] containing four GalNAc and four Gal moieties, acquired under the optimized sequencing conditions, gave rise to unambiguous determination of the glycosylation sites of the disaccharide units GalNAc-Gal to T^{225}, T^{228}, S^{230}, S^{232}, respectively (Fig. 17). Based on the accurate ESI-FTICR-MS1 mass determination, two of the glycopeptides were predicted as Gal-deficient glycopeptides. The AI-ECD FTICR-MS/MS of the [V^{216}–L^{246}] containing four GalNAc and three Gal residues, predicted within a mass accuracy of 3.6 ppm, is presented in Fig. 18. The product ions generated by the peptide bone cleavage uniquely localize three GalNAc-Gal residues at T^{225}, T^{228}, and S^{230}, and a GalNAc at S^{232} (Fig. 18).

8.3.2.4 **Complex ganglioside mixtures**

A recent study [48] was designed to assess the potential of the fully automated (−)nanoESIchip/QTOF-MS and automatic MS/MS for screening and sequencing of gangliosides extracted from normal human cerebellar gray matter. The highly complex ganglioside sample dissolved in 100% methanol was automatically infused under the premise of optimization of the ESI chip ionization yield for all single ganglioside components first at higher concentrations (15 pmol/μL) and elevated values of the ESI chip and sampling cone potentials. Such an approach has been already demonstrated [42, 44] to enhance the ionization yield of larger ganglioside species from human brain tissues. Under these solution and instrumental parameters major ganglioside variants were detected, but minor components and/or longer chains, assumed to be present in the mixture could, however, not be found. The conditions were further modified by diluting the sample to 2–3 pmol/μL and varying the cone potential values from 45 to 135 V and the ESI capillary potential from 1.45 to 1.67 kV (Fig. 19), which allowed high ionization yield and detection of an extended number of species. By this approach, a realistic representation upon the ganglioside mixture heterogeneity as compared to data provided by TLC analysis was reported. According to this set of data, it was concluded that by ESIchip/MS a high sensitivity for the analysis of ganglioside multicomponent mixture, without compromising the ionization of minor, biologically relevant components can be achieved. Moreover, in this work the capability of the

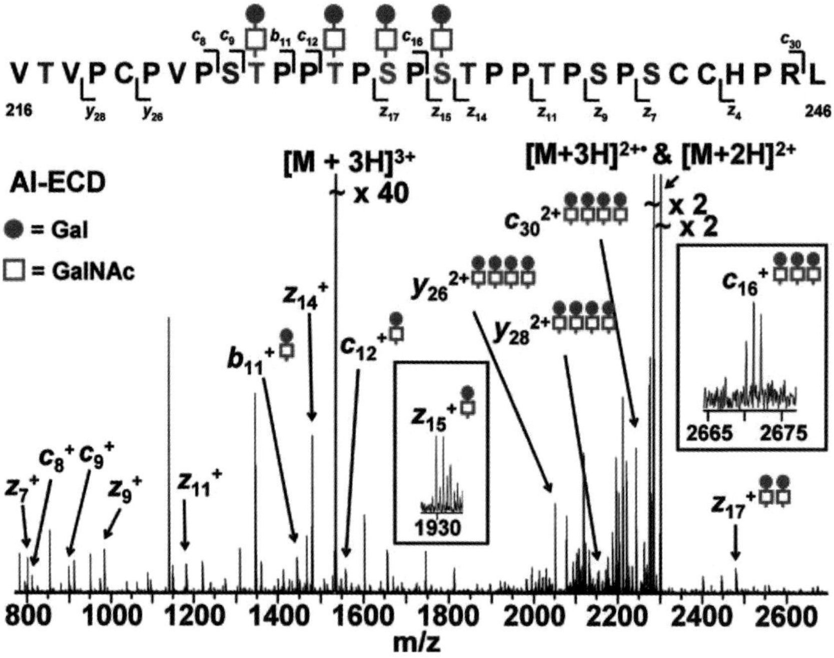

Fig. 17. (+)ESI/AI-ECD FTICR-MS/MS spectrum of a population of [^{216}VTVPCOVPSTP-PTPSPSTPPTPSPSCCHPRL ^{246} + 4GalNAc + 4Gal + 3H]$^{3+}$ ions from the *O*-glycosylated IgA1 HR peptide. Isolation of the precursor ions: by quadrupole and SWIFT. IRMPD event: 300 ms photon irradiation at 6.4 or 8 W laser power. ECD event: 10–20 ms electron irradiation event at 11 W cathode heating power, ~300 nA. Nomenclature for assignment of fragment ions is according to [66]. Reprinted by permission of The American Society for Biochemistry and Molecular Biology from [60].

NanoMate chipESI to minimize in-source fragmentation and significantly reduce the analysis time as compared to ganglioside MS screening by capillary Z-spray sample infusion [42, 44] was demonstrated.

In Tab. 2, the assignment of singly, doubly, and triply charged ions corresponding to 44 different ganglioside components identified by automated chip-based nanoESI-QTOF-MS is listed. From Tab. 2, it is obvious that the cerebellar gray matter is dominated by GD1 glycoforms. GM1, GM2 and GM3, GD2 and GD3, as well as GT1 and GQ1 expressing different ceramide portions and biologically important *O*-acetylated and fucosylated GD1, and GT1 and GQ1 exhibiting high heterogeneity in their ceramide motifs could also be detected and identified by the NanoMate approach.

In order to test the feasibility of an efficient ganglioside sequencing by automated MS to MS/MS mode switching available on the QTOF mass spectrometer, the triply charged ion at m/z 708.39 assigned to GT1 (d18:1/18:0) has been submitted to automatic chip MS infusion followed by automatic MS/MS fragmentation in data-dependent acquisition (DDA). In previous work on ganglioside frag-

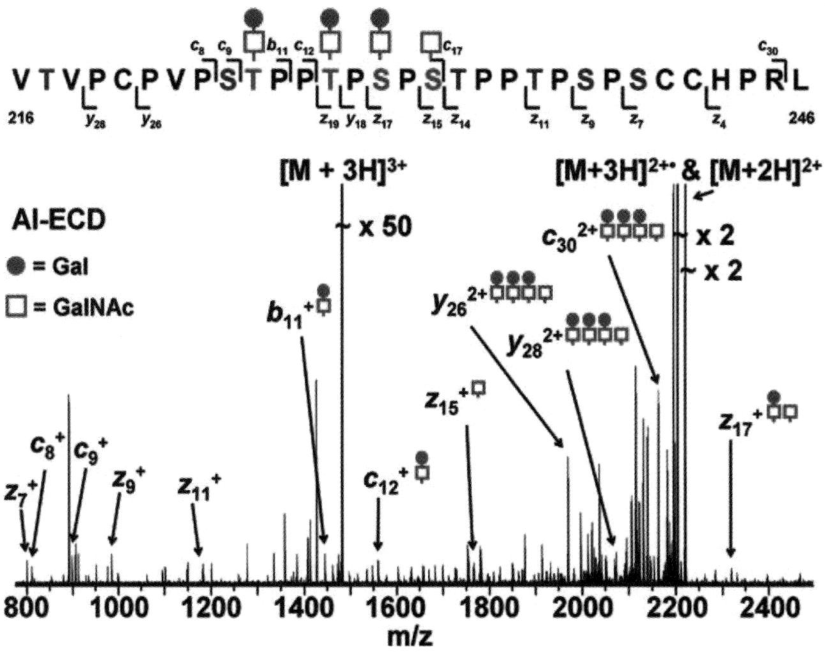

Fig. 18. (+)ESI/AI-ECD FTICR-MS/MS spectrum of a population of [^{216}VTVPCOVPSTP-PTPSPSTPPTPSPSCCHPRL ^{246} + 4GalNAc + 3Gal + 3H]$^{3+}$ ions from the *O*-glycosylated IgA1 HR peptide. Isolation of the precursor ions: by quadrupole and SWIFT. IRMPD event: 300 ms photon irradiation at 6.4 or 8 W laser power. ECD event: 10–20 ms electron irradiation event at 11 W cathode heating power, ~300 nA. Reprinted by permission of The American Society for Biochemistry and Molecular Biology from [60].

mentation using capillary-based (−)nanoESI-QTOF in Z-spray and manual selection for MS/MS [44], a long signal acquisition time with a sample consumption of about 50 pmol was necessary to obtain fairly high abundance of fragment ions diagnostic for the structure assignment of GT1 from a purified fraction. By Nano ESIchip-automatic fragmentation, within about 1 min of acquiring in MS/MS mode at a flow rate of 100 nL/min, approximately 0.5 pmol of material was sprayed and a sufficient set of fragment ions as fingerprints for structural elucidation of the molecule was obtained.

8.4
Conclusions and perspectives

Studies upon carbohydrate structure are indispensable to define complex life systems though they were generally considered as extremely difficult because of the lack of the basic efficient technologies commonly used for proteins or nucleic acids.

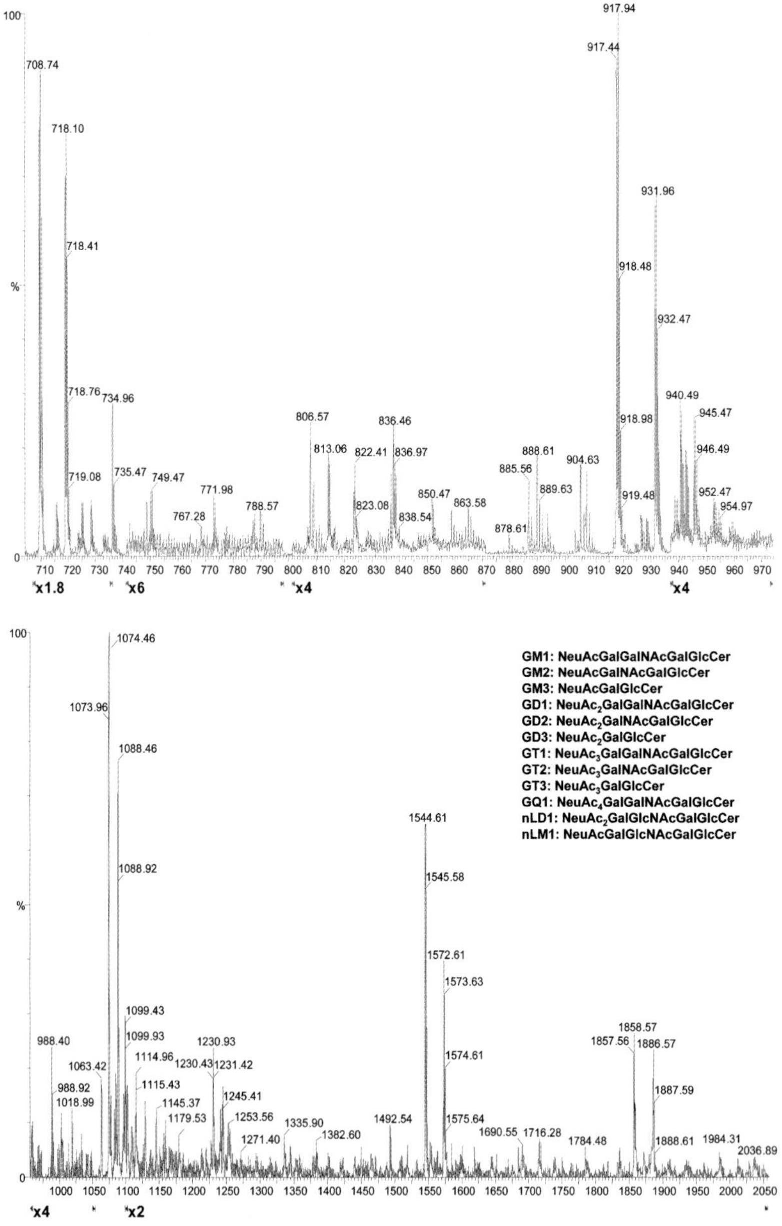

Fig. 19. Fully automated (−)nanoESIchip/QTOF-MS of the native ganglioside mixture from normal human cerebellar gray matter. Sample concentration: 2–3 pmol/μL in methanol; acquisition time: 3 min; number of scans: 90. Average sample consumption for MS experiment: 0.7 pmol, sampling cone potential: 45–135 V. n.a. = not assigned. (a) *m/z* (700–980). (b) (980–2050). Reprinted and adapted by permission of Elsevier [48].

Tab. 2 Composition of single components in the ganglioside mixture from gray matter of normal human cerebellum as detected by a fully automated (−)nanoESIchip/QTOF-MS.

Type of molecular ion	*m/z* (monoisotopic)		Assigned structure
	Detected	Calculated	
$[M + 2Na - 4H]^{2-}$	611.40	611.35	GM3 (d18:1/18:0)
$[M - H]^{-}$	1179.57	1179.74	
$[M - H]^{-}$	1382.60	1382.82	GM2 (d18:1/18:0)
$[M - 2H]^{2-}$	734.96	734.91	GD3 (d18:1/18:0)
$[M + Na - 2H]^{-}$	1492.78	1492.81	
$[M - 2H]^{2-}$	748.99	748.93	GD3 (d18:1/20:0)
$[M - H]^{-}$	1518.51	1518.85	GM1, nLM1, and/or LM1 (d18:0/16:0)
$[M - 2H]^{2-}$	771.98	771.93	GM1, nLM1, and/or LM1 (d18:1/18:0)
$[M - H]^{-}$	1544.61	1544.85	
$[M - 2H]^{2-}$	786.00	785.92	GM1, nLM1, and/or LM1 (d18:1/20:0)
$[M - H]^{-}$	1572.61	1572.85	
$[M - H]^{-}$	1690.55	1690.93	Fuc-GM1(d18:1/18:0)
$[M - H]^{-}$	1716.56	1716.94	Fuc-GM1(d18:1/20:1)
$[M - 2H]^{2-}$	836.46	836.45	GD2 (d18:1/18:0)
$[M - 2H]^{2-}$	850.47	850.47	GD2 (d18:1/20:0)
$[M - 2H]^{2-}$	917.44	917.48	GD1, nLD1, and/or LD1 (d18:1/18:0)
$[M + Na - 3H]^{2-}$	928.45	928.47	
$[M - H]^{-}$	1835.62	1835.96	
$[M + Na - 2H]^{-}$	1857.56	1857.95	
$[M - 2H]^{2-}$	926.44	926.48	GD1, nLD1, and/or LD1 (t18:0/18:0)
$[M - 2H]^{2-}$	924.44	924.49	GD1, nLD1, and/or LD1 (d18:1/19:0)
$[M - 2H]^{2-}$	931.46	931.49	GD1, nLD1, and/or LD1 (d18:1/20:0)
$[M + Na - 3H]^{2-}$	942.44	942.48	
$[M - H]^{-}$	1885.60	1885.98	
$[M - 2H]^{2-}$	940.49	940.50	GD1, nLD1, and/or LD1 (t18:0/20:0)
$[M - 2H]^{2-}$	938.44	938.50	GD1, nLD1, and/or LD1 (d18:1/21:0)
$[M - 2H]^{2-}$	945.47	945.51	GD1, nLD1, and/or LD1 (d18:1/22:0)
$[M - 2H]^{2-}$	954.46	954.51	GD1, nLD1, and/or LD1 (t18:0/22:0)
$[M - 2H]^{2-}$	952.47	952.52	GD1, nLD1, and/or LD1 (d18:1/23:0)
$[M - 2H]^{2-}$	958.46[a)]	958.52	GD1, nLD1, and/or LD1 (d18:1/24:1)
$[M - 2H]^{2-}$	966.44	966.53	GD1, nLD1, and/or LD1 (d18:1/25:0) or (d20:1/23:0)
$[M - 2H]^{2-}$	988.40	988.49	Fuc-GD1 (d18:1/18:2)
$[M - 2H]^{2-}$	990.40	990.51	Fuc-GD1 (d18:1/18:0)
$[M - 2H]^{2-}$	999.41[a)]	999.51	Fuc-GD1 (t18:0/18:0)
$[M - 2H]^{2-}$	1002.41	1002.51	Fuc-GD1 (d18:1/20:2)
$[M - 2H]^{2-}$	1004.42	1004.52	Fuc-GD1 (d18:1/20:0)
$[M - 2H]^{2-}$	1013.44[a)]	1013.53	Fuc-GD1 (t18:0/20:0)
$[M - 2H]^{2-}$	1018.99	1019.02	GalNAc-GD1 (d18:1/18:0)
$[M - 2H]^{2-}$	1032.93[a)]	1033.03	GalNAc-GD1 (d18:1/20:0)

Nomenclature for assignment of ions is according to [63–65]. Reprinted by permission of Elsevier from [48].

Tab. 2 Continued

Type of molecular ion	*m/z* (monoisotopic)		Assigned structure
	Detected	Calculated	
$[M-3H]^{3-}$	708.39	708.35	GT1 (d18:1/18:0)
$[M-2H]^{2-}$	1062.96	1063.03	
$[M+Na-3H]^{2-}$	1073.96	1074.02	
$[M+2Na-4H]^{2-}$	1084.93	1085.01	
$[M-3H]^{3-}$	714.41	714.35	GT1 (t18:0/18:0)
#$[M+Na-3H]^{2-}$	1082.92	1083.02	
$[M-3H]^{3-}$	717.75	717.69	GT1 (d18:1/20:0)
$[M-2H]^{2-}$	1076.97	1077.04	
$[M+Na-3H]^{2-}$	1087.95	1088.03	
$[M+2Na-4H]^{2-}$	1098.92	1099.02	
$[M-3H]^{3-}$	723.75	723.70	GT1 (t18:0/20:0)
$[M+Na-3H]^{2-}$	1096.93	1097.04	
$[M+Na-3H]^{2-}$	1094.95[a)]	1095.04	GT1 (d18:1/21:0)
$[M-3H]^{3-}$	727.11	727.04	GT1 (d18:1/22:0)
$[M+Na-3H]^{2-}$	1101.92	1102.05	
$[M+Na-3H]^{2-}$	1108.92[a)]	1109.06	GT1 (d18:1/23:0)
$[M+Na-3H]^{2-}$	1114.96	1115.06	GT1 (d18:1/24:1)
$[M-3H]^{3-}$	722.39	722.35	*O*-Ac-GT1 (d18:1/18:0)
$[M-3H]^{3-}$	731.74	731.70	*O*-Ac-GT1 (d18:1/20:0)
$[M-2H]^{2-}$	1128.95	1129.05	Fuc-GT1 (d18:1/17:0)
$[M-2H]^{2-}$	1144.89	1145.06	Fuc-GT1 (t18:0/18:0)
$[M-2H]^{2-}$	1159.89	1159.08	Fuc-GT1 (t18:0/20:0)
$[M-3H]^{3-}$	805.40	805.38	GQ1 (d18:1/18:0)
$[M+Na-4H]^{3-}$	812.73	812.71	
$[M+2Na-4H]^{2-}$	1230.43	1230.56	
$[M+3Na-5H]^{2-}$	1241.43	1241.55	
$[M-3H]^{3-}$	814.74	814.72	GQ1 (d18:1/20:0)
$[M+Na-4H]^{3-}$	822.07	822.05	
$[M+2Na-4H]^{2-}$	1244.42	1244.57	
$[M-3H]^{3-}$	819.38[a)]	819.38	*O*-Ac-GQ1 (d18:1/18:0)
$[M+Na-4H]^{3-}$	826.73[a)]	826.71	

d = dihydroxy sphingoid base; t = trihydroxy sphingoid base
a) Low intensity ions. Monoisotopic m/z values of ions are given.

In the past decade, due to the progress in systems biology and ascendant development of specific methods and techniques, glycomics could be introduced as a new concept to follow the genomic and proteomic ones.

MS is crucial for carbohydrate research, by contributing basic understanding of how posttranslational events such as glycosylation affect protein activities. In the past decade, capillary nanoESI-MS has developed as an effective means in glycomics. However, several disadvantages of the method were encountered including low sampling throughput, potential carryovers, and difficult reproducibility due to the variable shape of the spray tip.

The recent introduction of chip-based ESI-MS in glycomics is obviously driven by the higher performance, sensitivity, and reduced analysis time. To date, two different chipESI systems, a thin chip microsprayer and a fully automated chip-based nanoESI robot, have been coupled each to high performance mass spectrometers and tested for structural elucidation of glycoconjugates originating from various biological matrices.

The high sensitivity and ionization efficiency, provided at nano- and microscale level by these systems, will help to establish the new chip-based technology for applications that require identification of unknown, minor components in complex native mixtures of glycoconjugates. Finally, nano- and microchip ESI-MS methods are on the way to become the best suited tools for high-throughput structural elucidation of carbohydrates indicative of pathological states, allowing the chipMS-based glycomics perspectives for large-scale use in biomedical research and clinical diagnostics.

8.5 References

[1] Dahlin, A. P., Wetterhall, M., Liljegren, G., Bergstrom, S. K., Andren, P., Nyholm, L., Markides, K. E., Bergquist, J., *Analyst* 2005, *130*, 193–199.

[2] Leuthold, L. A., Grivet, C., Allen, M., Baumert, M., Hopfgartner, G., *Rapid Commun. Mass Spectrom.* 2004, *18*, 1995–2000.

[3] Overall, C. M., Tam, E. M., Kappelhoff, R., Connor, A., Ewart, T., Morrison, C. J., Puente, X., *et al.*, *Biol. Chem.* 2004, *385*, 493–504.

[4] Vlahou, A., Schellhammer, P. F., Wright, G. L. Jr., *Adv. Exp. Med. Biol.* 2003, *539*, 47–60.

[5] Sydor, J. R., Scalf, M., Sideris, S., Mao, G. D., Pandey, Y., Tan, M., Mariano, M., *et al.*, *Anal. Chem.* 2003, *75*, 6163–6170.

[6] Huikko, K., Kostiainen, R., Kotiaho, T., *Eur. J. Pharm. Sci.* 2003, *20*, 149–171.

[7] Schultz, G. A., Corso, T. N., Prosser, S. J., Zhang, S., *Anal. Chem.* 2000, *72*, 4058–4063.

[8] Zhang, S., Van Pelt, C. K., Henion, J. D., *Electrophoresis* 2003, *24*, 3620–3632.

[9] Dethy, J. M., Ackermann, B. L., Delatour, C., Henion, J. D., Schultz, G. A., *Anal. Chem.* 2003, *75*, 805–811.

[10] Kapron, J. T., Pace, E., Van Pelt, C. K., Henion, J. D., *Rapid Commun. Mass Spectrom.* 2003, *17*, 2019–2026.

[11] Van Pelt, C. K., Zhang, S., Fung, E., Chu, I. H., Liu, T. T., Li, C., Korfmacher, W. A., Henion, J., *Rapid Commun. Mass Spectrom.* 2003, *17*, 1573–1578.

[12] Xue, Q. F., Dunayevskiy, Y. M., Foret, F., Karger, B. L., *Rapid Commun. Mass Spectrom.* 1997, *11*, 1253–1256.

[13] Ramsey, R. S., Ramsey, J. M., *Anal. Chem.* 1997, *69*, 1174–1178.

[14] Rohner, T. C., Rossier, J. S., Girault, H. H., *Anal. Chem.* 2001, *73*, 5353–5357.

[15] Gobry, V., van Oostrum, J., Martinelli, M., Rohner, T., Rossier, J. S., Girault, H. H., *Proteomics* 2002, *2*, 405–412. Erratum in *Proteomics* 2002, *2*, 1474.

[16] Lion, N., Gellon, J. O., Jensen, H., Girault, H. H., *J. Chromatogr. A* 2003, *1003*, 11–19.

[17] Deng, Y., Henion, J., Li, J., Thibault, P., Wang, C., Harrison, D. J., *Anal. Chem.* 2001, *73*, 639–646.

[18] Dayon, L., Roussel, C., Prudent, M., Lion, N., Girault, H. H., *Electrophoresis* 2005, *26*, 238–247.

[19] Lion, N., Gellon, J. O., Girault, H. H., *Rapid Commun. Mass Spectrom.* 2004, *18*, 1614–1620.

[20] Lion, N., Rohner, T. C., Dayon, L., Arnaud, I. L., Damoc, E., Youhnovski, N., Wu, Z. Y., *et al.*, *Electrophoresis* 2003, *24*, 3533–3562.

[21] Marko-Varga, G., Nilsson, J., Laurell, T., *Electrophoresis* 2003, *24*, 3521–3532.

[22] Williams, J. G., Tomer, K. B., *J. Am. Soc. Mass Spectrom.* 2004, *15*, 1333–1340.

[23] Montreuil, J., Vliegenthart, J. F. G., Schachter, H., *Glycoproteins*, Elsevier, Science B. V. Amsterdam, The Netherlands 1995.

[24] Varki, A., Cummings, R., Esko, J., Freeze, H., Hart, G., Marth, J., *Essentials of Glycobiology*, Cold Spring Harbor Laboratory Press, New York 1999.

[25] Peter-Katalinić, J., *Mass Spectrom. Rev.* 1994, *13*, 77–96.

[26] Zaia, J., *Mass Spectrom. Rev.* 2004, *23*, 161–277.

[27] Zamfir, A., Lion, N., Vukelić, Ž., Bindila, L., Rossier, J., Girault, H. H., Peter-Katalinić, J., *Lab Chip* 2004, 2005, *5*, 298–307.

[28] Bindila, L., Froesch, M., Lion, N., Vukelić, Ž., Rossier, J., Girault, H. H., Peter-Katalinić, J., Zamfir, A., *Rapid Commun. Mass Spectrom.* 2004, *18*, 2913–2920.

[29] Rossier, J. S., Vollet, C., Carnal, A., Lagger, G., Gobry, V., Girault, H. H., Michel, P., Revmond, F., *Lab Chip* 2002, *2*, 145–150.

[30] Rossier, J. S., Youhnovski, N., Lion, N., Damoc, E., Reymond, F., Girault, H. H., Przybylski, M., *Angew. Chem. Int. Ed. Engl.* 2003, *42*, 53–58.

[31] Linden, H. U., Klein, R. A., Egge, H., Peter-Katalinić, J., Dabrowski, J., Schindler, D., *Biol. Chem. Hoppe Seyler* 1989, *370*, 661–672.

[32] Froesch, M., Bindila, L., Zamfir, A., Peter-Katalinić, J., *Rapid Commun. Mass Spectrom.* 2003, *17*, 2822–2832.

[33] Zamfir, A., Peter-Katalinić, J., *Electrophoresis* 2001, *22*, 2448–2457.

[34] Zamfir, A., Peter-Katalinić, J., *Electrophoresis* 2004, *25*, 1949–1963.

[35] Bindila, L., Peter-Katalinić, J., Zamfir, A., *Electrophoresis* 2005, *26*, 1488–1499.

[36] de Jong, J., van den Berg, C., Wijburg, H., Willemsen, R., van Diggelen, O., Schindler, D., Hoevenaars, F., Wevers, R., *J. Pediatr.* 1994, *125*, 385–391.

[37] Sakuraba, H., Matsuzawa, F., Aikawa, S., Doi, H., Kotani, M., Nakada, H., Fukushige, T., Kanzaki, T., *J. Hum. Genet.* 2004, *49*, 1–8.

[38] van Diggelen, O. P., Schindler, D., Kleijer, W. J., Huijmans, J. M. G., Galjaard, H., Linden, H. U., Peter-Katalinić, J., *et al.*, *Lancet* 1987, *2*, 804.

[39] Hakomori, S., *Curr. Opin. Hematol.* 2003, *10*, 16–24.

[40] Nagai, Y., *Behav. Brain Res.* 1995, *66*, 99–104.

[41] Kraćun, I., Rosner, H., Drnovšek, V., Vukelić, Ž., osović, C., Trbojević-Čepe, M., Kubat, M., *Neurochem. Int.* 1992, *20*, 421–431.

[42] Vukelić, Ž., Metelmann, W., Müthing, J., Kos, M., Peter-Katalinić, J., *Biol. Chem.* 2001, *382*, 259–274.

[43] Svennerholm, L., Boström, K., Jungbjer, B., Olsson, L., *J. Neurochem.* 1994, *63*, 1802–1811.

[44] Metelmann, W., Vukelić, Ž., Peter-Katalinić, J., *J. Mass Spectrom.* 2001, *36*, 21–29.

[45] Zamfir, A., Vukelić, Ž., Peter-Katalinić, J., *Electrophoresis* 2002, *23*, 2894–2903.

[46] Vukelić, Ž., Zamfir, A., Bindila, L., Froesch, M., Usuki, S., Yu, R. K., Peter-Katalinić, J., *J. Am. Soc. Mass Spectrom.* 2005, *16*, 571–580.

[47] Zamfir, A., Vakhrushev, S., Sterling, A., Niebel, H. J., Allen, M., Peter-Katalinić, J., *Anal. Chem.* 2004, *76*, 2046–2054.

[48] Zamfir, A., Vukelić, Ž., Bindila, L., Peter-Katalinić, J., Almeida, R., Sterling, A., Allen, M., *J. Am. Soc. Mass. Spectrom.* 2004, *15*, 1649–1657.

[49] Zhang, S., Chelius, D., *J. Biomol. Techn.* 2004, *15*, 120–133.

[50] Froesch, M., Bindila, L., Baykut, G., Allen, M., Peter-Katalinić, J., Zamfir, A., *Rapid Commun. Mass Spectrom.* 2004, *18*, 3084–3092.

[51] Zamfir, A., Seidler, D. G., Kresse, H., Peter-Katalinić, J., *Rapid Commun. Mass Spectrom.* 2002, *16*, 2015–2024. Erratum in *Rapid Commun. Mass Spectrom.* 2003, *17*, 265.

[52] Zamfir, A., Seidler, D. G., Kresse, H., Peter-Katalinić, J., *Glycobiology* 2003, *13*, 733–742.

[53] Zamfir, A., Konig, S., Althoff, J., Peter-Katalinić, J., *J. Chromatogr. A* 2000, *895*, 291–299.

[54] Zamfir, A., Vukelić, Ž., Peter-Katalinić, J., *Electrophoresis* 2002, *23*, 2894 – 2903.

[55] Bindila, L., Almeida, R., Sterling, A., Allen, M., Peter-Katalinić, J., Zamfir, A., *J. Mass Spectrom.* 2004, *39*, 1190–1201.

[56] Vakhrushev, S. Y., Zamfir, A., Peter-Katalinić, J., *J. Am. Soc. Mass Spectrom.* 2004, *15*, 1863–1868.

[57] Mechref, Y., Novotny, M. V., *Chem. Rev.* 2002, *2*, 321–369.

[58] Reinders, J., Lewandrowski, U., Moebius, J., Wagner, Y., Sickmann, A., *Proteomics* 2004, *4*, 3686–3703.

[59] Brooks, S. A., *Mol. Biotechnol.* 2004, *28*, 241–255.

[60] Renfrow, M. B., Cooper, H. J., Tomana, M., Kulhavy, R., Hiki, Y., Toma, K., Emmett, M. R., *et al.*, *J. Biol. Chem.* 2005, *280*, 19136–19145.

[61] Domon, B., Costello, C. E., *Glycoconj. J.* 1988, *5*, 397–409.

[62] Costello, C. E., Juhasz, P., Perreault, H., *Prog. Brain Res.* 1994, *101*, 45–61.

[63] Svennerholm, L., *Adv. Exp. Med. Biol.* 1980, *125*, 11.

[64] IUPAC-IUB Commission on Biochemical Nomenclature, *Eur. J. Biochem.* 1977, *79*, 11–21.

[65] IUPAC-IUB Joint Commission on Biochemical Nomenclature, *Eur. J. Biochem.* 1998, *257*, 293–298.

[66] Roepstorff, P., Fohlman, J., *Biomed. Mass Spectrom.* 1984, *17*, 601–609.

8.6 Appendix

Ganglioside abbreviations: **LacCer**, Galβ4Glcβ1Cer; **Gg$_3$Cer**, GalNAcβ4Galβ4Glcβ1-Cer; **Gg$_4$Cer**, Galβ3GalNAcβ4Galβ4Glcβ1Cer; **nLc$_4$Cer**, Galβ4GlcNAcβ3-Galβ4Glcβ1Cer; **GD3**, II$_3$-α-(Neu5Ac)$_2$-LacCer; **GT3**, II^3-α-(Neu5Ac)$_3$-LacCer; **GM2**, II^3-α-Neu5Ac-Gg$_3$Cer; **GD2**, II^3-α-(Neu5Ac)$_2$-Gg$_3$Cer; **GM1a**, II^3-α-Neu5Ac-Gg$_4$Cer; **GM1b**, IV^3-α-Neu5Ac-Gg$_4$Cer; **GD1a**, IV^3-α-Neu5Ac,II^3-α-Neu5Ac-Gg$_4$Cer; **GD1b**, II^3-α-(Neu5Ac)$_2$-Gg$_4$Cer; **GT1b**, IV^3-α-Neu5Ac,II^3-α-(Neu5Ac)$_2$-Gg$_4$Cer; **GQ1b**, IV^3-α-(Neu5Ac)$_2$,II^3-α-(Neu5Ac)$_2$-Gg$_4$Cer; **3'-nLM1 or nLM1**, IV^3-α-Neu5Ac-nLc$_4$Cer; **nLD1**, disialo-nLc$_4$Cer (IV^3-α-(Neu5Ac)$_2$-nLc$_4$Cer

9
Utility of lab-on-a-chip technology for high-throughput nucleic acid and protein analysis*

Paul Hawtin, Ian Hardern, Rainer Wittig, Jan Mollenhauer, AnneMarie Poustka, Ruediger Salowsky, Tanja Wulff, Christopher Rizzo, Bill Wilson

On-chip electrophoresis can provide size separations of nucleic acids and proteins similar to more traditional slab gel electrophoresis. Lab-on-a-chip (LoaC) systems utilize on-chip electrophoresis in conjunction with sizing calibration, sensitive detection schemes, and sophisticated data analysis to achieve rapid analysis times (< 120 s). This work describes the utility of LoaC systems to enable and augment systems biology investigations. RNA quality, as assessed by an RNA integrity number score, is compared to existing quality control (QC) measurements. High-throughput DNA analysis of multiplex PCR samples is used to stratify gene sets for disease discovery. Finally, the applicability of a high-throughput LoaC system for assessing protein purification is demonstrated. The improvements in workflow processes, speed of analysis, data accuracy and reproducibility, and automated data analysis are illustrated.

9.1
Introduction

Systems biology is typically described as the cyclical union of computational modeling and biological measurement [1, 2]. From the measurement perspective, the need for accurate, reproducible, and meaningful analyses is of paramount importance; otherwise, the modeling constructs proposed are without validation.

Microfluidic measurement systems in general provide advantages in the small sample volumes required, reduced reagent consumption, and enhanced mass detection [3, 4]. The lab-on-a-chip architecture (LoaC) employs capillary gel electrophoresis to separate biomolecules such as nucleic acids and proteins [5, 6]. The analytical figures of merit, such as band resolution, speed of analysis, sensitivity, and separation reproducibility for LoaC are typically superior to traditional slab gel elec-

* Originally published in Electrophoresis 2005, 26, 3674–3681

Microfluidic Applications in Biology. Edited by Niels Lion, Joël S. Rossier, and Hubert H. Girault

ISBN-10: 3-527-31761-9

trophoresis. In addition, the resultant data are already digitized, eliminating the need for scanning or photographic transfer prior to data analysis. Consequently, the data can be automatically processed at the conclusion of the separation.

Nucleic acid analysis and protein analysis are two critical pieces for developing a more comprehensive understanding of systems biology. Several examples are illustrative of this. For RNA, the widespread use of microarrays to determine gene expression levels places a premium on starting with high quality samples. Assessment of purity as well as degradation is required to ensure useful data [7–10]. The RNA integrity number (RIN), which is based on the electrophoretic separation of RNA samples, has quickly become a *de facto* standard for quality assessment, as it is clear that no single RNA quality control measurement can address all the potential issues in sample preparation for an array experiment [11]. In addition to the traditional DNA sizing measurement needs, some of the protocols in the burgeoning field of array comparative genomic hybridization (aCGH) require a quality control (QC) measurement after phi29 amplification and after digestion to validate the profile and amount of DNA [12, 13]. Furthermore, multiplex PCR analysis of reverse transcribed DNA shows promise as a means to screen for unique gene sets [14, 15]. To study the various protein interactions within even a single pathway, recombinant protein expression is an indispensable capability. Numerous measurements are required through the course of a single protein expression project to verify that the target has successfully been made [16].

With the current drive to characterize biological "systems" in a more comprehensive fashion, the need to incorporate automation is growing as well. Greater numbers of samples and replicates must be analyzed in order to build statistical relevance and confidence. In such a medium- to high-throughput environment, any analytical bottleneck confounds the productivity of the investigation. Manual analytical systems such as traditional gel electrophoresis can create such a bottleneck. Most biological sample handling formats are moving toward titer plates and parallel sample processing and analysis.

This work describes the utility of LoaC systems to enable and augment systems biology investigations. RNA quality, as assessed by an RIN score, is compared to existing QC measurements. High-throughput DNA analysis of multiplex PCR samples is used to stratify gene sets for disease discovery. Finally, the applicability of a high-throughput LoaC system for assessing protein purification is demonstrated. The improvements in workflow processes, speed of analysis, data accuracy and reproducibility, and automated data analysis are illustrated.

9.2 Materials and methods

9.2.1 RNA samples

HeLa cells were obtained from ATCC (Manassas, VA, USA). Approximately 1×10^6 cells were used *per* isolation. Cells were homogenized by resuspension in lysis solution at a concentration of $\leq 10^7$ cells/mL followed by vortexing for 1 min. Rat hearts were obtained from Pel-Freez (Rogers, AR, USA). Approximately 10 mg of fresh or frozen heart muscle was weighed and placed in lysis solution (20 μL of lysis solution *per* mg of tissue). The tissue was homogenized immediately using a conventional rotor-stator type homogenizer with a stainless steel probe for 1 min at 15 000 rpm (OMNI International TH homogenizer, Warrenton, VA, USA). The resulting lysate was pelleted, and the supernatant used without further treatment. Cellular RNA was isolated with either the total RNA isolation mini kit (Agilent Technologies, Palo Alto, CA, USA) or a commercially available silica-based kit, using the manufacturer's protocol.

9.2.2
DNA samples

All details for the sample source and preparation are given in [15]. Briefly, the 11 candidate genes (IFITM1, PEPP2, PDE3A, CYR61, PLAB, APOD, G1P3, G1P2, IL1B, CRYAB, and IFITM3) selected for the multiplex reverse transcription-PCR (mRT-PCR) assay differed in expression levels by a factor of at least 2 (upregulation) or less than 0.5 (downregulation) in drug-resistant melanoma cells when compared to a sensitive reference cell line. The UBB gene showed equal expression levels and therefore was selected as an internal control. Total RNA was isolated and measured for relative quantity and quality on an Agilent 2100 bioanalyzer. Four micrograms of RNA was denatured and reverse transcribed in a 50 μL reaction. The mRT-PCR was carried out in a volume of 50 μL for 30 cycles in a GeneAmp 9700 thermocycler (Applied Biosystems). The resulting solution was used for LoaC analysis without further manipulation. To evaluate the utility of the approach, drug-responsive MeWo cells were run for a reference case while three different populations of drug-resistant MeWoEto1 cells served as the test case. These two cell lines displayed prominent differences within the selected gene set [17].

9.2.3
Protein samples

Protein expression of proprietary target kinases was the objective of this experiment. Several constructs were prepared so that parallel expression could be conducted; however, the data shown reflects triplicate measurements from a single construct. Each of the constructs was designed with a hexahistidine tag to allow affinity purification of the resulting lysates. Ultimately, the best constructs from these experiments along with the most productive conditions were used for a larger volume production process. The resulting expression lysates were purified by an automated multiple column chromatographic approach (Akta 3-D purification system, GE-Amersham). During method development, the system was run in a single column mode. In this single dimension, an Ni-NTA column was used to isolate the hexahistidine tagged proteins. The flow through, wash, and the eluted fractions were collected for subsequent analysis on the LoaC system. The subsequent chromatographic separation utilized gel filtration as a purification and buffer exchange step to obtain material of sufficient purity for crystallography studies.

9.2.4
RNA sample analysis

Total RNA was quantified by measuring A_{260}/A_{280} on an ND-1000 spectrophotometer (NanoDrop Technologies, Rockland, DE, USA). Samples also were analyzed on an Agilent 2100 bioanalyzer using the RNA 6000 Nano reagent kit and the Eukaryote total RNA assay. All reagents and consumables were used from the reagent kit. Data processing and RIN scoring was done with the default parameters.

9.2.5
DNA and protein analysis

DNA samples were analyzed on an Agilent 5100 Automated LoaC Platform using the DNA 1000 HT-4 chip, reagent kit, and protocol. This chip allows four samples to be processed in parallel. Reagents such as fresh gel, sizing ladder, and markers are loaded in resource plates and utilized in an automated fashion. This affords high-throughput measurements in an unattended mode. Protein samples were analyzed on a similar system with the Protein 200 HT-2 assay kit and protocol. For this chip, two samples are processed simultaneously. All assay specific reagents and consumables were used from the kit according to the manufacturer directions. Samples were prepared in Eppendorf PCR twin plates and sealed with a Remp plate sealer.

For comparison to the protein purification analysis, slab gel electrophoresis was also conducted. Electrophoresis was carried out on a 10 cm × 12 well Criterion 10% Bis-Tris gel (BioRad, Hercules, CA, USA) at 200 V (constant voltage) for 40 min. The gel was stained for 2 h with 0.2% CBB and destained overnight. Protein size calibration was done with the Seeblue (Invitrogen, Carlsbad, CA, USA) ladder standard (18, 28, 38, 49, 64, and 98 kDa).

9.3
Results and discussion

9.3.1
RNA analysis and results

Twelve replicates of a HeLa cell extract were analyzed to determine the RIN value. Fig. 1 shows a typical electropherogram for an RNA extract with a RIN number of 10 (RIN range 1–10 where 1 is low integrity, 10 is high integrity). Note the strong bands for the 18S and 28S ribosomal RNA and the small, almost undetected peak for the 5S ribosomal RNA. This indicates that the RNA sample has undergone almost no degradation and that it is not contaminated with DNA. Low-molecular-weight DNA contamination would be visible in the "fast" region close to the 5S ribosomal RNA, while high-molecular-weight genomic DNA would be visible in the "inter" region between 18S and 28S ribosomal RNA. The calculation of the RIN score and the effects of various types of contamination are given elsewhere [18].

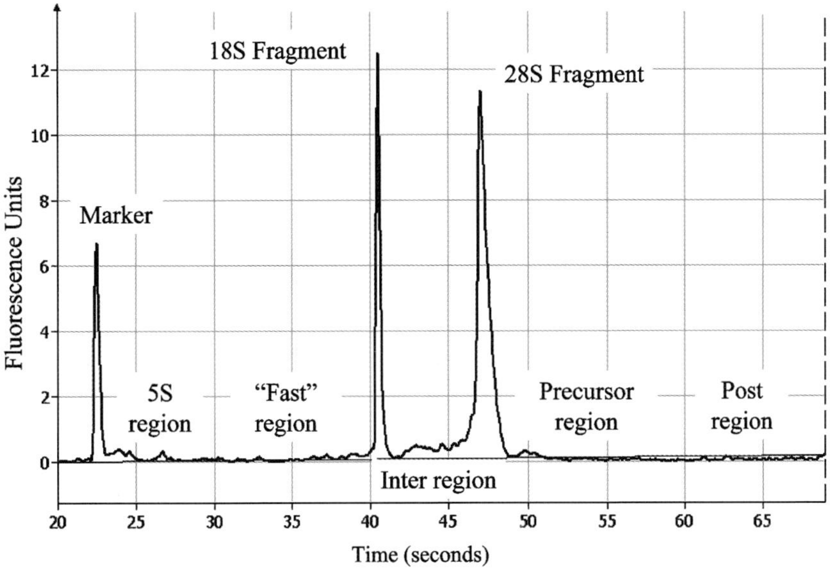

Fig. 1 RIN measurement regions.

Fig. 2 shows a comparison of RNA quality after different extraction protocols. In Fig. 2A, the sample is very clean and has a high-integrity RIN number of 9. The electropherogram for the silica-based extraction is shown in Fig. 2B. Note the large amount of genomic DNA contamination that occurs between the 18S and 28S rRNA peaks. This RNA trace is sufficiently distorted by the DNA present that the algorithm cannot calculate a RIN value. The algorithm is user-adjustable to permit calculation of RIN values even with this distortion; however, the user must set the peak thresholds for the detected anomaly peaks. In this case, any peaks detected in the predefined regions from Fig. 1 can be set to impact the RIN calculation [18]. For comparison, Tab. 1 shows the RNA QC measurements performed on these two extracts. With respect to the UV measurement, the samples are nearly the same. In addition, the rRNA ratios are also very similar. Using only these two QC measurements, the extraction protocols look to be about equal. However, the electropherogram clearly shows substantial contamination in the silica-based preparation. In addition to assessing the RNA from the sample, the LoaC technology also provides invaluable information about cDNA production and cRNA formation for a broad range of gene microarray platforms [11].

9.3.2
DNA analysis and results

Recently, it was shown that the combination of on-chip electrophoresis with multiplex RT-PCR resulted in a powerful screening tool for the rapid and reliable semiquantitative expression screening of candidate gene sets. The good correlation

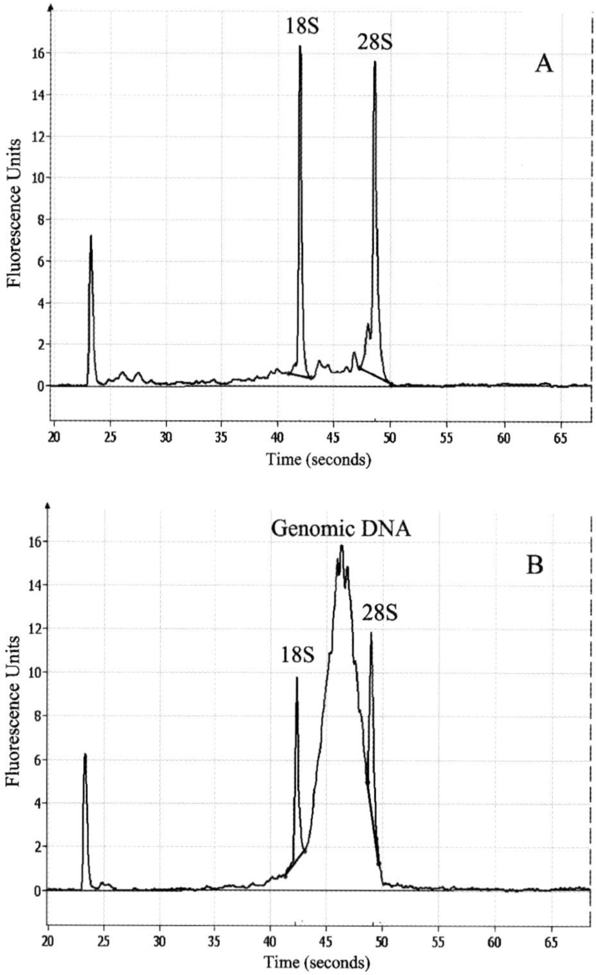

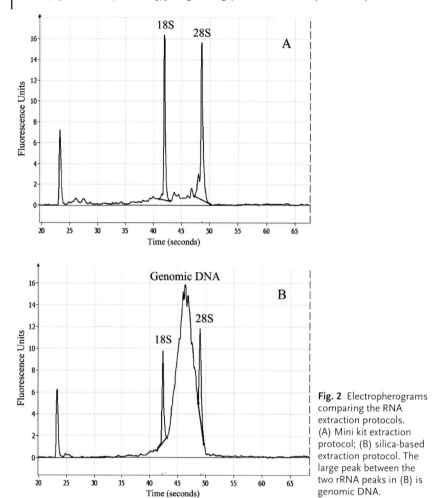

Fig. 2 Electropherograms comparing the RNA extraction protocols. (A) Mini kit extraction protocol; (B) silica-based extraction protocol. The large peak between the two rRNA peaks in (B) is genomic DNA.

Tab. 1 RNA QC results

RNA extraction method	A_{260}/A_{280}	28S/18S ratio	RIN score
Mini kit	2.03	1.2	9
Silica-based kit	1.90	1.0	N/A

of array data, Northern data, and the LoaC data provided preliminary validation for this approach [15, 17]. To extend the approach toward the screening of a larger sample set, the analysis was tested on an automated microfluidic platform. Fig. 3 shows the electropherogram and peak identification of the PCR products from the automated LoaC separation.

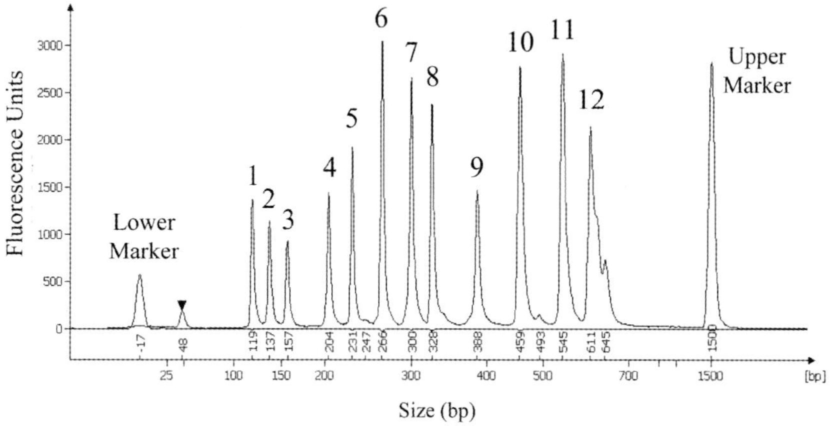

Fig. 3 High-throughput electropherogram of gene screening set. Peak identification: (1) IFITM3, (2) UBB, (3) CRYAB, (4) IL1B, (5) G1P2, (6) G1P3, (7) APOD, (8) PLAB, (9) CYR61, (10) PDE3A, (11) PEPP2, and (12) IFITM1. Note the multiple peaks migrating for (12), illustrating the slightly higher resolution afforded by the 5100 platform.

Tab. 2 HT mRT-PCR analysis performance

Amplicon	**Target size, bp**	**Average measured size, bp**	**Size deviation, %**	**Average measured amount, ng/μL**	**CV measured amount, %**
IFITM3	116	119	2.2	9.1	3.6
UBB	134	137	1.9	7.8	3.1
CRYAB	153	156	1.8	5.7	2.6
IL1B	196	203	3.7	8.1	1.2
G1P2	225	230	2.1	9.2	2.9
G1P3	263	265	0.6	15	4.5
APOD	303	299	−1.2	14	2.6
PLAB	330	325	−1.6	12	4.4
CYR61	386	387	0.1	7.7	0.8
PDE3A	450	458	1.7	13	2.9
PEPP2	498	545	9.4	13	1.7
IFITM1	582	607	4.3	7.8	4.1

To verify the performance of the platform in terms of reproducibility and accuracy, 24 mRT-PCR samples were analyzed in quadruplicate. RT-PCR amplicons were derived from different melanoma cell lines [15]. The positive control containing the amplicons of all 12 target genes was used to test the reproducibility and accuracy of sizing results within a single lane (Tab. 2). The average sizing accuracy was less than 10% deviation with all but one fragment below 5% deviation. The sizing of all

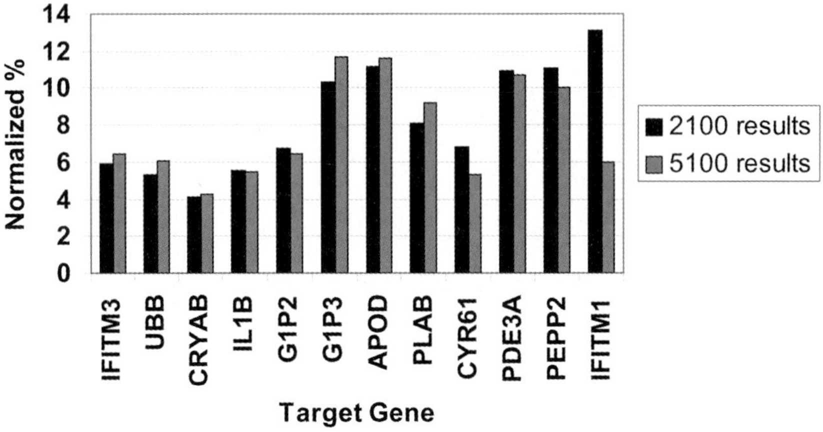

Fig. 4 Comparison of 2100 Bioanalyzer gene screen profile *versus* 5100 ALP profile. Major difference between the profiles is for IFITM1. This is a result of the higher resolution on the 5100 system. Summing the multiple peaks gives essentially equivalent results between the manual and automated platforms.

12 fragments showed a high reproducibility with deviations of $< 1.1\%$ CV for the four replicate runs. In terms of quantitation, the measured amount precision was very good at less than 5% CV.

The correlation of the 2100 Bioanalyzer data to the 5100 automated LoaC is shown in Fig. 4. The calculated amount of PCR product observed in both systems was normalized. The 12 amplicons show the same general trend except for IFITM1. The normalized percent measured on the 2100 is about twice that seen on the 5100. From the electropherogram in Fig. 3, the band for IFITM1 tails and shows several small bands migrating on the tail. The 2100 system analyzes this as a single component; however, the higher resolution of the 5100 separates these small fragments. By summing the three smaller bands into a single composite band, akin to the 2100, the 5100 gives a normalized percent of 12.1, which is very comparable to the 2100 profile.

Fig. 5 shows the gel-like images for the mRT-PCR samples after electrophoretic separation using the automated LoaC platform. The first lane is the ladder standard for system calibration. The 12th lane shows the positive control containing all 12 target genes. The sixth lane shows the absence of IL1B (~200 bp) and IFITM1 (~600 bp) with the PEPP2 signal (~500 bp) showing up prominently. This clearly contrasts with the ninth lane as IL1B and IFITM1 are clearly visible and the level of PEPP2 is similar to IFITM1. These profiles correctly reflect the gene expression differences noted previously [15]. Using predictive gene sets, mRT-PCR followed by automated LoaC analysis therefore can be used to stratify the cell's drug resistance or responsiveness, and this work is currently under more extensive validation. The high analysis speed and automation possible with the LoaC approaches create significant benefit for developing these gene-screening protocols.

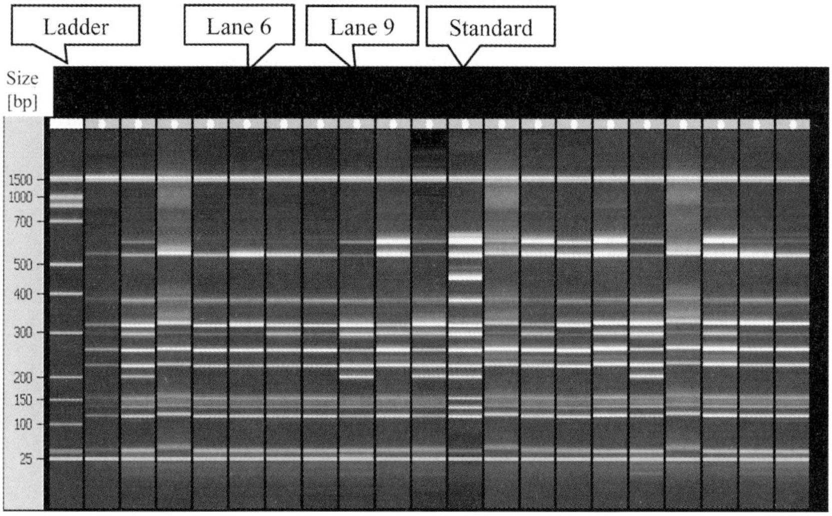

Fig. 5 Gel-like image of gene-screening analyses. The first lane is the sizing ladder while the 12th lane is the positive control standard. Lane 6 and lane 9 illustrate the gene expression differences between two cell lines.

9.3.3 Protein analysis and results

Fig. 6 shows the gel-like image from the automated LoaC system and a comparable traditional gel electrophoresis analysis for a ~46 kDa kinase construct. In both pictures, triplicate results from a single expression constructs are analyzed. Lanes 1–3 correspond to the original cell lysates. Lanes 4–6 correspond to the affinity column flow through. Lanes 7–9 show the affinity column wash and lanes 10–12 show the eluted column fractions. Although the images are very similar, note the approximate size for the kinase construct is greater than 50 kDa from the slab gel, while the LoaC analysis more accurately sized the protein at 46 kDa. In addition, the anomalous migration behavior seen on the slab gel is not seen on the gel-like image. By using marker components in every analysis, the LoaC system can compensate for variations in migration behavior from run to run.

While both measurement approaches produce viable data, the manual nature of the traditional slab gel system becomes problematic for parallel protein expression. The 12 samples analyzed were run on a 10 cm gel at 200 V for 40 min. The gel was then stained for 2 h and destained overnight. The length of destaining is dependent on the desired clarity of the gel and sensitivity. Assuming that destaining for 1 h is sufficient that sets the cycle time for 12 samples at almost 4 h. Clearly, multiple gels could be run in parallel to improve the throughput; however, these will require more operator interaction as well as a greater amount of reagents. Imaging of the resultant gels and data collection and reduction also entail more operator interaction.

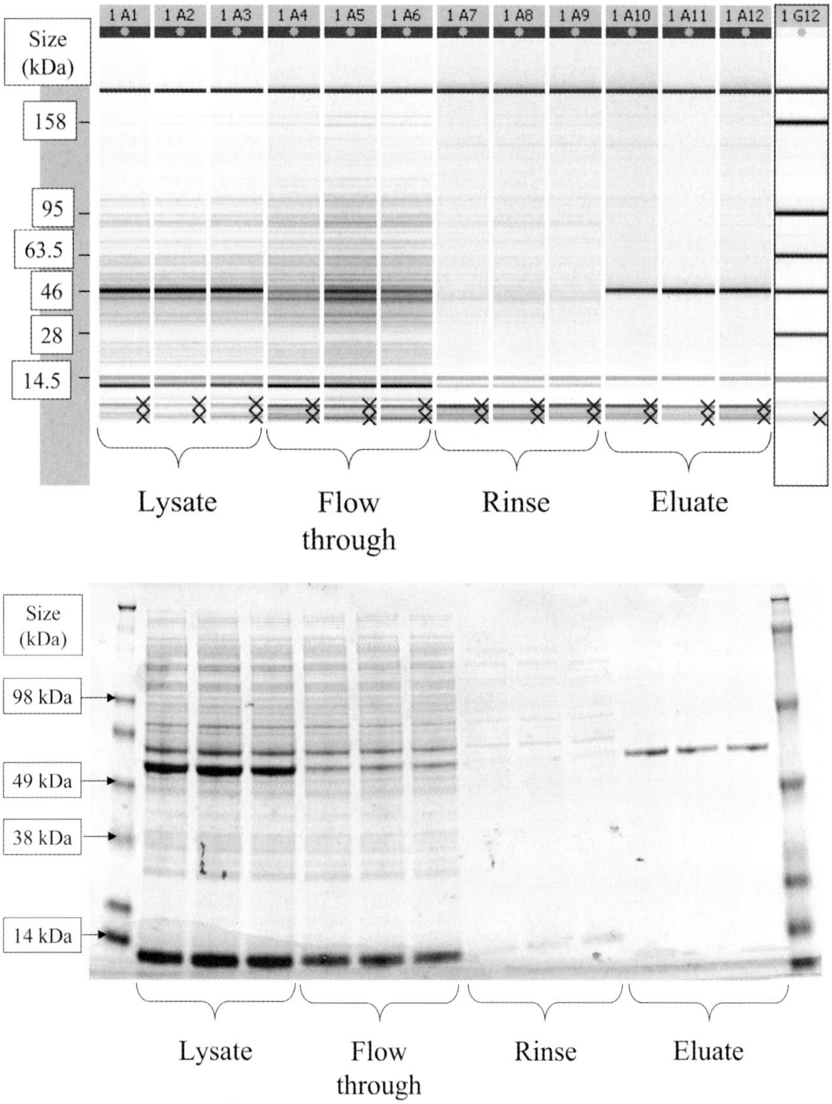

Fig. 6 Comparison of gel-like image (A) and standard 1-D SDS-PAGE (B) for assessing protein purification. Lanes 1–3 correspond to the original cell lysates, lanes 4–6 correspond to the Ni-NTA affinity column flow through fraction, lanes 7–9 show the affinity column wash, and lanes 10–12 show the eluted column fractions.

For the automated LoaC system, the instrument automatically loads the chip with reagents initially and replenishes the chip as scheduled in the assay. The initial chip preparation requires approximately 90 min. The actual analysis time for 96 samples plus sizing ladders is also approximately 90 min. The instrument then

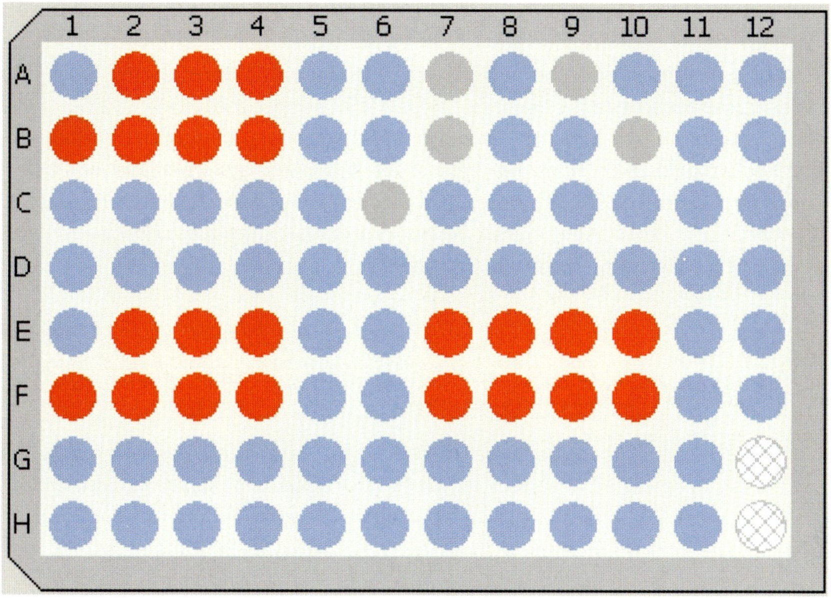

Fig. 7 Protein purification results display. Criteria are the presence of a 27 kDa (±5%) protein and that it constitutes >50% of the total protein content. Red circles, criteria met; blue circles, criteria not met; hashed circles, protein ladder; gray circles, no peaks detected.

requires approximately 30 min to prepare the chip for storage and clean the instrument. This entire process is termed a "job" and constitutes the total time for analysis. Once the job is started, the instrument requires no further operator interaction until all the analyses are completed. Thus, in the same time that the slab gel system needed for 12 samples, 96 were analyzed with the LoaC system.

The initial 90 min and the final 30 min are constant regardless of the number of sample plates analyzed (up to 10 × 384 wells); consequently, the overall time *per* analysis decreases as more samples are analyzed. To run 96 samples, the total run time is ~210 or 2.2 min/run. For 192 samples, the total run time is ~300 or 1.6 min/run. A similar situation is seen for DNA analyses by the automated LoaC system [19].

With the growing interest in structural biology and protein function, the need for pure and correctly folded target protein is of primary importance. A parallel protein expression and purification approach enables the investigator to conduct optimization experiments on a small scale before expending the time and resources to produce the desired quantity on a larger scale. In protein purification, the research goal is to develop the separation conditions needed to meet the purity and quality goals while optimizing throughput and cost. Even with established purification protocols, overall protein purity and recovery measurements are needed. With the digital data possible from the LoaC technology, a purity assessment can be incorporated into the data analysis. Using a collection of rules-based decisions, the measurement results can be flagged at a titer plate level. In a separate expression experiment, the protease construct targeted was 27 kDa. Fig. 7 shows a titer plate view of a protein purification optimization experiment. The results are flagged to indicate the presence of a band at 27 kDa (±5%) and constituting >50% of the total protein content. Red wells meet the aforementioned requirements while the blue wells failed these rules. The gray wells show analyses where no significant peaks were detected. Finally, the hashed marked wells indicate the protein ladder standards. This level of automated data interpretation simplifies subsequent sample manipulation and purification optimization.

9.4 Concluding remarks

Systems biology will require numerous analytical measurements to unravel the complexity of even simple systems. Microfluidics offers benefits in miniaturization, speed of analysis, small sample and reagent consumption, and automation capabilities. Chip electrophoresis, such as the LoaC format, allow rapid separation, sizing, and detection of nucleic acids and proteins. RNA quality measurement, gene-screening analysis, and protein purification measurements, exemplified in this paper, illustrate the power and figures of merit for such an analytical approach. Furthermore, the resulting digital data is well suited for medium- and high-throughput biological studies and will allow statistically valid studies to be conducted.

The authors would like to thank the guest editors (Dr. Lion, Dr. Rossier, and Dr. Girault) for their invitation to contribute to this special issue.

9.5 References

[1] Kitano, H., *Nature* 2002, *420*, 206–210.

[2] Nurse, P., *Nature* 2003, *424*, 883.

[3] Bousse, L., Mouradian, S., Minalla, A., Yee, H., *et al.*, *Anal. Chem.* 2001, *73*, 1207–1212.

[4] Yin, H., Killeen, K., Brennen, R., Sobek, D., *et al.*, *Anal. Chem.* 2005, *77*, 527–533.

[5] Kuschel, M., Neumann, T., Barthmaier, P., Kratzmeier, M., *J. Biomol. Tech.* 2002, *13*, 172–178.

[6] Mueller, O., Hahnenberger, K., Dittman, M., Yee, H., *et al.*, *Electrophoresis* 2000, *21*, 128–134.

[7] Hlaing, M., Spitz, P., Padmanabhan, K., Cabezas, B., *J. Biol. Chem.* 2004, *279*, 43625–43633.

[8] Sanchez-Carbayo, M., Saint, F., Lozano, J. J., Viale, A., *et al.*, *Clin. Chem.* 2004, *49*, 2096–2100.

[9] Naderi, A., Ahmed, A. A., Barbosa-Morais, N. L., Aparacio, S., *et al.*, *BMC Genomics* 2004, *5*, 9.

[10] Auer, H., Lyianarachchi, S., Newsom, D., Klisovic, M., *et al.*, *Nat. Genet.* 2003, *35*, 292–293.

[11] Dumur, C. I., Nasim, S., Best, A. M., Archer, K. J., *et al.*, *Clin. Chem.* 2004, *50*, 1994–2002.

[12] Baldus, C. D., Liyanarchchi, S., Mrozek, K., Auer, H., *et al.*, *PNAS* 2004, *101*, 3915–3920.

[13] Barrett, M. T., Scheffer, A., Ben-Dor, A., Sampas, N., *et al.*, *PNAS* 2004, *101*, 17765–17770.

[14] Dehainault, C., Lauge, A., Caux-Moncoutier, V., Pages-Berhouet, S., *et al.*, *Nucleic Acids Res.* 2004, *32*, e139.

[15] Wittig, R., Salowsky, R., Blaich, S., Lyer, S., *et al.*, *Electrophoresis* 2005, *26*, 1687 – 1691.

[16] Doyle, S., Murphy, M., Massi, J., Richardson, P., *J. Proteom. Res.* 2002, *1*, 531–536.

[17] Wittig, R., Nessling, M., Will, R. D., Mollenhauer, J., *et al.*, *Cancer Res* 2002, *62*, 6698–6705.

[18] Mueller, O., Lightfoot, S., Schroeder, A., Agilent Application Note, Pub # 5989-1165EN, May 2004.

[19] Pike, K., Plona, T., Stewart, C., Kim, C., *et al.*, Agilent Application Note, Pub # 5989-1870EN, November 2004.

10
Analysis of amino acids and proteins using a poly(methyl methacrylate) microfluidic system*

Masaru Kato, Yukari Gyoten, Kumiko Sakai-Kato, Tohru Nakajima, Toshimasa Toyo'oka

Plastic microchips are very promising analytical devices for the high-speed analysis of biological compounds. However, due to its hydrophobicity, their surface strongly interacts with nonpolar analytes or species containing hydrophobic domains, resulting in a significant uncontrolled adsorption on the channel walls. This paper describes the migration of fluorescence-labeled amino acids and proteins using the poly(methyl methacrylate) microchip. A cationic starch derivative significantly decreases the adsorption of analytes on the channel walls. The migration time of the analytes was related to their molecular weight and net charge or p*I* of the analytes. FITC-BSA migrated within 2 min, and the theoretical plate number of the peak reached 480 000 plates/m. Furthermore, proteins with a wide range of p*I* values and molecular weights migrated within 1 min using the microchip.

10.1
Introduction

The sequencing of the human genome [1, 2] has ushered in the era of system biology with its goal of studying the function and control of a biological system *via* the systematic and quantitative analysis of the system's components. One of the important techniques in system biology is the quantitative analysis of gene expression changes due to external or internal perturbations. The high-speed and comprehensive analysis of intracellular biological compounds in living organisms can reveal the connection of biochemical networks and provide a system-level understanding of living organisms [3].

The area of integrated chemistry is rapidly growing and now many analytical systems have been integrated into microfabricated devices [4–15]. The integrated analytical systems were used in many fields because the integration provided small

* Originally published in Electrophoresis 2005, 26, 3682–3688

Microfluidic Applications in Biology. Edited by Niels Lion, Joël S. Rossier, and Hubert H. Girault

ISBN-10: 3-527-31761-9

volumes, faster responses, highly parallel analyses, and minimal cross contamination [4–10]. Therefore, these microfluidic analytical systems are suitable analytical techniques for use in the field of system biology.

The plastic microchip is suitable for disposable use because it can be cheap to manufacture in large volumes and many analytical systems have now been integrated into a plastic microchip [16–23]. However, the sample was sometimes adsorbed onto the polymer surface and peak-broadening occurred during the analysis of hydrophobic compounds, such as fluorescence-labeled amino acids and proteins. A commonly reported solution to this unfavorable separation performance is the covalent or dynamic coating modification of the polymer surface [24–33].

We recently found that a cationic starch derivative, 2-hydroxy-3-(triethylammonio)propyl ether, chloride, 2-hydroxypropyl ether starch, was a good additive for the suppression of the adsorption by an analyte on the poly-(methyl methacrylate) (PMMA) microchip surface [34], and successfully separated some fluorescently labeled amino acids and peptides. In this study, we applied the cationic starch derivative for protein analysis using the microchip. Although protein separation is the main analytical target in proteomic research, only a few papers have reported protein separation using a plastic microchip [35, 36]. Furthermore, we tried to clarify the separation mechanism of analytes using the cationic starch derivative by comparing the migration times of the analytes.

10.2 Materials and methods

10.2.1 Materials and chemicals

4-Fluoro-7-nitro-2,1,3-benzoxadiazole (NBD-F), a fluorescein derivative (FITC-BSA), and ribonuclease were purchased from Tokyo Kasei (Tokyo, Japan). FITC-BSA with a different derivatization ratio was purchased from Molecular Probes (Eugene, OR, USA). Amino acids, acid glycoprotein, trypsin inhibitor, β-lactoglobulin, carbonic anhydrase II, myoglobin, and lysozyme were purchased from Sigma-Aldrich (Milwaukee, WI, USA). α-Chymotrypsinogen was purchased from Wako Pure Chemicals (Osaka, Japan). Bradykinin was from Peptide Institute (Osaka, Japan). 2-Hydroxy-3-(trimethylammonio)propyl ether, chloride, 2-hydroxypropyl ether starch (cationic starch) (Fig. 1a) were supplied by Nippon Starch Chemical (Osaka, Japan). Water was purified by a Milli-Q apparatus (Millipore, Bedford, MA, USA).

10.2.2 Apparatus

The PMMA chip, *i*-chip 3 DNA (Hitachi, Tokyo, Japan), was used [29, 31, 32]. The diagram of the chip is shown in Fig. 1b. This microchip has dimensions of 85 mm × 50 mm with three simple cross channels of 100 μm in width and 30 μm

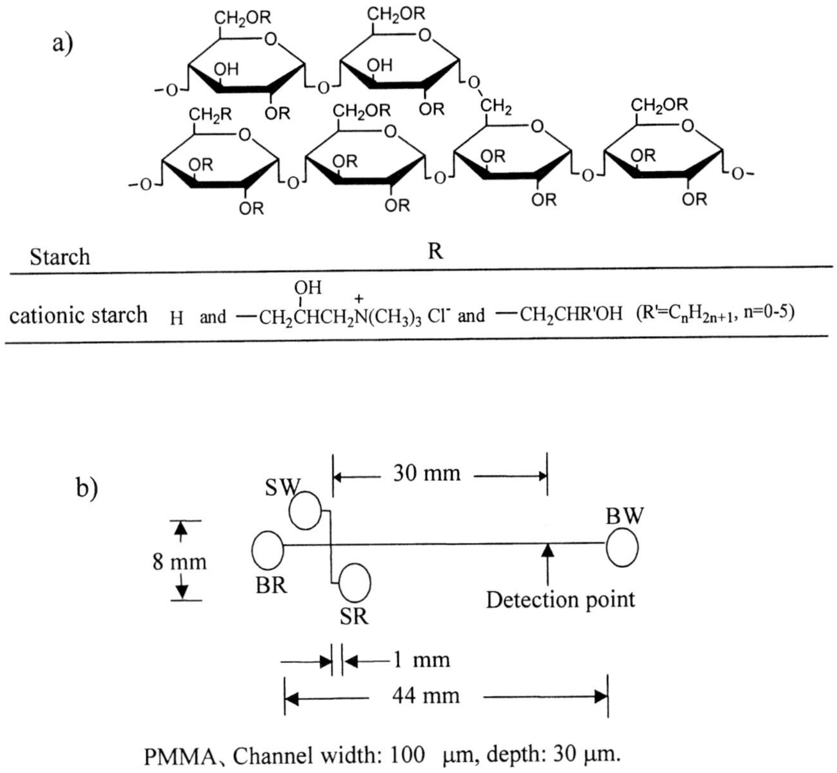

Fig. 1 (a) Chemical structure of starch derivative and (b) diagram of the microfluidic system.

in depth. The distance between the sample reservoir (SR) and the sample waste (SW) was 10 mm, whereas the distance from the buffer reservoir (BR) to the buffer waste (BW) was 44 mm. All experiments were performed using a Hitachi SV1100 microchip electrophoresis instrument (Tokyo, Japan), and the detail of the instrument was shown in our previous report [34]. Sample introduction and separation were controlled through manipulation of the electric field strengths. Phosphate or Tris-HCl buffers at the concentration of 10 mM were used as the running buffer. The program of applied voltage for each well is summarized in Tab. 1. A high-voltage supply (LabSmith HVS448, LabSmith, Livermore, CA, USA) was used to apply the voltage.

The running buffer solution was heated at 90°C for dissolving the starch derivatives. The width at half height of a peak was used for calculation of the efficiencies measurement.

Tab. 1 List of applied voltages for each well

		SR	SW	BR	BW
Positive voltage mode	Injection	0	300	0	0
	Separation	130	130	0	750
Negative voltage mode	Injection	0	−300	0	0
	Separation	−130	−130	0	−750
High voltage mode	Injection	0	300	0	0
	Separation	260	260	0	1500

Voltages are given in V.

10.2.3
Derivatization of samples with NBD-F

Ten microliters of 50 mM NBD-F in ACN and 30 μL of 100 mM borate buffer (pH 8.5) were added to 10 μL sample solution dissolved in 100 mM borate buffer (pH 8.5). The mixture was heated at 60°C. The derivatization time of amino acids and proteins was 5 and 20 min, respectively. A volume of 100 μL of the running buffer was added to the reaction solution, followed by filtration with a 0.22-μm filter (Millipore, Bedford, MA, USA) and degassing by ultrasonication for 20 s. The filtrate was applied to SR.

10.2.4
EOF measurement

The velocity of the EOF was measured using an NBD-derivatized *N,O*-dimethyloxazolidine. This compound was found to be an effective EOF marker in our previous study [34].

10.3
Results and discussion

10.3.1
Migration of amino acids using cationic starch derivatives

Amino acids are typical metabolites in living organisms. Twenty DNA-coded amino acids were derivatized with NBD-F and these derivatives migrated through the microchip using a buffer solution containing a 3% cationic starch. NBD-F reacts with the first and secondary amines. We used the positive voltage mode, Tab. 1, for the migration of the NBD-amino acids. Tab. 2 shows the migration results and physical properties of each NBD-amino acid. The EOF without the starch was 1.4×10^{-4} cm^2/V × s [32], and its direction was reversed with the addition of the starch derivatives. The mobility reached 4.9×10^{-5} cm^2/V × s when the concentra-

Tab. 2 Migration results and properties of amino acids

Amino acid	M_r	Migration time, s	Electrophoretic mobility, $\times 10^{-4}$ cm^2/V $\times$ s	Thoretical plate number, plates/m	Charge	NBD group
Asp	133	54.6	3.0	380 000	−2	1
Glu	147	57.6	2.8	370 000	−2	1
Gly	75	73.0	2.1	290 000	−1	1
Ala	89	75.8	2.0	200 000	−1	1
Ser	105	76.5	2.0	210 000	−1	1
Pro	115	76.4	2.0	140 000	−1	1
Val	117	84.5	1.8	190 000	−1	1
Thr	119	82.0	1.9	230 000	−1	1
Cys	121	83.0	1.8	140 000	−1	1
Leu	131	91.1	1.6	220 000	−1	1
Ile	131	89.9	1.7	230 000	−1	1
Asn	132	79.3	1.9	230 000	−1	1
Gln	146	86.6	1.7	250 000	−1	1
Met	149	90.8	1.6	270 000	−1	1
Phe	165	102.5	1.4	140 000	−1	1
Tyr	181	106.2	1.3	130 000	−1	1
Lys	146	197.4	0.5	130 000	−1	2
His	155	164.4	0.7	100 000	−1	2

tion of the starch was 3%. The acidic amino acids, such as Asp and Glu, migrated faster than the other amino acids, because these amino acids have two negative charges on the molecule under the given separation condition (pH 7.0). Among the neutral amino acids, those amino acids with a low molecular weight migrated faster than those with the higher molecular weights. Two basic amino acids, Lys (M_r 146.19) and His (M_r 155.16), migrated later than Gln (M_r 146.15) and Met (M_r 149.21), which had molecular weights similar to Lys and His. This can be explained by the fact that these basic amino acids reacted with two NBD-Fs and the molecular weights were then double those of the other amino acids. Another basic amino acid, Arg, did not migrate within 5 min. Arg has a guanidine group, which did not react with NBD-F. Although both the guanidine and carboxyl groups of Arg were ionized, the net charge of NBD-Arg was neutralized under the migration condition (pH 7.0). Therefore, the migration time of NBD-Arg was slower than those of the other NBD-amino acids, which have a negative charge and migrated toward the anode (BW). Because the fluorescence of the Trp derivatives was quenched under our detection conditions, we could not determine the migration time of NBD-Trp. Fig. 2 shows the relationship between the m/z of the NBD-amino acids and their migration times ("m" and "z" were indicated "molecular weight" and "charge", respectively). The good correlation coefficient of $r^2 = 0.87$ was obtained. However, Asp, Glu, and Lys had poorer correlations. This result indicated that the separation mechanism of the NBD-amino acids using the cationic starch was mainly based on the m/z of the analytes, and the interaction between cationic starch and NBD-amino acid was negligible.

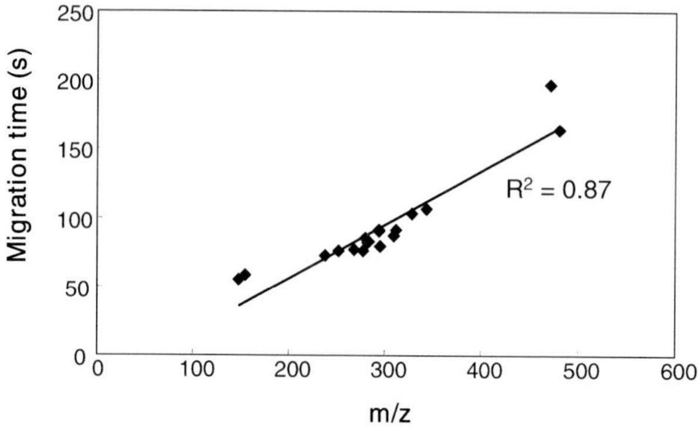

Fig. 2 Relationship between the migration time of NBD-amino acids and their *m/z* values. Conditions: running buffer, 10 mM phosphate buffer (pH 7.0) with 3% cationic starch derivative.

The RSD of the within-day and day-to-day reproducibilities of the migration times was about 1 and 4%, respectively, which seems to indicate acceptable reproducibilities of this method.

10.3.2
Migration of Arg using cationic starch derivative

NBD-Arg did not migrate using the buffer solution at pH 7.0. NBD-Arg has free carboxyl and guanidine groups, and their pK_a values were about 2.2 (carboxyl group) and 12.5 (guanidine group), respectively. NBD-Arg has a positive charge at pH 2.2 or lower, no charge between pH 2.2 and 12.5, and a negative charge at pH 12.5 or higher pH. This means that NBD-Arg migrated toward the cathode at the pHs lower than 2.2. Therefore, the applied voltage was changed to the negative voltage mode (Tab. 1), which was a reversed polarity condition compared with that in the previous study. Although the direction of the electrophoresis of NBD-Arg and that of the EOF mobility were the opposite, NBD-Arg migrated in about 180 s using the buffer solution at pH 1.85. This result showed that the EOF mobility was slower than the electrophoretic mobility of NBD-Arg.

These results showed that the cationic starch was a suitable additive for the migration of both the positive and negative charged compounds. Therefore, we used the cationic starch for further studies.

10.3.3
Migration of FITC-BSA using cationic starch derivatives

For proteomic research, a high-speed and efficient analysis method of proteins in living organisms is strongly desired. Protein analysis using a microchip is a promising technique for this high-speed analysis. The cationic starch was used for the adsorption suppression of the protein sample on the microchip surface. Because the SV 1100 microchip electrophoresis instrument has a fluorescence detection system, FITC-labeled BSA was used as the model protein sample. BSA consisted of 583 amino acids and its molecular weight was about 66 400. Theoretically,

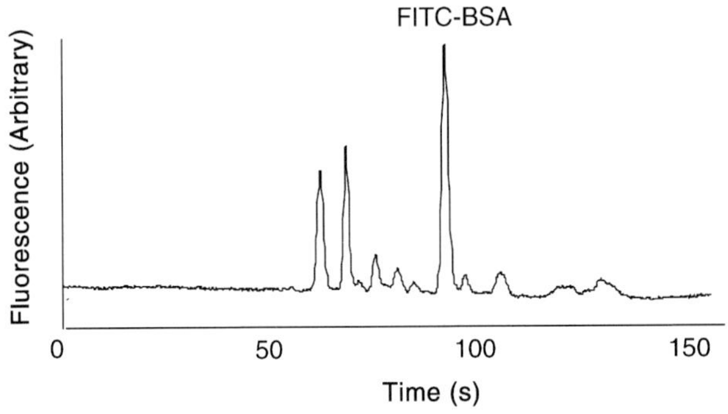

Fig. 3 Electropherogram of FITC-BSA. Conditions: running buffer, 10 mM Tris/HCl buffer (pH 8.3) with 3% cationic starch derivative.

less than 60 FITC can react with BSA, because BSA has 59 Lys residues and 1 *N*-terminal. We first used a BSA sample with 11 FITCs on average. The voltage applied to each well was doubled in the high voltage mode (Tab. 1), because the protein was a larger molecule than the amino acid, and it was expected that the protein took a long time to migrate. Fig. 3 shows the electropherogram of FITC-BSA using a 3% cationic starch as the additive. Although some peaks migrated within 2 min, these peaks were supposed to be derived from FITC-BSA with a different quantity of the FITC. We used another FITC-BSA sample, which has a lower quantity of the FITC. The lower FITC sample migrated faster than the higher FITC BSA sample under the same conditions. This result showed that the migration time became longer by increasing the quantity of the FITC group. The theoretical plate number of the peak, which migrated at 95.0 s in Fig. 3, was 480 000 plates/m. This value was similar to that reported by other researchers using a microchip [35–37]. This developed method was very simple, by only dissolving the cationic starch in the buffer solution. Therefore, the cationic starch derivative was an effective additive for protein analysis.

10.3.4 Separation of proteins using cationic starch derivative

Finally, we tried to separate protein samples. Before analyses, all protein samples were derivatized with a fluorogenic reagent, NBD-F. Nine proteins (trypsin inhibitor, β-lactoglobulin, acid glycoprotein, carbonic anhydrase II, myoglobin, α-chymotrypsinogen, lysozyme, ribonuclease, and bradykin) were used for analysis. The model proteins represented a wide range of p*I* values and molecular weights. These p*I* and molecular weights are summarized in Tab. 3. These NBD-proteins migrated within a minute when using the high voltage mode, which was faster than the migration time of FITC-BSA (Fig. 3). Fig. 4 shows the electropherogram of mixture sample of four proteins. Although the reaction time of protein and NBD-F was increased compared with that of amino acid and NBD-F, reaction efficiency was not enough and a peak derived from the derivatization reagent appeared at about 17 and 32 s. Fig. 5 shows the relationship between the migration time and molecular weight of the protein sample. The migration times of the protein with p*I* value lower than pH value of the running buffer (pH 7.0) tended to be short, but

Tab. 3 Migration results and properties of proteins

Protein	p*I* [a]	Molecular weight[a], kDa	Migration time, s	Thoretical plate number, plates/m
Trypsin inhibitor	4.6	20.1	19.6	120 000
β-Lactoglobulin	4.8	18.3	22.2	110 000
Acid glycoprotein	5.0	21.6	29.0	400 000
Carbonic anhydrase II	6.9	29.1	24.2	110 000
Myoglobin	7.0	16.9	20.0	110 000
α-Chymotrypsinogen	8.5	25.7	24.2	150 000
Lysozyme	9.3	14.7	25.8	130 000
Ribonuclease	9.6	13.7	36.6	120 000
Bradykinin	12.0	1.06	48.8	140 000

a) Value was calculated by ProtParam tool (http://krexpasy.org/tools/protparam.html).

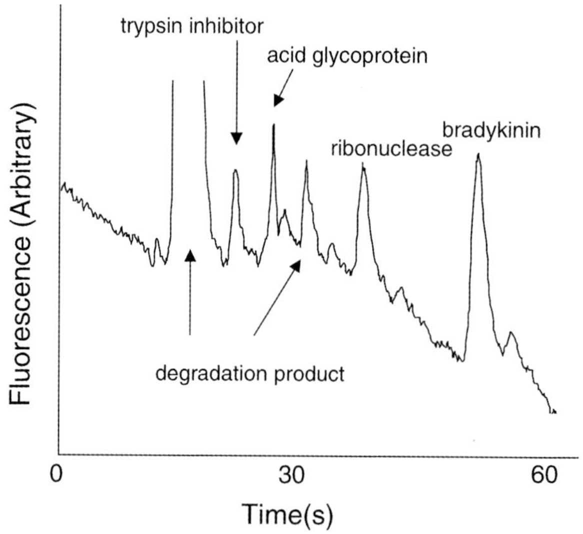

Fig. 4 Electropherogram of NBD-derivatized protein mixture. Conditions: running buffer, 10 mM phosphate buffer (pH 7.0) with 3% cationic starch derivative.

the migration time of the protein with p*I* value higher than 7.0 tended to be long. Furthermore, in the case of the protein with low p*I* value, the small-molecular-weight protein migrated fast, while in the case of the protein with high p*I* value, the large-molecular-weight protein migrated fast. These results indicated that the EOF occurred from BR to BW and the direction of the electrophoresis of the protein with low p*I* value was the same as the direction of the EOF. In this case, the small-molecular-protein, which showed a fast electrophoretic mobility, migrated faster compared with the high-molecular-weight protein. On the contrary, for the protein with a high p*I* value, the direction of the electrophoresis of the protein and that of the

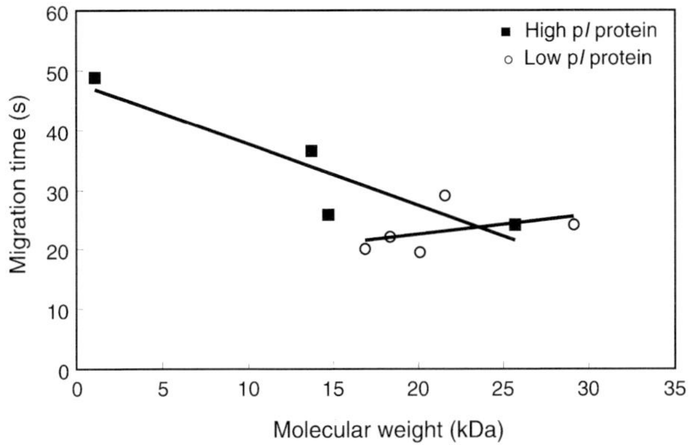

Fig. 5 Relationship between the migration time of NBD-proteins and their molecular weights.

EOF was opposite. Therefore, net mobilities of these proteins were very slow. In this case, the increase in the electrophoretic mobility produced a long migration time. Based on this result, the small protein migrated later than the large protein. The mobility of the NBD-bradykinin was thought to be fast, because the molecular weight was about 1000. This result showed that the mobility of EOF allowed the migration of a wide range of compounds.

The net charge (z) of the derivatized protein was not clear, because we do not know how many NBD-Fs reacted with each protein. We used the molecular weight instead of the

m/z and examined the relationship between the migration time and molecular weight of the protein (Fig. 5). Although the migration time was very fast, the correlation between the migration time and molecular weight was good. Acid glycoprotein did not fit to the correlation coefficient, because we did not calculate the molecular weight of glycan. The separation mechanism was supposed to be based on the molecular weight and p*I* (net charge) of the protein.

These results showed the possibility that a wide range of biological compounds could be analyzed by the microchip within a very short time. By combination of the microchip and MS, we will develop a very powerful tool for the analysis of biological compounds in living organisms.

10.4 Concluding remarks

We found that the dynamic coating of the PMMA channel with a cationic starch derivative could dramatically improve the analysis of amino acids and proteins derivatized with a fluorogenic reagent, NBD-F, and FITC. Using a 3% solution in 10 mM phosphate buffer (pH 7.0), FITC-BSA migrated within 2 min. The separation efficiency was 480 000 plates/m, which is similar to the reported values. Because the high-speed analytical method of biological compounds in living organisms is a critical technique in the research field of system biology, we believe that these microchip separation systems are capable of providing a solution to this requirement.

This work was supported by Goto Research Grant from University of Shizuoka and a grant from the Ministry of Education, Culture, Sports, Science and Technology (MEXT).

10.5
References

[1] Lander, E. S., Linton, L. M., Birren, B., Nusbaum, C., Zody, M. C., Baldwin, J., Devon, K., *et al.*, *Nature* 2001, *409*, 860–921.

[2] Venter, J. C, Adams, M. D., Myers, E. W., Li, P. W., Mural, R. J., Sutton, G. G., Smith, H. O., *et al.*, *Science* 2001, *291*, 1304–1351.

[3] Ideker, T., Thorsson, V., Ranish, J. A., Christmas, R., Buhler, J., Eng, J. K, Bumgarner, R., *Science* 2001, *292*, 929–934.

[4] Harrison, D. J., Manz, A., Fan, Z., Lüdi, H., Widmer, H. M., *Anal. Chem.* 1992, *64*, 1926–1932.

[5] Harrison, D. J., Fluki, F., Seiler, K., Fan, Z., Effenhauser, C. S., Manz, A., *Science* 1993, *261*, 895–897.

[6] Verpoorte, E., *Electrophoresis* 2002, *23*, 677–712.

[7] Salimi-Momosavi, H., Tang, T., Harrison, D. J., *J. Am. Chem. Soc.* 1997, *119*, 8716–8717.

[8] Kopp, M. U., de Mello, A. J., Manz, A., *Science* 1998, *280*, 1046–1048.

[9] Schilling, E. A., Kamholz, E., Yager, P., *Anal. Chem.* 2002, *74*, 1798–1804.

[10] Olsen, K. G., Ross, D. J., Tarlov, M. J., *Anal. Chem.* 2002, *74*, 1436–1441.

[11] Hadd, A. G., Jacobson, S. C., Ramsey, J. M., *Anal. Chem.* 1999, *71*, 5206–5212.

[12] Chiem, N. H., Harrison, D. J., *Clin. Chem.* 1998, *44*, 591–598.

[13] Sato, K., Tokeshi, M., Odake, T., Kimura, H., Ooi, T., Nakao, M., Kitamori, T., *Anal. Chem.* 2000, *72*, 1144–1147.

[14] Reyes, D. R., Iossifidis, D., Auroux, P.-A., Manz, A., *Anal. Chem.* 2002, *74*, 2623–2636.

[15] Auroux, P.-A., Iossifidis, D., Reyes, D. R., Manz, A., *Anal. Chem.* 2002, *74*, 2637–2652.

[16] Soper, S. A., Ford, S. M., Qi, S., McCarley, R. L., Kelly, K., Murphy, M. C., *Anal. Chem.* 2000, *72*, 643A–651A.

[17] Boone, T. D., Hugh Fan, Z., Hooper, H. H., Ricco, A. J., Tan, H., Williams, S. J., *Anal. Chem.* 2002, *74*, 78A–86A.

[18] Effenhauser, C. S., Bruin, G. J. M., Paulus, A., *Anal. Chem.* 1997, *69*, 3451–3457.

[19] Duffy, D. C., McDonald, J. C., Schueller, O. J. A., Whitesides, G. M., *Anal. Chem.* 1998, *70*, 4974–4984.

[20] Ford, S. M., Kar, B., McWhorter, S., Davies, J., Soper, S. A., Klopf, M., Calderon, G., Saile, V., *J. Microcol. Sep.* 1998, *10*, 413–422.

[21] Henry, A. C., Tutt, T. J., Galloway, M., Davidson, Y. Y., McWhorter, C. S., Soper, S. A., McCarley, R. L., *Anal. Chem.* 2000, *72*, 5331–5337.

[22] McCormick, R. M., Nelson, R. J., Alonso-Amigo, M. G., Benvegnu, D. J., Hooper, H. H., *Anal. Chem.* 1997, *69*, 2626–2630.

[23] Sassi, A. P., Paulus, A., Cruzado, I. D., Bjornson, T., Hooper, H. H., *J. Chromatogr. A* 2000, *894*, 203–217.

[24] Slentz, B. E., Penner, N. A., Lugowska, E., Regnier, F., *Electrophoresis* 2001, *22*, 3736–3743.

[25] Hu, S. W., Ren, X. Q., Bachman, M., Sims, C. E., Li, G. P., Allbritton, N., *Anal. Chem.* 2002, *74*, 4117–4123.

[26] Liu, Y., Fanguy, J. C., Bledsoe, J. M., Henry, C. S., *Anal. Chem.* 2000, *72*, 5939–5944.

[27] Barker, S. L. R., Ross, D., Tarlov, M. J., Gaitan, M., Locascio, L. E., *Anal. Chem.* 2000, *72*, 5925–5929.

[28] Barker, S. L. R., Tarlov, M. J., Canavan, H., Hickman, J. J., Locascio, L. E., *Anal. Chem.* 2000, *72*, 4899–4903.

[29] Kato, M., Gyoten, Y., Sakai-Kato, K., Toyo'oka, T., *J. Chromatogr. A* 2003, *1013*, 183–189.

[30] Wang, S. C., Perso, C. E., Morris, M. D., *Anal. Chem.* 2000, *72*, 1704–1706.

[31] Dang, F., Zhang, L., Jabasini, M., Kaji, N., Baba, Y., *Anal. Chem.* 2003, *75*, 2433–2439.

[32] Sakai-Kato, K., Kato, M., Toyo'oka, T., *Anal. Chem.* 2003, *75*, 388–393.

[33] Belder, D., Ludwig, M., *Electrophoresis* 2003, *24*, 3595–3606.

[34] Kato, M., Gyoten, Y., Sakai-Kato, K., Nakajima, T., Toyo'oka, T., *Anal. Chem.* 2004, *76*, 6792–6796.

[35] Tabuchi, M., Kuramitsu, Y., Nakamura, K., Baba, Y., *Anal. Chem.* 2003, *75*, 3799–3805.

[36] Wu, D., Luo, Y., Zhou, X., Dai, Z., Lin, B., *Electrophoresis* 2005, *26*, 211–218.

[37] Herr, A. E., Singh, A. K., *Anal. Chem.* 2004, *76*, 4727–4733.

11
Single cell manipulation, analytics, and label-free protein detection in microfluidic devices for systems nanobiology*

Wibke Hellmich, Christoph Pelargus, Kai Leffhalm, Alexandra Ros, Dario Anselmetti

Single cell analytics for proteomic analysis is considered a key method in the framework of systems nanobiology which allows a novel proteomics without being subjected to ensemble-averaging, cell-cycle, or cell-population effects. We are currently developing a single cell analytical method for protein fingerprinting combining a structured microfluidic device with latest optical laser technology for single cell manipulation (trapping and steering), free-solution electrophoretical protein separation, and (label-free) protein detection. In this paper we report on first results of this novel analytical device focusing on three main issues. First, single biological cells were trapped, injected, steered, and deposited by means of optical tweezers in a poly(dimethylsiloxane) microfluidic device and consecutively lysed with SDS at a predefined position. Second, separation and detection of fluorescent dyes, amino acids, and proteins were achieved with LIF detection in the visible (VIS) (488 nm) as well as in the deep UV (266 nm) spectral range for label-free, native protein detection. Minute concentrations of 100 fM injected fluorescein could be detected in the VIS and a first protein separation and label-free detection could be achieved in the UV spectral range. Third, first analytical experiments with single Sf9 insect cells (*Spodoptera frugiperda*) in a tailored microfluidic device exhibiting distinct electropherograms of a green fluorescent protein-construct proved the validity of the concept. Thus, the presented microfluidic concept allows novel and fascinating single cell experiments for systems nanobiology in the future.

11.1
Introduction

In systems biology [1], a multitude of different disciplines from biology, chemistry, physics, material science, micro- and nanoengineering, and (bio)informatics aim to link quantitative molecular structural and functional (proteomic) information to the

* Originally published in Electrophoresis 2005, 26, 3689–3696

Microfluidic Applications in Biology. Edited by Niels Lion, Joël S. Rossier, and Hubert H. Girault

ISBN-10: 3-527-31761-9

different genetically programmed and regulated networks in a living cellular organism. To date, proteomes are analyzed at the level of 10^5–10^6 cells accessing functional information only on the basis of that probed cellular ensemble. Averaging effects from cell-cycle-dependent states, the different and inhomogeneous cellular response to an external stimulus, or the introduction of genomic and proteomic variabilities during eucaryotic cell proliferation are completely neglected. The analysis of smallest analyte quantities and the hunt for low-abundant proteins at the single cell level, however, requires new techniques for efficient and sensitive separation, detection, and analysis. As a rule of thumb, the typical protein content of a cell is about 15% proteins w/w which equals 75 pg, 2 fmol, or 10^8 protein molecules, assuming an average molecular mass of 40 kDa. A low-abundant protein in a cell at μM concentration runs at the amol level which is equivalent to 10^5 molecules.

In systems nanobiology [2], microfabrication and nanotechnology offer novel tools to detect, measure, analyze, steer, and manipulate individual molecules and cells. Such tools allow more detailed insights into the interplay of genomic information and functional peculiarity at the single molecule or single cell level. Micro total analysis systems (μTAS) or lab-on-a-chip systems [3, 4] offer the possibility to handle minute volumes down to the pL and even fL range. In recent years, protein electrophoresis, one of the most efficient techniques for protein separation, has been investigated in microfluidic systems [3]. Standard capillary separation techniques were transferred to the microchip format and LIF detection systems for the ultrasensitive detection of labeled proteins were developed. Using LIF detection in the visible (VIS) range, separation of covalently labeled proteins [5, 6] and peptides [7, 8] as well as postcolumn labeling for protein chip electrophoresis have been demonstrated [9, 10]. Recently, a label-free interferometric backscatter method has been described by Wang *et al.* [11], where proteins could be detected in the nM range. It is worth noting that the achievable detection limit in microfluidic devices upon injecting nL volumes at nM concentrations lies in the amol range.

In contrast to microfluidic systems, native LIF detection, based on the fluorescence of the aromatic amino acids tryptophan (Trp), tyrosine (Tyr), and phenylalanine (Phe), has found widespread use in conventional CE. In 1992, Yeung and co-workers [12] pioneered a LIF detection method with 275 nm excitation light provided by an Ar^+-laser with pM detection limits. Since then this method has served for the exocytose monitoring of single mast cells [13] as well as for the separation of hemoglobin variants in red blood cells [14]. Exploiting alternative laser systems, nM detection limits for Trp could be achieved with a metal vapor laser [15] or an excimer laser [16]. With solid state lasers, nM detection limits of peptides [17] and proteins [18, 19], as well as a pM detection limit for carbonic anhydrase [19] were reported. In contrast to these conventional CE methods protein separation with native UV detection on a quartz microchip with μM detection limit has only been demonstrated in the very recent past [20]. CE could thus be a method of choice for label-free single cell analysis.

First single cell fingerprinting with capillary sieving electrophoresis in 1-D [21, 22] and 2-D format using a protein stain with LIF detection in the VIS range was pioneered by the group of Dovichi and co-workers [23]. Recently, microfluidic devices have been explored for separation and detection of fluorescent dyes [24, 25] and a specific small peptide [26] or vitamin [27] from single cells.

In this work, we focus on three main issues: (i) we describe the trapping, steering, and deposition of a single target cell out of a cell culture by means of optical tweezers (OTs) and consecutive lysis in a microfluidic device. (ii) The separation and detection of native amino acids and proteins by VIS and UV-LIF in a microfluidic device is demonstrated. (iii) The monitoring of a first single cell electropherogram of a fluorescent protein (a green fluorescent protein (GFP)-construct) in a microfluidic device is presented. The manipulation and lysis of single cells and the incorporation of a detection system are based on a poly(dimethylsiloxane) (PDMS) microfluidic chip adapted to an inverted microscope with tailored UV optics and single photoncounting detection. Such microfluidic devices will be a future tool for single cell proteomic fingerprinting.

11.2 Materials and methods

11.2.1 Chemicals and reagents

PDMS (Sylgard 184) was purchased from Dow Corning (Midland, MI, USA). Quartz slides were from SPI Supplies (USA), glass microscope slides from Menzel (Germany). Avidin and Pullulan were obtained from Sigma (Deisenhofen, Germany) and lysocyme C from Serva (Heidelberg, Germany). Trp, PBS tablets, CHES, and Tris were purchased from Fluka (Buchs, Switzerland). Triblock copolymer Pluronics F-108 was a generous gift from BASF (Ludwigshafen, Germany). SDS was from Merck (Darmstadt, Germany). For all solutions deionized water from a Milli-Q biocel (Millipore, Bedford, MA, USA) was used.

11.2.2 Cells

Sf9 insect cells (*Spodoptera frugiperda*) from Novagen (Madison, WI, USA) transfected with pIEx4-vector (Novagen) containing the gene for the fusion protein were used in these studies. The GFP-Sf9 cells expressed a GFP-labeled "loss-of-function" mutant of the cytoplasmic G-protein ArF1 of *Medicago truncatula*. Transfection of cells resulted in maximal 50% efficiency for live cells containing the GFP-construct protein (T31N-GFP, molecular mass 49.5 kDa, p*I* 5.6). Portions of this cell culture (200 μL) in BacVector insect cell medium (Novagen) were washed with PBS-buffer (10 mM phosphate, 137 mM NaCl, 2.7 mM KCl, titrated to the optimal cell medium pH 6.4) *via* centrifugation and subsequent buffer addition. An estimated concentration of 10^5 cells/mL was used for the single cell manipulation and lysis experiments.

11.2.3 Fabrication of the PDMS device

A master with the inverted relief of the microstructure was fabricated *via* spincoating a photoresist (SU-8) onto an Si-wafer, UV-exposing through a chromium mask, and developing in a developer bath. The detailed fabrication procedure was recently published by Duong *et al.* [28]. The polymer Sylgard 184 and its curing agent were mixed in a 10:1 ratio and poured over the microstructured wafer. After curing at 85°C for 4.5 h the cross-linked polymer was easily peeled off the wafer and the reservoir holes were punched through the structured side. The PDMS slab and a clean quartz (for UV) or glass (for VIS detection) slide were oxidized in a UVO-Cleaner (Model 42–220, Jelight, USA) for 3 min. Afterwards, the PDMS slab was placed onto the quartz or glass slide forming an irreversible seal. The microfluidic channels had a typical cross section of $20 \times 20\ \mu m^2$ and channel walls were coated with a triblock copolymer F-108 [25] significantly reducing unwanted cell adhesion during cell steering.

11.2.4 OTs and microfluidic liquid handling

Individual Sf9 cells were trapped, injected, and steered in the microfluidic channel by a home-made single-beam OT system. The OT was incorporated into an inverted optical microscope [29] with additional standard fluorescence microscopy capabilities in the VIS range. A self-constructed x/y-stage allowed long range positioning of cells with the OTs within the microchannel for maximal 25 mm with a precision of 1–2 μm. By this means, an individual cell can be captured in the reservoir and navigated to the injection cross. Cell movement and lysis was observed using a CCD camera (DMK 3002-IR/C, The Imaging Source, Germany) grabbing images at a rate of 25 frames/s.

11.2.5 LIF detection in the VIS spectral range

The experimental setup was mounted on an inverted microscope (Axiovert 100, Zeiss, Jena, Germany), which additionally served as a platform for the confocal LIF detection system (see Fig. 2). Laser light (488 nm) from an Ar^+-ion laser (25 mW, Omnichrome, Germany or 2 mW, Spectraphysics, Germany) was coupled into the rear port of the microscope *via* two mirrors (New Focus, USA). The excitation light was reflected by a dichroic mirror and focused by a 20× objective (Zeiss) into the microchannel. The detection window was adjusted with the x/y-stage along the separation channel at various distances from the injector (usually several mm). The emitted fluorescence light was collected by the objective and passed through the dichroic mirror and a longpass filter (520 nm). The tube lens focused the emission light through an x/y adjustable 400 μm wide pinhole (unless otherwise stated) onto the photon-counting photomultiplier (Hamamatsu H6240, Japan).

11.2.6
LIF detection in the UV spectral range

For sensitive detection in the UV range, excitation light from a frequency quadrupled Nd:YAG laser (266 nm, Nanolase, France) was coupled into the rear port of the microscope (see Fig. 2). The wavelength was adjusted to the autofluorescence excitation maximum of Trp. Switching between the two excitation modes was achieved *via* a flipper equipped with a silver-coated mirror (New Focus).

For UV excitation and detection, a dichroic mirror (Laseroptik, Germany) with high reflectivity at 266 nm and high transmission >300 nm passed the excitation light through the 52 × reflective objective (Ealing, USA), which focused the laser light on the microchannel. Emitted fluorescence was collected with the reflective objective and focused with a high UV transmission tube lens (Zeiss) through the pinhole onto the photomultiplier. Fluorescence emission spectra of PDMS excited at 266 nm demonstrated a reduction to approximately 30% from maximal PDMS background fluorescence for wavelengths above 325 nm (data not shown). An interference emission filter with high transmission at 360 nm (50%) was thus chosen (360/50, Analysentechnik, Germany).

11.2.7
Chip operations

Initial filling of the microchannels was performed by pipetting the buffer into one reservoir and by applying a vacuum to the other reservoirs. Subsequent flows were either initialized by hydrostatic pressure due to different droplet volumes on the corresponding reservoirs or through electrokinetic pumping. Electrical connection to the microchip device was achieved with four platinum electrodes which were dipped into the reservoirs. Voltage was applied using power supplies from FUG (Modell HCN 14–12500 and HCN 7E-12500, Germany). Instrumental control and data acquisition were performed with software programmed in LabView (National Instruments, Austin, TX, USA). For protein or dye analysis from reservoir 1 the floating method was used according to [30]. Separation buffer for VIS-LIF and UV-LIF separation and detection was Tris-buffer (10 mM, pH 8.2). The single cell lysate was injected into the separation channel 2 by applying a positive voltage to reservoir 4 filled with the separation buffer (100 mM Tris, 100 mM CHES, 8% Pullulan, 0.1% SDS, pH 8.6).

11.3
Results and discussion

11.3.1
Single cell trapping, steering, and lysis

Diluted insect cells in PBS buffer were pipetted into the buffer reservoir of channel 1 of our microfluidic device (Fig. 1a) where an individual cell was optically selected,

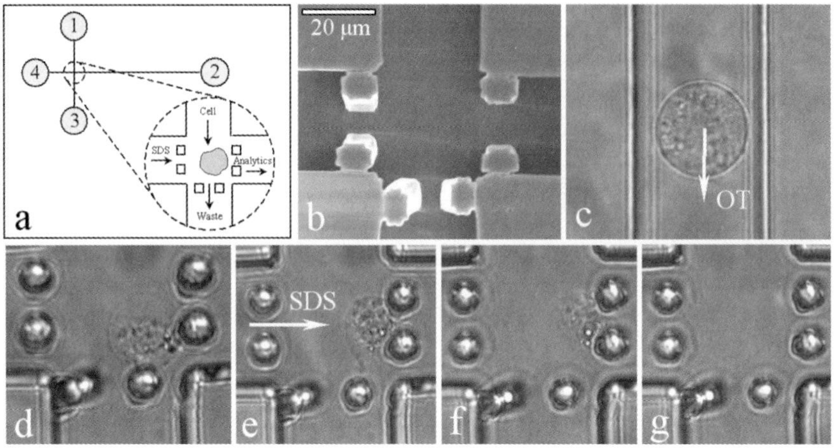

Fig. 1 (a) Scheme of our PDMS microfluidic device. Inset: channel crossing with the cell trap composed of microstructured obstacles, (b) scanning electron micrograph of the cell trap, (c) single cell in a channel navigated by OTs in the microchannel, and (d–g) optical micrographs of a single cell at the injection position during SDS lysis. SDS flow is from channel 4 through the cell trap into channel 2.

trapped, and injected into the microfluidic channel with our OT setup. The injection and steering of the cell (Fig. 1c) was realized with the dedicated *x*/*y*-stage. The individual cells were transferred to the crossing of our microfluidic device which was microstructured by vertical posts in order to act as a physical cell trap (Figs. 1b and d). Once the cell was navigated into this position the optical trap was switched off and the cell was allowed to adhere to the microchannel wall. Consecutively, cell lysis was performed by flushing a 0.5% SDS solution in PBS by hydrostatic pressure into the perpendicular channel (from channel 4 to 2). The cell lysis was visualized and controlled by optical bright-field microscopy. Figures 1d–g demonstrate a sequence of snapshots from a single cell lysis at the injection point close to the entrance of channel 2, in which subsequent analytics will be performed. Complete cell lysis was typically achieved within 6 s.

11.3.2
LIF setup and VIS-LIF detection

The detection system for VIS and native UV LIF of proteins, which will be applied to the lysed cells in the future, is schematically depicted in Fig. 2. It is based on an inverted microscope providing a robust and versatile setup for microchip inspection, especially for the optimization of microchip injection and the placement of the detection window in the microchannel. The microfluidic device is constructed as PDMS/glass or PDMS/quartz hybrid in order to minimize fluorescence in the chosen wavelength range. Molded PDMS microchips were thus sealed with a quartz or glass objective slide after oxidative UV treatment of the two components.

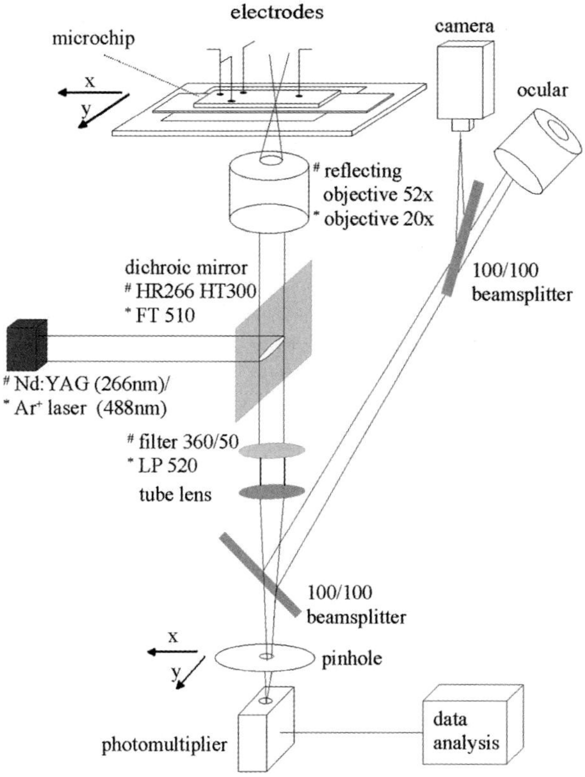

Fig. 2 Scheme of the setup for the UV (#) and VIS (*) LIF detection systems realized on an inverted optical microscope.

This UV treatment provided a tight and irreversible seal of the PDMS device both for glass and quartz slides, which was also reported by plasma treatment of PDMS and glass surfaces [31]. The microscope slides served thus as the bottom of the microstructure through which the excitation light passed the microchip.

In order to estimate the sensitivity limit in our microchips, the floating method [32, 33] was chosen to thoroughly study fluorescent dye detection in the VIS range. Thus, extremely small analyte amounts (fM range) could be detected in our PDMS microfluidic devices [30]. Fig. 3a shows the electropherogram of an electrokinetically injected 100 fM fluorescein solution. With respect of the detection volume (defined by the pinhole and the microchannel dimensions) this corresponds roughly to 50–100 molecules. This is well below the anticipated number of low-abundant protein copies in a single cell. Furthermore this is, to our knowledge, the smallest dye concentration which was electrokinetically injected and detected in a microfluidic device. For the fluorescently labeled protein avidin, an electropherogram resulting from the injection of 84 nM sample concentration is demonstrated in Fig. 3b. The required amol detection sensitivity for low abundant proteins can thus be achieved already with a standard protein in our detection setup in the VIS range.

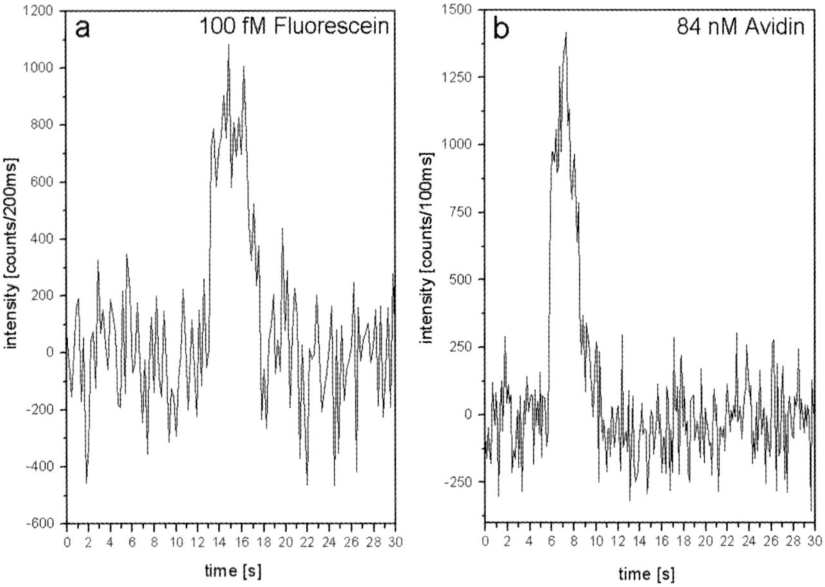

Fig. 3 (a) Electropherogram for the VIS LIF detection of 100 fM fluorescein injected into the PDMS microchannel device (separation electric field 1250 V/cm). Baseline correction and background substraction were applied to the original data. (b) Electropherogram for the VIS LIF detection of 84 nM fluorescein-avidin injected into a PDMS device with a separation field strength of 1500 V/cm. Laser power was 25 mW for both electropherograms.

11.3.3
UV-LIF amino acid detection

The UV fluorescence setup was optimized for the detection of Trp with an excitation and emission maxima at 295 and 353 nm, respectively [34]. With the present fluorescence detection system, fluorescence of the amino acids Phe and Tyr could also be detected. However, a smaller contribution to the overall signal could be expected due to the different emission and excitation maxima and in addition to the smaller quantum yield of Phe [34]. We have thus successively investigated injection and UV-LIF detection of Trp. The inset of Fig. 4a demonstrates a typical electropherogram obtained with the UV detection system for the injection of 50 μM Trp in Tris buffer. The detection window was placed at 10 mm from the cross injector. Furthermore, a plot of peak heights of injected Trp *versus* concentration is shown in Fig. 4 (see figure legend for detailed separation conditions). The linear regression with a regression factor of $r = 0.998$ demonstrates an excellent linearity for Trp concentrations from 50 to 1000 μM. For the smallest concentration, an S/N of 9 has been determined, so that the theoretical detection limit is 17 μM for an S/N of 3. Thus, proteins with at least one Trp in their amino acid composition should be detectable with this setup in the low μM range, if fluorescence quenching is absent.

For 50 μM injected Trp the number of theoretical plates (*N*) resulted in 449, corresponding to a height equivalent of a theoretical plate (*H*) of 22.3 μm. This is an order of magnitude higher than obtained for fluorescently labeled amino acids [7, 35]. However, it has been reported that the separation efficiency in PDMS devices is often lower compared to glass devices, it can be satisfactory by careful control of separation parameters [7]. Ocvirk *et al.* [36] report that only with voltage control on all four reservoirs injections for the dye fluorescein with $N > 15\,000$ could be obtained in PDMS. In contrast to this work, we used a positive voltage at the buffer reservoir during separation while all other reservoirs were grounded. This implies a pushback flow of analyte to the sample and sample waste reservoir during separation, so that analyte leakage into the separation channel was prevented. This fact is underlined by the high reproducibility of Trp injections with SDs smaller than 5% for the peak heights (except the 100 μM injections with 9.2%). Furthermore, we expect that peak-broadening could significantly be reduced in the future with a change of the injection method, *e.g.*, pinched injection or narrow sample channel injectors [32]. However, the actual sensitivity of our UV-LIF detection system applied to the chosen hybrid quartz/PDMS microchip device compares well with a previously reported detection of 2 μM Trp in a full body quartz microfluidic chip [20]. The outstanding advantage of PDMS as microstructure material is the relatively low production cost compared to quartz microchips and also its versati-

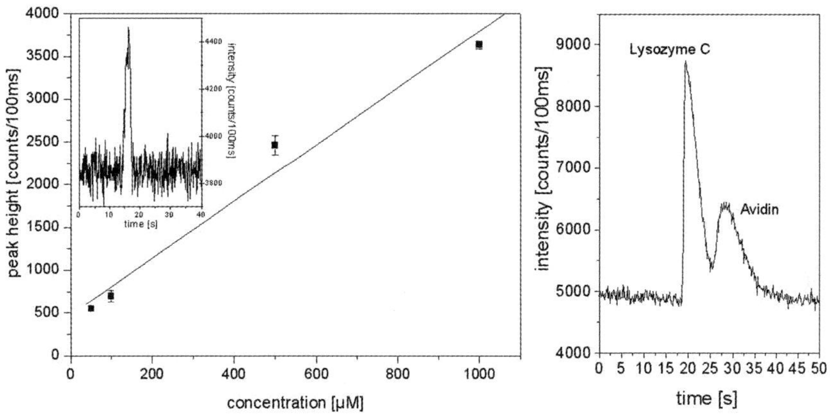

Fig. 4 Separation and detection in the UV. (a) Peak heights *versus* Trp concentration for injections in a PDMS microchannel: the line represents a linear regression with $r = 0.998$ indicating a theoretical detection limit of 17 μM Trp with S/N = 3. Inset: Electropherogram of 50 μM injected Trp in 10 mM Tris with the detection window placed at 10 mm after the cross injector. An electrical field of 350 V/cm was used during the loading phase of the injection and of 325 V/cm during the separation phase, respectively. (b) Electropherogram for the separation of lysocyme C and avidin. Sample and running buffer was 10 mM Tris and concentrations of avidin and lysocyme C were 125 and 500 μM, respectively. For the injection, the analyte was loading the cross injector at 350 V/cm and a separation electrical field of 770 V/cm was used for separation. Separation length was 12 mm.

lity. This is of special interest for single cell handling in a microfluidic device, where structured parts or obstacles can easily be incorporated in order to serve as physical barriers for cell trapping.

11.3.4 Protein separation and UV-LIF detection

Two standard proteins, lysocyme C (6 Trp) and avidin (16 Trp), were selected. Fig. 4 shows an electropherogram for the separation of a mixture of 125 µM avidin and 500 µM lysocyme C. The protein peaks could clearly be identified by a comparison to single injections of the respective protein (data not shown). After a separation distance of only 12 mm the proteins were separated nearly to baseline resolution.

It is worth noting that protein separation in PDMS microdevices is strongly influenced by the surface properties so that the detection limit of 17 µM for Tyr can only be achieved with adequate control of protein–surface interaction. However, lysocyme C and avidin could be separated without further treatment of the PDMS channel surface. A significant improvement of the detection limit should arise by controlling PDMS protein interaction to circumvent unspecific adhesion. For that reason we are currently investigating PDMS surface coatings with poly(oxyethylene) compounds [37].

11.3.5 Single cell electropherograms

In order to prove the general concept of our single cell analytics approach, we monitored the electropherogram of a single Sf9 cell which was modified by a GFP-variant of a cytoplasmatic G-protein T31N-GFP (Fig. 5). The GFP-cell was optically selected in the PBS buffer reservoir from a solution containing less than 30% live GFP-cells and navigated with OT to the microdevice crossing. There, the OT was switched off and the cell was allowed to adhere (Fig. 5 inset). Subsequently, the separation buffer containing a sieving additive and 0.1% SDS was electrokinetically injected from reservoir 4 to the captured cell. The electropherogram of the single cell lysate was recorded by VIS LIF detection at a distance of 1000 µm from the initial cell position in the microchannel at an electrical field of 100 V/cm. In the single cell electropherogram (Fig. 5) one distinct peak can be identified at 334 s with an S/N of 400 as expected from a single component analyte, originating from the only fluorescent GFP-construct protein. This peak is characterized by a full width of half maximum of 0.8 s and a baseline width of only 5 s. This baseline width corresponds well with the lysis time recorded for a cell lysis sequence (Figs. 1d–g) indicating minute diffusion dispersion from the injection process. Thus, the electropherogram denotes transport of the whole lysis fraction to the detection point and reflects the injection of the complete protein content from a single cell cytoplasm. This is to our knowledge the first electropherogram of a single protein compound from a single cell in a microfluidic device.

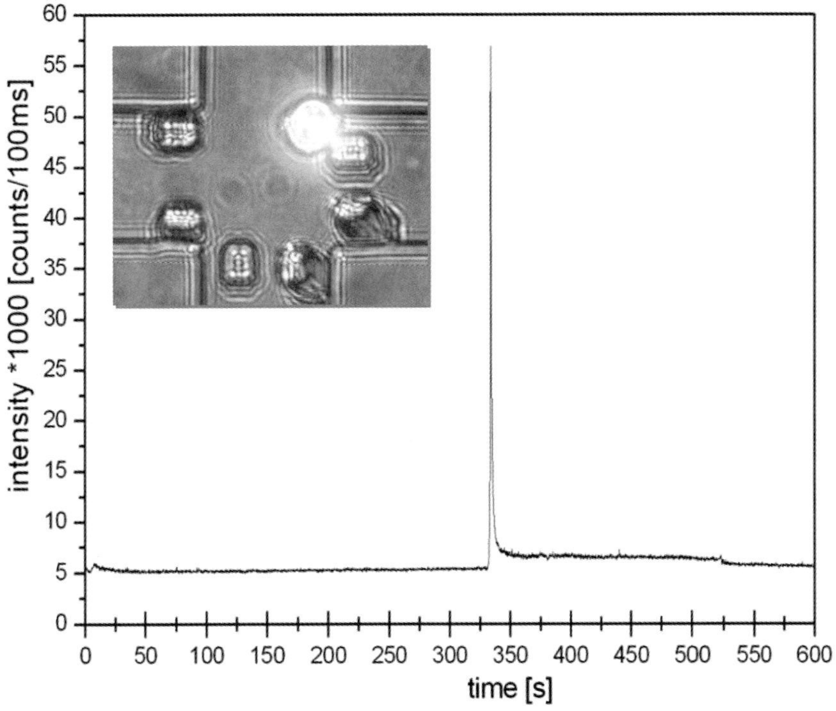

Fig. 5 Electropherogram of a single GFP-Sf9 insect cell with a distinct single component peak of the 49.5 kDa T31N-GFP variant. Starting point (0 s) of the electropherogram is defined by the application of the separation voltage to the horizontal channel and thus does not represent the initiation of cell lysis. Power of the incident laser light (at 488 nm) was 2 mW and the pinhole size 200 μm. Inset: Fluorescence micrograph of a single GFP-Sf9 insect cell captured at the injection position (for details see text).

11.4 Concluding remarks

We are currently developing a microfluidic chip platform which combines single cell trapping, steering, deposition, lysis, and subsequent electrophoretical protein separation, and LIF-detection in the VIS and in the UV spectral range. We have demonstrated the manipulation and controlled lysis of single Sf9 insect cells as well as the separation of proteins with native, label-free UV-LIF detection in a microfluidic PDMS device. Avidin and lysocyme C were separated with nearly baseline resolution within less than 40 s. Injections of the amino acid Trp resulted in a theoretical detection limit of 17 μM which should be also applicable for proteins. We have also demonstrated that our detection setup in the VIS spectral range is capable of detecting minute concentrations necessary for the analysis of low-abundant proteins from a single cell.

In addition, we injected and lysed single Sf9 insect cells expressing a single GFP-construct protein in our microfluidic device and monitored the corresponding single cell electropherogram in the visual spectral range. These first single cell analytical experiments exhibited the distinct feature of a single fluorescent component, in agreement with the probed cell species. Further optimization of the UV optical detection setup will significantly improve separation efficiencies and detection sensitivity, and facilitate a label-free microchip based single cell device for systems nanobiology in the future.

Sf9 cells were generously donated by Nickels Jensen and Professor Karsten Niehaus from the Genetics Department of Bielefeld University. Financial support from the Deutsche Forschungsgemeinschaft (Project: Single Cell Analytics – An 370/1–2) and the generous Pluronics sample donation by BASF (Dr. Stephan Altmann and Dr. Kati Schmidt) is gratefully acknowledged. We thank Andy Sischka, Dr. Katja Tönsing, and Thanh Tu Duong from the Experimental Biophysics Group for technical assistance in OT manipulation and microfabrication.

11.5
References

[1] Ideker, T., Galitski, T., Hood, L., *Annu. Rev. Genomics Hum. Genet.* 2001, *2*, 343–372.

[2] Heath, J. R., Phelps, M. E., Hood, L., *Mol. Imaging Biol.* 2003, *5*, 312–325.

[3] Auroux, P. A., Iossifidis, D., Reyes, D. R., Manz, A., *Anal. Chem.* 2002, *74*, 2637–2652.

[4] Reyes, D. R., Iossifidis, D., Auroux, P. A., Manz, A., *Anal. Chem.* 2002, *74*, 2623–2636.

[5] Liu, Y., Foote, R. S., Culbertson, C. T., Jacobson, K. B., Ramsey, R. S., Ramsey, J. M., *J. Microcol. Sep.* 2000, *12*, 407–411.

[6] Bousse, L., Mouradian, S., Minalla, A., Yee, H., Williams, K., Dubrow, R., *Anal. Chem.* 2001, *73*, 1207–1212.

[7] Lacher, N. A., de Rooij, N. F., Verpoorte, E., Lunte, S. M., *J. Chromatogr. A* 2003, *1004*, 225–235.

[8] Chiem, N., Harrison, D. J., *Anal. Chem.* 1997, *69*, 373–378.

[9] Liu, Y., Foote, R. S., Jacobson, S. C., Ramsey, R. S., Ramsey, J. M., *Anal. Chem.* 2000, *72*, 4608–4613.

[10] Colyer, C. L., Mangru, S. D., Harrison, D. J., *J. Chromatogr. A* 1997, *781*, 271–276.

[11] Wang, Z., Swinney, K., Bornhop, D. J., *Electrophoresis* 2003, *24*, 865–873.

[12] Lee, T. T., Yeung, E. S., *J. Chromatogr.* 1992, *595*, 319–325.

[13] Lillard, S. J., Yeung, E. S., McCloskey, M. A., *Anal. Chem.* 1996, *68*, 2897–2904.

[14] Lillard, S. J., Yeung, E. S., Lautamo, R. M. A., Mao, D. T., *J. Chromatogr. A* 1995, *718*, 397–404.

[15] Zhang, X., Sweedler, J. V., *Anal. Chem.* 2001, *73*, 5620–5624.

[16] Paquette, D. M., Song, R., Banks, P. R., Waldron, K. C., *J. Chromatogr. A* 1998, *714*, 47–57.

[17] Kuijt, J., van Teylingen, R., Nijbacker, T., Ariese, F., Brinkman, U. A. Th., Gooijer, C., *Anal. Chem.* 2001, *73*, 5026–5029.

[18] Chan, K. C., Muschik, G. M., Issaq, H. J., *Electrophoresis* 2000, *21*, 2062–2066.

[19] Tseng, W.-L., Chang, H.-T., *Anal. Chem.* 2000, *72*, 4805–4811.

[20] Schulze, P., Ludwig, M., Kohler, F., Belder, D., *Anal. Chem.* 2005, *77*, 1325–1329.

[21] Hu, S., Zhang, L., Krylow, S., Dovichi, N. J., *Anal. Chem.* 2003, *75*, 3495–3501.

[22] Hu, S., Zhang, L., Newitt, R., Aebersold, R., Kraly, J. R., Jones, M., Dovichi, N. J., *Anal. Chem.* 2003, *75*, 3502–3505.

[23] Hu, S., Michels, D. A., Abu Fazal, M., Ratisoontorn, Ch., Cunningham, M. L., Dovichi, N. J., *Anal. Chem.* 2004, *76*, 4044–4049.

[24] McClain, M. A., Culbertson, C. T., Jacobson, S. C., Albritton, N. L., Sims, C. E., Ramsey, J. M., *Anal. Chem.* 2003, *75*, 5646–5655.

[25] Munce, N. R., Li, J., Herman, P. R., Lilge, L., *Anal. Chem.* 2004, *76*, 4983–4989.

[26] Gao, J., Yin, X. F., Fang, Z.-L., *Lab Chip* 2004, *4*, 47–52.

[27] Xia, F., Jin, W., Yin, X., Fang, Z.-L., *J. Chromatogr. A* 2005, *1063*, 227–233.

[28] Duong, T., Kim, G., Ros, R., Streek, M., Schmid, F., Brugger, J., Ros, A., Anselmetti, D., *Microelectr. Eng.* 2003, 67–68, 905–912.

[29] Sischka, A., Eckel, R., Toensing, K., Ros, R., Anselmetti, D., *Rev. Sci. Instrum.* 2003, *74*, 4827–4831.

[30] Ros, A., Hellmich, W., Duong, T., Anselmetti, D., *J. Biotechnol.* 2004, *122*, 65–67.

[31] Duffy, D. C., Cooper McDonald, J., Schueller, O. J. A., Whitesides, G. M., *Anal. Chem.* 1998, *70*, 4974–4984.

[32] Zhang, C.-X., Manz, A., *Anal. Chem.* 2001, *73*, 2656–2662.

[33] Effenhauser, C. S., Bruin, G. J. M., Paulus, A., Ehrat, M., *Anal. Chem.* 1997, *69*, 3451–3457.

[34] Lakowicz, J. R., *Principles of Fluorescence Spectroscopy*, Kluwer Academic/Plenum Publisher, New York 1999.

[35] Seiler, K., Harrison, D. J., Manz, A., *Anal. Chem.* 1993, *65*, 1481–1488.

[36] Ocvirk, G., Munroe, M., Tang, T., Oleschuk, R., Westra, K., Harrison, D. J., *Electrophoresis* 2000, *21*, 107–115.

[37] Hellmich, W., Regtmeier, J., Duong, T., Ros, R., Anselmetti, D., Ros, A., Langmuir 2005, *21*, 7551–7557.

12
Fast immobilization of probe beads by dielectrophoresis-controlled adhesion in a versatile microfluidic platform for affinity assay*

Janko Auerswald, David Widmer, Nico F. de Rooij, André Sigrist, Thomas Staubli, Thomas Stöckli, Helmut F. Knapp

The use of probe beads for lab-on-chip affinity assays is very interesting from a practical point of view. It is easier to handle and trap beads than molecules in microfluidic systems. We present a method for the immobilization of probe beads at defined areas on a chip using dielectrophoresis (DEP)-controlled adhesion. The method is fast, *i.e.*, it takes between 10 and 120 s – depending on the protocol – to functionalize a chip surface at defined areas. The method is versatile, *i.e.*, it works for beads with different types of probe molecule coatings. The immobilization is irreversible, *i.e.*, the retained beads are able to withstand high flow velocities in a flow-through device even after the DEP voltage is turned off, thus allowing the use of conventional high-conductivity analyte buffers in the following assay procedure. We demonstrate the on-chip immobilization of fluorescent beads coated with biotin, protein A, and goat–antimouse immunoglobulin G (IgG). The number of immobilized beads at an electrode array can be determined from their fluorescence signal. Further, we use this method to demonstrate the detection of streptavidin and mouse IgG. Finally, we demonstrate the feasibility of the parallel detection of different analyte molecules on the same chip.

12.1
Introduction

In the life sciences research and in medical diagnostics, lab-on-chip devices offer advantages such as less consumption of sample and expensive reagents, considerably shorter assay time because of shorter diffusion paths, and controlled handling of microparticles and nanoparticles, *e.g.*, cells, beads, or molecules. Lab-

* Originally published in Electrophoresis 2005, 26, 3697–3705

Microfluidic Applications in Biology. Edited by Niels Lion, Joël S. Rossier, and Hubert H. Girault

ISBN-10: 3-527-31761-9

on-a-chip systems often exploit a specific biochemical reaction between probe molecules and the detected analyte molecules (affinity assay). Microfluidic flow-through devices for affinity assay with immobilized probe molecules on the chip surface offer the advantage of a high detection sensitivity. By maintaining a constant flow of analyte solution and using the small diffusion distances in microfluidic systems, even very low concentrations of analyte molecules can be detected by adjusting the flow velocity. In many applications, the probe molecules have to be immobilized on a chip surface in a complex and expensive procedure.

Bead-based microfluidic systems or microarrays have become particularly interesting in affinity assay development, as several recent review articles on this topic show [1, 2]. Bead-based assays can be an interesting alternative to direct probe molecule deposition on the chip surface or on microfluidic channel walls. The probe molecules can be attached to bead surfaces by specialized suppliers in cost-efficient large-scale production. The beads are introduced into the lab-on-chip system upon demand. Bead-based platforms offer a high degree of versatility because many kinds of beads with a variety of functionalizations are available. Different approaches are possible to immobilize beads on a chip, *e.g.*, mechanical barriers [3, 4], magnetic contraptions [5], EOF and pressure-driven counterflow for contactless trapping of beads in vortices generated at diffuser-nozzle elements [6], and retention by dielectrophoresis (DEP) either in contactless field cages using negative DEP [7] or at electrode edges using positive DEP [8]. Walt *et al.* [9] presented a bead-based assay concept using color-encoded beads similar to the ones known from the Luminex multiplexed flow cytometry assays [10]. They immobilized these encoded beads on an array of optical fibers. Though the array fabrication is rather complex, the concept allows a highly parallel assay of analyte molecules, *e.g.*, different oligonucleotides.

The retention of beads by positive DEP offers several advantages, *e.g.*, simple chip layout with planar electrodes, good compatibility with miniaturization, and the option to separate different beads according to their dielectric properties or sizes. However, in fluids with a conductivity of above 100 mS/m positive DEP alone becomes too weak to effectively retain the beads in a flow-through device [11]. Typical analyte solutions, *e.g.*, physiological buffers and body fluids, usually have conductivities up to 2000 mS/m. Therefore, we developed a bead immobilization method using positive DEP in a frequency range of 10–20 kHz and a superimposed pressure-driven flow to immobilize functionalized beads on a chemically conditioned chip surface with electrodes. The method is fast and versatile, *i.e.*, it is suitable for beads with different types of probe molecule coatings. After immobilization of the beads at the electrodes, the DEP voltage can be turned off and the functionalized beads serve as substrates for the detection of analyte molecules in typical high-conductivity analyte solutions.

In this paper, we demonstrate the fast on-chip immobilization of beads coated with biotin, protein A, and goat–antimouse immunoglobulin G (IgG). Further, we discuss the concept and quantitative aspects of the bead immobilization method. The detection of streptavidin and IgG is shown as demonstration assays based on the new bead immobilization method. Finally, we discuss the feasibility of a parallel assay concept.

12.2 Materials and methods

12.2.1 Experimental setup

The experimental setup is shown in Fig. 1. A microfluidic channel, microprecision-milled into poly(methyl methacrylate) (PMMA) with a height of 30 μm and a width of 4 mm, was placed on top of a Pyrex chip (20 × 13.5 mm^2) with 100 nm thin platinum microelectrode pads made by lift-off technology. The electrode pad layout is depicted in Fig. 2. The reversible sealing between Pyrex chip and PMMA microfluidic cover was accomplished by a home-built clamping system made of austenitic stainless steel. After each experiment, the chip with the adhering beads is discarded. The PMMA microfluidic channel is washed and reused. On the inlet side, three vial reservoirs were connected to the microfluidic setup. Fluid pumping was accomplished using a syringe pump (Kloehn, USA) in suction mode on the outlet side. A Zeiss Axioscope (Zeiss, Jena, Germany) optical microscope with an Epiplan 10 × objective lens (numerical aperture of 0.20) was used in the dark field (DF) and fluorescence mode (FL) for observation and fluorescence detection, respectively. The yellow-green fluorescence module comprised a 450–490 nm excitation filter and an emission filter for wavelengths greater than 515 nm. The red fluorescence module comprised a 665/45 nm excitation filter and a 725/50 nm emission filter. All pictures were taken with the ChemiLum Cooled Image System GM320 CC 220 (Scion, USA). Quantitative measurements of the fluorescence were performed using the Scion Image software (Scion). The software for electric field and affinity assay simulations was CFD-ACE$^+$ (ESI, France).

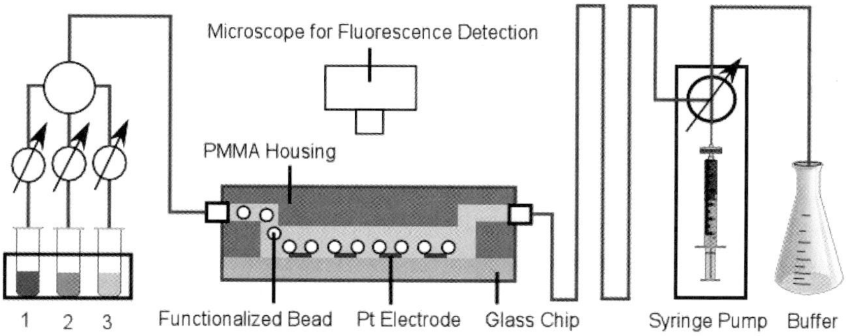

Fig. 1 Scheme of the experimental setup (1, bead suspension; 2, analyte solution; 3, buffer), not to scale. Dimensions of the fluidic chamber are 10 mm in length, 4 mm in width, and 30 μm in height. Platinum electrodes are 100 nm thin.

12.2.2
Beads, chemicals, and chip surface conditioning

Yellow-green dyed Fluoresbrite polystyrene beads (excitation maximum at 490 nm, emission maximum at 520 nm) of 2 μm in diameter coated with biotin and of 1 μm in diameter coated with protein A or goat–antimouse IgG, respectively, were purchased from Polysciences (Warrington, PA, USA). Yellow-green dyed polystyrene beads of 1 μm in diameter coated with biotin, streptavidin labeled with Alexa Fluor 680 (excitation maximum at 679 nm, emission maximum at 702 nm), and Zenon Alexa Fluor 680 mouse IgG1 labeling kit were supplied by Invitrogen (Carlsbad, CA, USA). Monoclonal antialbumin mouse IgG1 was obtained from Fluka Chemie (Buchs, Switzerland). The emitted yellow-green fluorescence of the beads did not excite the fluorescence of the Alexa Fluor 680 analyte label. The concentration of the beads in the original suspension was 1.25% solids. Before use, the beads were washed in deionized (DI) water, centrifuged at 2000 × g for 5 min, and resuspended by vortexing and ultrasound in DI water three times, the final concentration being 10% v/v of the original concentration. Alexa Fluor 680 streptavidin and mouse IgG were dissolved in PBS buffer (pH 7.4, 200 mM Na^+, conductivity of 2070 mS/m) at various concentrations and centrifuged at 2000 × g for 5 min. Only the supernatant was used for the experiments to get rid of possibly agglomerated protein. The Zenon Alexa Fluor mouse IgG label was used in a 0.1 μM concentration in PBS buffer.

A polyelectrolyte multilayer (PEM) coating was used to condition the chip surface. It consisted of a self-assembled monolayer (SAM) of 11-mercapto-undecanoic acid, a layer of hexadimethrine bromide (Polybrene, PB), and a layer of dextran sulfate (DS). The deposition process of this PEM and its ability to protect the chip from uncontrolled adhesion of polystyrene beads without interfering with their DEP-controlled adhesion have been reported earlier [12]. The PEM also protected the chip surface from uncontrolled adhesion of beads coated with goat–antimouse IgG, protein A, and biotin, and did not interfere with the bead immobilization mechanism described in this paper. Fig. 2 shows that the beads adhere only to the electrodes after the DEP voltage was applied, but not to the rest of the chip surface. PEM coatings [13] and other dextran-based coatings [14] are reported to passivate surfaces against proteins in solution. This is important for the analyte molecule detection. Further, PEM enhances the wetting of hydrophobic plastic materials resulting in smoother filling of microfluidic channels [15]. The PEM coating process is suitable for automation.

12.2.3
Fluorescence measurement

Yellow-green fluorescent beads of 1 μm in diameter were immobilized at the electrode pads by DEP-controlled adhesion. After flushing away excess beads, two fluorescence images were taken. The first image measured the yellow-green fluorescence (*A*) of the beads at the electrode pads. The background noise (*a*) without

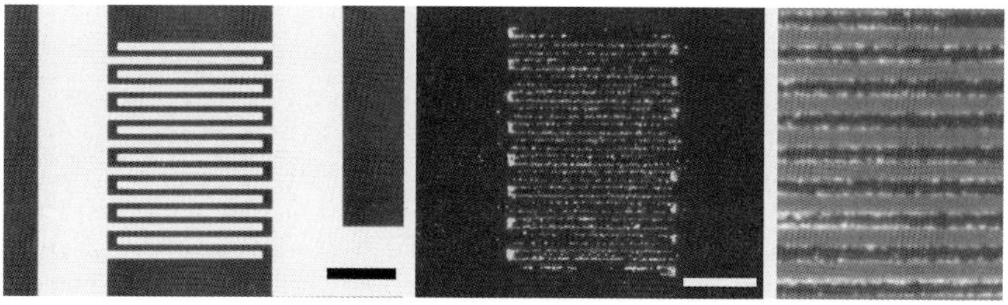

Fig. 2 Experimental setup, electrodes. Left: Bright field image of an interdigitated electrode pad with 10 μm wide fingers and 10 μm interelectrode gaps (scale bar: 100 μm). Electrodes have bright contrast. Middle: Fluorescence image of immobilized yellow-green beads of 1 μm in diameter coated with goat–antimouse IgG (scale bar: 100 μm). Right: DF image showing that beads of 1 μm in diameter adhere to the edges of the interdigitated electrodes after immobilization by DEP-controlled adhesion. The 10 μm micron wide electrodes have dark contrast.

beads measured left and right of the electrode pad was subtracted from the measured intensity. The thus obtained signal is proportional to the number of immobilized beads and hence to the number of available binding sites. The second image measured the red fluorescence intensity to verify that no red signal was present.

Then the analyte solution was introduced into the system at a constant flow rate and analyte molecules bound to the probe beads. The red fluorescence of the analyte label was measured by integrating 50 frames on the camera chip. In order to measure only the intensity coming from the analyte bound to the probe bead surfaces, the background noise (*b*) coming from the nonbound streptavidin solution was measured left and right of the electrode pad and subtracted from the intensity (*B*) measured above the electrode pad. Finally, the streptavidin signal was corrected by a factor (*F*) representing the number of available probe beads (*i.e.*, binding sites) according to the following procedure:

$$F_x = \frac{A_x - a_x}{A_1 - a_1} \quad (1)$$

$$S_x = \frac{(B_x, - b_x)}{F_x} \quad (2)$$

Here, F_x is the correction factor for the *x*-th measurement, A_x is the green fluorescence signal coming from the beads immobilized on the electrode pad, and a_x is the corresponding background measured left and right of the electrode pad (no beads). A_x is the green fluorescence signal and a_x the corresponding background of a reference measurement which ensures that the correction factor is around 1 for convenience. This reference measurement and the resulting correction factor (F_x) are important to adjust for the varying number of immobilized beads from pad to pad. This ensures interpad and interchip reproducibility of the analyte concentration measurement. B_x is the red fluorescence signal coming from the label of bound and unbound analyte on the electrode pad during the dynamic binding measurement, and b_x is the corresponding background of the label of unbound analyte left and right of the pad. S_x is the corrected signal of bound analyte, the measurement result.

12.3 Results and discussion

12.3.1 Bead immobilization on the chip

We named the immobilization mechanism DEP-controlled adhesion, because positive DEP was used to bring the beads into contact with the electrodes triggering the immobilization by nonspecific adhesion. This immobilization mechanism works best with beads of 1 μm or less in diameter. To avoid nonspecific adhesion of the beads to the rest of the chip surface, the chip was conditioned with the surface chemistry described in Section 2.2, and the bead suspension was kept in perpetual flow above the chip surface during the immobilization process. Only the beads drawn to the electrodes by the positive DEP force came in close contact with the electrode edges.

DEP is based on polarizability and requires a strong inhomogeneous electric field. If the beads are more polarizable than the surrounding liquid medium, the DEP force moves them toward areas with high electric field gradient (positive DEP). In the present setup, these areas correspond to the electrode edges. The opposite case is called negative DEP. Positive and negative DEP are confined to certain frequency ranges separated by the cross-over frequency. The DEP force is proportional to the field gradient and the electric field strength [8, 33]. A positive DEP frequency range of 10–20 kHz was used. At these frequencies, the bead immobilization was most efficient, *i.e.*, a homogeneous occupation of the electrode edges could be achieved within seconds. In this range, alternating-current (AC) EOF is also present. Under AC EOF condition, the beads are levitated above the electrodes by electrohydrodynamic forces without the pressure-driven flow [16–18]. However, the maintained pressure-driven flow disrupts the AC EOF flow pattern and the beads go to the electrodes edges due to positive DEP.

Positive DEP alone, although very effective at the bead suspension conductivity of 1 mS/m, was not suitable for bead immobilization. The reason is that in a flow-through device, the positive DEP force cannot retain protein-coated beads of 1 μm in diameter at the electrodes at high flow rates and high analyte buffer conductivities above 100 mS/m (conservative limit, *i.e.*, no positive DEP is observed above this value; the actual limit depends on the bead type and size and is usually found at values above 10 mS/m). Typical analyte buffer or body fluid conductivities in the life sciences are one order of magnitude larger than this. Therefore, we took advantage of the stronger adhesion forces in the microworld. We used the positive DEP force to bring the probe beads in contact with the electrodes. As a result, a certain number of beads made a strong irreversible contact with the electrodes by adhesion. DEP-controlled adhesion immobilizes only a fraction of the beads available in suspension, until a full and homogeneous occupation of the electrodes is achieved. This can be accomplished within a short period of time using the appropriate bead size, DEP frequency, and chip surface conditioning. The unused beads, which do not stick to the electrodes by adhesion and are washed away after the DEP voltage is turned off, can be collected at the device outlet and reused, if desired.

Functionalizing a chip surface with probe beads by DEP-controlled adhesion has several advantages. The method is fast. Depending on the bead type and protocol, it takes only between a few seconds and 2 min to immobilize the beads. It is versatile, *i.e.*, it works for beads with different probe molecule coatings. It immobilizes the beads irreversibly, *i.e.*, the beads withstand high flow rates (at least up to 6.3 μL/s or 52 mm/s with our device geometry) even in conventional high-conductivity buffers, after the DEP voltage is turned off. Finally, it requires a simple chip layout with planar electrodes.

The finally chosen bead immobilization protocol for the biotin-coated and protein-A-coated beads of 1 μm in diameter comprised two steps: first, applying a voltage of 17 $V_{p\text{-}p}$ with a frequency of 20 kHz at the electrodes at a flow rate of 0.1 μL/s for 1 min, and second, applying a voltage of 17 $V_{p\text{-}p}$ with a frequency of 10 kHz at the electrodes at a flow rate of 0.1 μL/s for 1 min. When the 20 kHz step was omitted, the biotin-coated beads tended to form large nonuniform agglomerations at the 90° angles at the ends of the interdigitated electrode fingers.

The finally chosen bead immobilization protocol for the goat–antimouse IgG-coated beads of 1 μm in diameter comprised one step. Applying a voltage of 20 $V_{p\text{-}p}$ with a frequency of 10 kHz at the electrodes at a flow rate of 0.2 μL/s for only 10 s was sufficient to obtain a homogeneous occupation of the electrode edges with beads.

The good adhesion of beads at the electrodes under the described conditions can be explained using the DLVO theory (DLVO – Derjaguin, Landau, Verwey, Overbeek) [19, 20], which was originally developed for colloidal stability in suspensions and later adapted to particle–particle and particle–surface interactions in aqueous solutions [21]. The DLVO theory describes that a suspended particle usually has to overcome a potential of repulsive energy (*e.g.*, double-layer repulsion) before it feels the strong attractive van der Waals interaction with another surface. During the DEP-controlled adhesion of beads, this repulsive barrier is overcome by the positive DEP force. The chip surface passivated with DS possesses a negative surface charge at pH 7.4. This is also the case for all probe bead coatings – biotin [22], protein A [23], and antimouse IgG [24] – *i.e.*, they have anionic character and their p*I*s are at pH values lower than those of the aqueous suspensions and solutions used here.

The positive DEP force is in the picoNewton range for low-conductivity fluids (for our bead suspension with 1 mS/m). However, the positive DEP force gets significantly weaker when the conductivity of the aqueous solution increases above 100 mS/m, which was the case for our analyte PBS buffer with a conductivity of 2070 mS/m. Stokes friction forces, acting on the micrometer-sized beads as a result of the viscous drag of the fluid flow in the microfluidic channel, are typically also in the picoNewton range. They range from 16 pN during the incubation flow velocity up to 490 pN at the flow velocities which we used during the washing steps. Therefore, the DEP force alone is too weak to permanently retain the protein-coated beads at the electrodes against the Stokes force, especially if a typical buffered analyte with a conductivity of greater than 100 mS/m is introduced into the microfluidic device. It is the noncovalent interactions, such as van der Waals forces which are very strong in the microworld, that lead to the irreversible bead adhesion at the electrode edges. Attractive van der Waals interactions exert forces in the nanoNewton range on micrometer-sized beads when they come in contact with the surface. They are stronger than the Stokes friction forces. Therefore, the beads adhering to the electrodes, after overcoming the repulsive barrier described in the DLVO theory by applying a positive DEP force, are able to withstand the Stokes force of the fluid flow in the microfluidic channel even after the DEP voltage is turned off.

Besides the Coulomb and van der Waals forces in the DLVO theory, other forces contribute to the bead–chip surface interaction, like hydrophobic interaction or steric interaction. A theoretical discussion of each individual contribution would go beyond the scope of this work.

The chip surface functionalization with probe beads by DEP-controlled adhesion was very fast and worked for different bead types. It was demonstrated for polystyrene beads coated with different kinds of probe molecules, namely, with biotin, protein A, and goat–antimouse IgG. The bead immobilization process required just 10 s for the goat–antimouse IgG-coated beads and 2 min for the biotin- and protein-A-coated beads.

Interdigitated electrodes were used for further quantitative bead immobilization experiments. One important point was the orientation of the interdigitated electrodes. If the electrodes ran across the channel perpendicular to the fluid flow, the beads were retained at the electrode edges facing the fluid flow (luff or upstream side). The lee side of the electrode was occupied with much less beads. If the electrodes ran along the channel parallel to the fluid flow, beads were retained more homogeneously at both edges of the electrodes, making the immobilization procedure more effective. Therefore, the parallel orientation was chosen.

DEP-controlled adhesion works best for bead sizes of 1 μm in diameter or smaller since a homogeneous occupation of the electrode edges is achieved for these bead sizes. Quantitative bead immobilization experiments, however, were performed using 2 μm beads because their number and size still allowed for individual counting on the microscopy images. The beads were yellow-green fluorescently dyed and coated with biotin molecules. The goal was to find out if and how the fluorescent signal coming from immobilized beads correlated with their number. The bead number is proportional to the number of available binding sites. Beads of 2 μm in diameter immobilized at interdigitated electrode arrays were counted individually in a window of 200 by 300 pixels. On an average, 205 beads (SD of 35 or 17%) were immobilized in such a window for a given set of immobilization parameters. In addition, the fluorescence signal of the counted beads in these windows was measured. As expected, there was a linear correlation between the bead number in the window and their fluorescence signal. The precision of the measured fluorescence signal coming from the same number of beads was better than 2%. This is important for the affinity assay procedure and the interpretation of the results using the correction factor (*F*) as described in Section 2.3.

12.3.2 Streptavidin demonstration assay

Biotin-coated yellow-green fluorescent beads of 1 μm in diameter were immobilized at the electrode pads by DEP-controlled adhesion. Subsequently, the analyte solution was introduced into the system and the dynamic binding detection of streptavidin labeled with a red fluorescence marker started. The beads were perfused with streptavidin solution at a pump rate of 0.2 μL/s. A red fluorescence picture was taken every 15 s to monitor the dynamic binding.

The biotin–streptavidin assay is an affinity assay by its nature. Affinity binding reaches an equilibrium state of saturation when all available binding sites are occupied. The streptavidin–biotin affinity assay equation can be written as [25, 26]

$$c_{SB}(t) = B_0\lfloor 1 - \exp(-k_f c_S t)\rfloor \qquad (3)$$

where c_{SB} is the concentration of the formed streptavidin–biotin complex, t is the time, B_0 is the concentration of available biotin binding sites, k_f is the forward reaction rate constant which is much larger than the reverse reaction rate constant for the biotin–streptavidin reaction [26, 27], and c_S is the streptavidin concentration.

The number of available unbound streptavidin molecules at a given time determines if and how fast the state of saturation is reached. The number of available molecules at a given time increases with their concentration, diffusion constant, and depends on whether a constant flow of fresh streptavidin molecules is maintained over the chip surface. Therefore, no universal data for the streptavidin assay can be given here. However, using the setup and the assay conditions described above, the LOD was reached for a streptavidin concentration of 0.5 μg/mL (8 nM). A concentration of 0.1 μg/mL (1.7 nM) could also be detected but did not satisfy the LOD criterion of signal being three times that of the noise. Fig. 3A shows dynamic binding curves for various concentrations at the specified flow velocity of 1.7 mm/s. After 180 s, fluorescence quenching of the Alexa Fluor 680 dye started to affect the measurements. Under the assay conditions chosen here, 120 s was a good incubation time to distinguish between different streptavidin concentrations (Fig. 3B, dose–response curve). The precision of the streptavidin assays was 14%. The major source of error was quenching of the fluorescent dye of the analyte. Accuracy was inherent because analyte solutions with known concentrations were used for the demonstration assay. No additional calibration had to be done.

Fig. 4 shows that the dynamic biotin–streptavidin binding can be accelerated by increasing the analyte flow rate thus making the measurement more sensitive. For example, an increase of the analyte flow velocity from 0.1 up to 5.1 mm/s would result in an increase of bound streptavidin (and its measured signal) by a factor of 3.6 after 120 s of incubation time. The LOD under the demonstration assay condi-

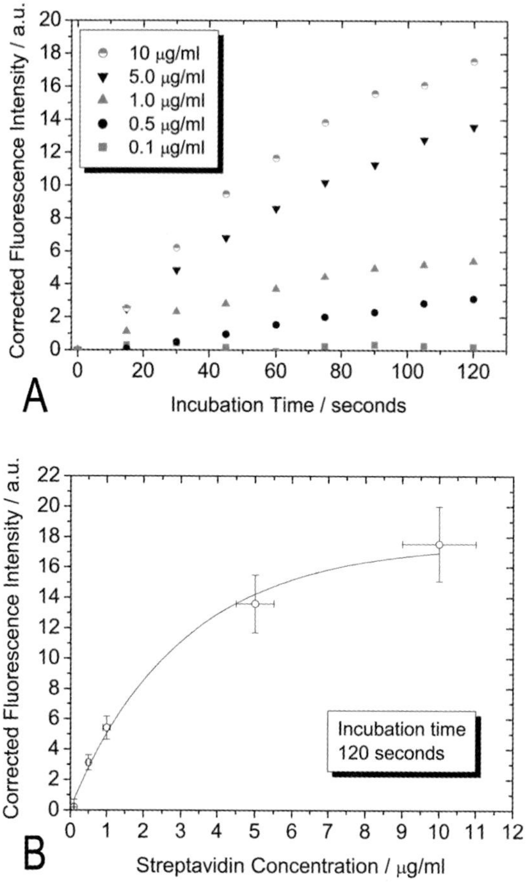

Fig. 3 (A) Dynamic streptavidin–biotin binding curves for various streptavidin concentrations. Fluorescence signal of bound streptavidin *versus* incubation time. (B) Dose–response curve constructed from (A). Fluorescence signal of bound streptavidin *versus* streptavidin concentration after 120 s of incubation time. Line curve is not a theoretical fit curve but emphasizes the saturation of available binding sites at higher concentrations.

tions can therefore be shifted to lower streptavidin concentrations by adjusting the assay parameters, particularly the flow velocity in the microchannel during incubation and, alternatively, by increasing the incubation time. The dynamic binding curves were computed by finite element method taking into account the flow dynamics (Navier–Stokes equation), the concentration profile (mass transport equation), and the reversible affinity binding (reaction equation) [28]. Input parameters for the simulations were 17 nM (1 μg/mL) streptavidin with a molecular weight of 60 kg/mol (value provided by supplier Invitrogen, includes molecular weight of streptavidin, of four Alexa Fluor 680 labels *per* molecule and of the corresponding four spacers), a mobility of 1×10^{-11} m^2/V × s and a diffusivity of 7.4×10^{-11} m^2/s [29] in PBS buffer with a pH of 7.4, conductivity of 2.07 S/m, viscosity of 0.001 Pa × s, and density of 1000 kg/m^3. The biotin–streptavidin forward reaction rate constant was 10^7/M × s, the reverse reaction rate constant was 10^{-8}/s, resulting in an association constant K_a of 10^{15}/M [26, 27]. The dynamic binding was simulated on immobilized biotin-coated beads of 1 μm in diameter with a biotin surface density of 4.4×10^{-7} mol/m^2 (data provided by bead manufacturer Poly-

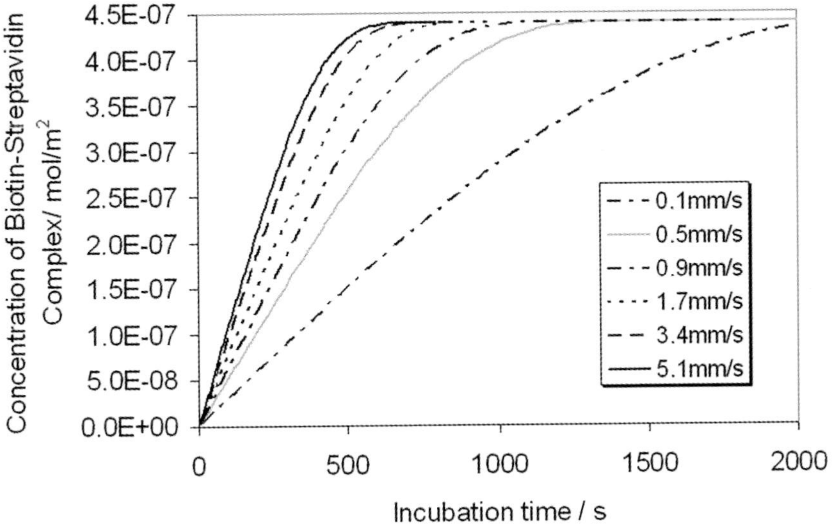

Fig. 4 Simulated streptavidin–biotin dynamic binding curves for various flow velocities of a streptavidin solution with a concentration of 1 μg/mL. Measurement sensitivity and the dynamic range can be shifted by changing the analyte flow velocity during the dynamic binding detection. For example, an increase of the analyte flow velocity from 0.1 up to 5.1 mm/s would result in an increase of bound streptavidin and hence of the detected fluorescence signal by a factor of 3.6 after 120 s of incubation time. For comparison: In the dynamic binding experiments of Fig. 3, the analyte flow velocity was 1.7 mm/s and the incubation time 120 s. Simulation does not account for the quenching of the Alexa Fluor 680 fluorescent label which is present in the dynamic binding experiments.

sciences). The simulation model comprised 220 beads in a row on the bottom in the middle of a 30 μm high and 5 μm wide channel. This corresponds to the length of one electrode finger edge of the interdigitated electrodes where the beads are immobilized. Only the upper semisphere of the beads was active resulting in 6.9×10^{-19} mol of biotin molecules (available binding sites) *per* bead. The total number of biotin molecules for all 220 beads was 1.5×10^{-16} mol (available binding capacity in the simulation).

12.3.3
Discussion of a parallel assay concept

To discuss the feasibility of detection of two different components in a multianalyte solution on the same chip (parallel assay), IgG was chosen as the second analyte component besides streptavidin. First, the LOD for IgG had to be determined. Second, both streptavidin and IgG were detected at concentrations above the LOD on the same chip in the one multianalyte solution.

12.3.3.1 LOD for IgG

The goat–antimouse IgG beads were immobilized by DEP-controlled adhesion. After flushing away excess beads, the immobilized beads were perfused with mouse IgG containing solution at a flow rate of 0.1 μL/min for 16 min. After flushing away nonbound IgG, the bound IgG was labeled by perfusing it with 0.1 μM Zenon Alexa Fluor 680 label at a flow rate of 0.1 μL/min for 16 min. After that, excess label was flushed away. A picture averaging 50 frames on the camera chip was taken to detect the red fluorescence signal of the label. The assay procedure is not optimized yet and certainly has potential for assay time reduction.

The LOD under the chosen demonstration assay conditions was at 1.6 μg/mL (10 nM) mouse IgG with the standard detection scheme described in Section 2.3. It can be improved by using a better detection system instead of a cooled CCD camera. Typical LODs for on-chip immunoassays are summarized in [30]. According to that review, typical limits for on-chip fluorescence detection of IgG are between 1 and 50 nM IgG. The LOD of a comparable on-chip mouse IgG bead-based sandwich assay using magnetic beads, but amperometric detection at integrated electrodes rather than fluorescence, was reported to be at 0.1 μg/mL (0.6 nM) after 40 min analysis time [31]. While our not yet optimized system is less sensitive at the moment, it does not require the integration of magnetic contraptions for bead immobilization in addition to the integrated electrodes. Lower LODs are also reported, but not always necessary depending on the application. The detection sensitivity of microfluidic systems does not need to be pushed down very low for all applications. For example, typical immunoglobulin concentrations in healthy human blood serum are IgG 12 mg/mL (80 μM), IgA 3 mg/mL (6–17 μM), IgM 1 mg/mL (1 μM), IgD 100 μg/mL (600 nM), and IgE 1 μg/mL (5 nM) [32].

12.3.3.2 Proof of principle of the parallel assay concept

In principle, DEP would offer the option to immobilize different bead types on different electrode pads by applying different frequencies according to the bead size [33–35]. However, as discussed in Section 3.1, the bead immobilization had to take place at 10–20 kHz. All tested protein-coated beads are under positive DEP in this frequency range and can therefore not be separated from each other.

The following procedure was chosen to demonstrate the detection of two different analytes–streptavidin and mouse IgG–in the same solution on the same chip. In a first step, goat–antimouse IgG-coated beads were immobilized on electrode array number 1 (actually, on a set of six electrode arrays simultaneously to have better statistics). In a second step, biotin-coated beads were immobilized on electrode array number 2 by DEP-controlled adhesion (again, a set of six electrode arrays was used). After each bead immobilization step, the excess beads were flushed away. The green fluorescence signal from the beads immobilized at the electrode arrays was measured. In a third step, the immobilized beads were perfused with the analyte solution containing both unlabeled mouse IgG 81 μg/mL (0.5 μM) and red fluorescently labeled streptavidin 10 μg/mL (0.17 μM) at a flow rate of 0.1 μL/min for 16 min. (The relatively high concentrations were chosen to better see the

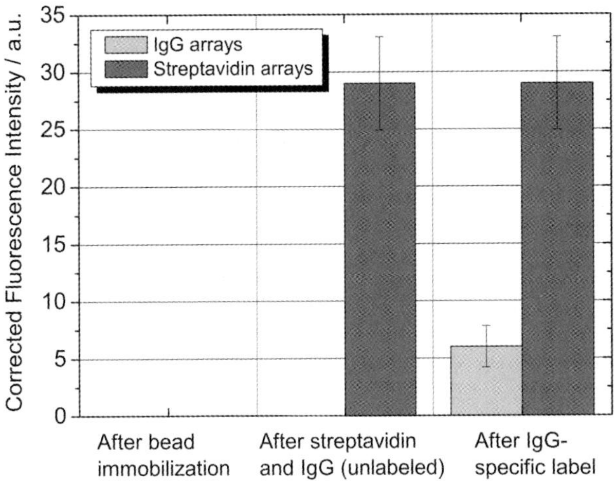

Fig. 5 Demonstration of the parallel assay concept. Red fluorescence signal from IgG detection site and from streptavidin detection site after the sequential immobilization of goat–antimouse IgG-coated beads and biotin-coated beads, after 16 min of incubation with multi-analyte solution containing unlabeled mouse IgG and labeled streptavidin and washing, and after labeling the bound mouse IgG with an IgG-specific label and washing.

cross reactivity which is discussed below.) After flushing away excess analyte solution, red fluorescence images of both electrode arrays were taken. Since only the streptavidin was labeled with a red fluorescent marker at this point, only array number 2 delivered a signal whereas array number 1 was dark. In a fourth step, the chip was perfused with the red fluorescent mouse IgG-specific label with a concentration of 5.0 μg/mL (0.1 μM) for 16 min. After flushing away excess IgG label, red fluorescence images of both electrode arrays were taken again. This time, both arrays delivered a signal (Fig. 5). Again, the red fluorescence signal of the labeled analyte was corrected by the number of available beads (green signal).

There are two major sources of cross reactivity (cross talk) in a bead-based assay: (i) if bead type A is immobilized on the detection site of bead type B and *vice versa*, and (ii) in the case of nonspecific binding of analyte. The first case occurs because of the sequential immobilization of bead type A and B in the same microchannel. However, if the bead plugs are well separated in time, this type of cross talk does not cause a problem. Cross reactivity due to unspecific analyte binding cannot be avoided if a multianalyte solution is used. Therefore, it is important to know if a measured signal comes from specific affinity assay binding (true signal) or from unspecific cross reactivity.

The probe beads specific to the multianalyte component a_i are immobilized at electrode array i. Due to nonspecific binding, the other analyte components contained in the multianalyte solution will also bind to the beads at electrode array i. Thus specific as well as nonspecific binding events will contribute to the fluorescence signal S_i measured at electrode array i. We quantitated the cross reactivity signal contribution in experiments without the binding site-specific analyte component a_i by measuring the fluorescence signal S_i (without a_i) at electrode array i. If the measured signal S_i (with a_i) at an electrode array was in the same range as the calibrated cross reactivity signal contribution S_i (without a_i), S_i did not qualify as a measure for the concentration of the binding site-specific analyte component a_i.

The cross reactivity criterion is described in detail in [36]. The criterion can be adapted to all multianalyte solutions which are measured on our platform.

12.4 Concluding remarks

DEP-controlled adhesion of beads is a promising method for the functionalization of chips at defined areas with probe molecules. The proposed bead immobilization mechanism is very fast, versatile, *i.e.*, it works for beads with different probe molecule coatings, and the chip design is relatively simple. Beads coated with biotin, protein A, and goat–antimouse IgG were immobilized on microelectrodes using a combination of a slight pressure-driven flow and positive DEP at 10–20 kHz, resulting in irreversible adhesion of the beads on the electrodes. Once immobilized, the beads remain at the electrodes after the DEP voltage is turned off, even at high flow rates which typically occur during the washing steps and even at high analyte solution conductivities.

The detection of streptavidin and mouse IgG was shown in demonstration assays using the concept of DEP-controlled adhesion of the corresponding probe beads on the chip. The demonstration assay comprised two steps. In the first step, the number of immobilized probe beads was measured using their yellow-green fluorescence signal. In the second step, the amount of bound analyte was measured using the red fluorescence of the analyte label and corrected by the intensity coming from unbound molecules still in solution and the number of available beads. Simulations showed that the LOD and the dynamic range can be offset by adjusting the flow velocity (pump rate) of the analyte solution during dynamic binding.

The parallel detection of two different analyte molecules in the same solution on the same chip was demonstrated. The probe beads were immobilized sequentially by subsequent DEP-controlled adhesion at different electrode arrays.

The authors would like to thank E. Casartelli for discussions, the CSEM – University of Neuchâtel IMT ComLab for fabricating the chips, Roche Instrument Center for making the microfluidic structure, and Micro Center Central Switzerland AG, HTA Luzern as well as the CTI Bern (Project-number 6061.1 KTS-NM) for financing this project.

12.5 References

[1] Verpoorte, E., *Lab Chip* 2003, *3*, 60–68.

[2] Templin, M. F., Stoll, D., Bachmann, J., Joos, T. O., *Comb. Chem. High Throughput Screening* 2004, *7*, 223–229.

[3] Grumann, M., Dobmeier, M., Schippers, P., Brenner, T., Zengerle, R., Ducrée, J., *Proceedings of the NSTI Nanotechnology Conference and Trade Show*, Boston, USA, 2004, *Vol. 2*, pp. 383–386.

[4] L'Hostis, E., Michel, P. E., Fiaccabrino, G., Strike, D. J., de Rooij, N. F., Koudelka, M., *Sens. Actuators B* 2000, *64*, 156–162.

[5] Fan, Z. H., Mangru, S., Granzow, R., Heaney, P., Ho, W., Dong, Q. P., Kumar, R., *Anal. Chem.* 1999, *71*, 4851–4859.

[6] Lettieri, G.-L., Dodge, A., Boer, G., de Rooij, N. F., Verpoorte, E., *Lab Chip* 2003, *3*, 34–39.

[7] Fuhr, G., Schnelle, T., *Phys. Blätter* 2001, *57*, 49–52.

[8] Pohl, H. A., *Dielectrophoresis*, Cambridge University Press, Cambridge 1978.

[9] Walt, D. R., *Science* 2000, *287*, 451–452.

[10] Fulton, R. J., McDade, R. L., Smith, P. L., Kienker, L. J., Kettman, J. R., *Clin. Chem.* 1997, *443*, 1749–1756.

[11] Auerswald, J., Knapp, H. F., *Microelectron. Eng.* 2003, 67–68, 879–886.

[12] Auerswald, J., Linder, V., Knapp, H. F., *Microelectron. Eng.* 2004, 73–74, 822–829.

[13] Katayama, H., Ishihama, Y., Asakawa, N., *Anal. Chem.* 1998, *70*, 5272–5277.

[14] Linder, V., Verpoorte, E., Thormann, W., de Rooij, N. F., Sigrist, H., *Anal. Chem.* 2001, *73*, 4181–4189.

[15] Barker, S. L. R., Tarlov, M. J., Canavan, H., Hickman, J. J., Locascio, L. E., *Anal. Chem.* 2000, *72*, 4899–4903.

[16] Ramos, A., Morgan, H., Green, N. G., Castellanos, A., *J. Phys. D* 1998, *31*, 2338–2353.

[17] Ajdari, A., *Phys. Rev. E* 2000, *61*, R45–R48.

[18] Mpholo, M., Smith, C. G., Brown, A. B. D., *Sens. Actuators B* 2003, *92*, 262–268.

[19] Derjaguin, B. V., Landau, L., *Acta Physiochim. URSS* 1941, *14*, 633–662.

[20] Verwey, E. J. W., Overbeek, J. T. G., *Theory of Stability of Lyophobic Colloids*, Elsevier, Amsterdam 1948.

[21] Israelachvili, J., *Intermolecular and Surface Forces*, 2nd Edn., Academic Press, London 1995.

[22] Zempleni, J., Green, G. M., Spannagel, A. W., Mock, D. M., *J. Nutr.* 1997, *127*, 1496–1500.

[23] Floros, J., *Swiss Med. Weekly* 2001, *131*, 87–90.

[24] van Lent, P. L., van den Berg, W. B., Schalkwijk, J., van de Putte, L. B., van den Bersselaar, L., *J. Rheumatol.* 1987, *14*, 798–805.

[25] Dodge, A., Fluri, K., Verpoorte, E., de Rooij, N. F., *Anal. Chem.* 2001, *73*, 3400–3409.

[26] Huang, S.-C., Stump, M. D., Weiss, R., Caldwell, K. D., *Anal. Biochem.* 1996, *237*, 115–122.

[27] Livnah, O., Bayer, E. A., Wilchek, M., Sussman, J. L., *Proc. Natl. Acad. Sci. USA* 1993, *90*, 5076–5080.

[28] The simulation software containing these equations was CFD-ACE, www.cfdrc.com.

[29] Spinke, J., Liley, M., Schmitt, F. J., Guder, H. J., Angermaier, L., Knoll, W., *J. Chem. Phys.* 1993, *99*, 7012–7019.

[30] Verpoorte, E., *Electrophoresis* 2002, *23*, 677–712.

[31] Choi, J. W., Oh, K. W., Han, A., Wijayawardhana, C. A., Lannes, C., Bhansali, S., Schlueter, K. T., Heineman, W. R., Halsall, H. B., Nevin, J. H., Helmicki, A. J., Henderson, H. T., Ahn, C. H., *Biomed. Microdevices* 2001, *3*, 191–200.

[32] Berg, J. M., Tymoczko, J. L., Stryer, L., *Biochemistry*, 5th Edn., Freeman and Co., New York 2002.

[33] Hughes, M. P., *Nanotechnology* 2000, *11*, 124–132.

[34] Pethig, R., *Automation in Biotechnology*, Proceedings of 4th Toyota Conference, Elsevier, Aichi, Japan 1991, pp. 159–185.

[35] Green, N. G., Morgan, H., *J. Phys. D: Appl. Phys.* 1998, *31*, L25–L30.

[36] Widmer, D., Demonstration of a lab-on-chip protein assay concept, Diploma thesis, École Polytechnique Fédérale de Lausanne (EPFL), Institut de Microélectronique et des Microsystèmes (IMM), Faculté des Sciences et Techniques de l'Ingénieur (STI), 1015 Lausanne, Switzerland, 2005.

13
Droplet fusion by alternating current (AC) field electrocoalescence in microchannels*

Max Chabert, Kevin D. Dorfman, Jean-Louis Viovy

We present a system for the electrocoalescence of microfluidic droplets immersed in an immiscible solvent, where the undeformed droplet diameters are comparable to the channel diameter. The electrodes are not in direct contact with the carrier liquid or the droplets, thereby minimizing the risk of cross-contamination between different coalescence events. Results are presented for the coalescence of buffered aqueous droplets in both quiescent and flowing fluorocarbon streams, and on-flight coalescence is demonstrated. The capillary-based system presented here is readily amenable to further miniaturization to any lab-on-a-chip application where the conductivity of the droplets is much greater than the conductivity of the stream containing them, and should aid in the further application of droplet microreactors to biological analyses.

13.1
Introduction

13.1.1
General aspects

The further advance of highly integrated analyses of biological systems, and systems biology in general, requires the confluence of high-throughput analyses and the means to analyze the massive data generated by them. While several high-throughput methods exist, most notably automated CE and related sequencing applications, furthering data acquisition capabilities for systems biology applications invariably requires new analysis technologies. Miniaturization, for example by using capillaries and labs-on-a-chip, is certain to play an important role in future developments in this field.

* Originally published in Electrophoresis 2005, 26, 3706–3715

Microfluidic Applications in Biology. Edited by Niels Lion, Joël S. Rossier, and Hubert H. Girault

ISBN-10: 3-527-31761-9

Microfluidic droplet reactors are one particularly promising route for developing high-throughput analyses. In these protocols, droplets containing the biomolecules to be analyzed are formed either from discrete samples [1] or by varying the contents of a multicomponent inlet stream [2]. The droplets typically contain a unique reaction mixture, and cross-contamination between droplets represents a serious problem, especially for clinical and high-throughput applications. The literature includes numerous applications of droplet microreactors, such as PCR [1] and the measurement of millisecond RNAase kinetics [2]. With automated sample injection and suitable control of interfacial chemistry, PCR systems will be capable of tens of thousands of PCR reactions *per* day without any contamination between samples [3]. The integration of the latter system with online fluorescence detection (for quantitative PCR) or CE (for conventional analysis) promises to greatly increase the throughput of genomic analyses.

As analysis speeds increase, the bottleneck in high-throughput analyses is moving toward sample preparation, in particular online (or real-time) manipulations. For PCR systems, even with robotics, the tens of thousands of offline reagent pipetting steps required to prepare each of the individual PCR samples is an overwhelming task. In addition, open systems based on pipetting and microtiter plates are limited to volumes no less than several microliters, owing to evaporation, surface tension, and pipetting reproducibility. In contrast, the fluorescence detection methods used in CE or quantitative PCR (Q-PCR) are actually able to work on much smaller sample quantities, typically tens of nanoliters. Similar interfacing concerns are ubiquitous in the high-throughput analysis community, and extend far beyond our particular example. Indeed, variability in sample preparation and the wait time before analysis is especially important when the results are analyzed by automated data mining tools. With the latter, it is essential to minimize the number of potential experimental artifacts, in order to arrive at significant biological results.

For high-throughput droplet-based systems, a key step in sample preparation is controlling the coalescence of two droplets in order to mix their contents. In the context of droplet-based PCR, for instance, an online coalescence system will allow us to prepare reagents (such as a PCR ready mix) in bulk, and then add precise volumes of reagents to each sample droplet online immediately prior to amplification. However, droplet fusion in miniaturized systems is still difficult. Moreover, it is essential that the fusion mechanism presents no opportunities for contamination between droplets.

The simplest method for coalescing droplets would be to force them to arrive simultaneously at a T intersection. However, in confined channels, one drop simply follows the second one into the T without coalescing (Link, D. L., Weitz, D. A., unpublished results). For different sized droplets, it is possible to coalesce a smaller droplet when it trails a larger one, since the smaller droplet moves with a higher average velocity [4]. However, this is not a rationale strategy in microfluidics, since the time for film drainage between the droplets is very slow, leading to coalescence distances between 30 and 100 tube diameters [4]. Moreover, under certain conditions (relative droplet sizes, viscosities, *etc.*) coalescence is impossible on a reasonable time scale [4]. Even more so, one can envision applications that involve equal sized droplets, where this coalescence mechanism is not active.

The inherent difficulties in droplet coalescence have led researchers to search for a method to fuse two droplets in a controlled manner. An ideal protocol would satisfy two criteria: (i) ease of implementation and (ii) minimal potential for contamination between droplets. Criterion (i) is clearly desirable for any application, while criterion (ii) is especially important for analyzing discrete samples in high-throughput applications.

The most straightforward method forcing coalescence is to form droplets with opposite charges by placing electrodes in the two different streams (or reservoirs) used to form the droplets [5]. While this certainly satisfies criterion (i), the direct contact with the electrodes should result in significant contamination between coalescence events. Another option is to alter the channel geometry to favor coalescence. After forming droplets from alternating inlet streams by using "push–pull" droplet formation at a double-T intersection, one droplet from each of the inlet streams can be fused by having them enter a diverging channel [6]. After coalescence, the droplets exit through a "pore" at the end of the diverging channel that presumably leads to the rest of the microfluidic device. From the standpoint of implementation, this technique is ideal in many circumstances, since it only requires fabricating a specific (nonrectangular) channel geometry. Likewise, since the droplets are never in contact with the walls, the potential for cross-contamination is minimal. However, this technique requires precise synchronization between the two droplets so that they coalesce before arriving at the pore. For applications involving the generation of vesicles and emulsions, the push–pull strategy [6] provides the requisite synchronization. However, in the highly integrated systems envisioned for systems biology, analyte droplets will need to coalesce at various points with different reagent droplets, and it is not clear how to produce the desired synchronization when the droplets are entrained in an immiscible, flowing stream.

The issues of synchronization and coalescence can be resolved by using so-called "digital" systems, where the droplets are manipulated using complex electric field patterns. The most common approach is to do away with the entraining oil entirely and use electrowetting-based actuation (also known as electrowetting on dielectric, or EWOD) [7, 8] of droplets in direct contact with the surface. This technique relies upon electrically induced changes in solid–liquid surface tension that arise when different potentials are applied to a surface pattern of electrodes. Similar types of actuation can also be realized by using photoresponsive surfaces [9] or thermocapillary forces [10]. When two droplets are driven to the same location on the pattern (corresponding to the minimum in potential energy), they coalesce instantaneously. From the standpoint of contamination, these strategies are not ideal, since the droplets are in direct contact with the surface. However, surface contamination can be attenuated by altering the solution pH and the electrode polarity [11]. Alternatively, the droplets floating on a fluorinated oil can be actuated by dielectrophoresis *via* a submerged electrode pattern [12]. Even if the potential for contamination is eliminated, this method still requires complex control systems [8], making its implementation less straightforward than the closed systems discussed thus far. Indeed, one of the often cited advantages of this system [13] is that it directly mimics traditional bench-top protocols, which means that it is subject to similar limitations when converted into a high-throughput application.

Consequently, there exists a clear need for a method of forcing droplet coalescence in a closed microfluidic system that does not rely upon precise synchronization. Moreover, the coalescence system must be able to fuse similar sized droplets. Even more so, the system must work when the undeformed droplet diameter is slightly larger than the size of the channel, corresponding to the typical situation in PCR applications [1, 3].

We present such a technique here, based upon the principle of electrocoalescence [14]. Electrocoalescence occurs when high conductivity droplets are immersed in a low conductivity medium and subjected to an electric field. These conditions are readily realized in droplet systems for biological applications, since the aqueous droplets are usually buffered at a relatively high salt concentration, whereas the entraining solvent possesses very low conductivity. The system presented here satisfies both of the generic criteria for a coalescence mechanism: (i) it only requires two external electrodes connected to an alternating current (AC) power source, making implementation straightforward; and (ii) the electrodes never come into contact with the droplets or the entraining fluid, thereby minimizing the potential for contamination between coalescence events. Even more so, the forces involved in electrocoalescence can be sufficiently strong to arrest the motion of the leading droplet in a flowing stream, making precise synchronization between the droplets unnecessary.

In Section 1.2, we recall the key aspects of electrocoalescence in unbounded systems and describe our electrocoalescence setup. We then present results for coalescence of droplets in both quiescent and flowing solvents. We conclude with some perspectives on how our capillary-based system can be further miniaturized.

13.1.2
Principle of electrocoalescence

Electrocoalescence in unbounded systems (*i.e.*, when the droplets are very far from the walls or in the absence of walls) is a well-studied phenomenon. It is important in atmospheric physics, where electric fields strongly affect the coalescence and dissolution of falling raindrops [15]. The petrochemical industry has also utilized electrocoalescence for over 50 years as a bulk process for removing dispersed water from crude oil [16, 17] by forcing the coalescence of small water drops and then collecting the condensed droplets by gravity settling. Inasmuch as we believe the system presented here is the first application of electrocoalescence in the context of microfluidics or systems biology, and because the relevant literature is scattered amongst a number of different sources, it behooves us to recall the basic physics of the phenomenon.

Our discussion is based primarily on the exposition by Atten [14]. Electrocoalescence occurs when the conductivity of the droplets is much higher than that of the surrounding medium. The electric field vanishes inside the droplets, whereupon the effect of the droplets on the field inside the oil phase can be approximated by dipolar disturbances situated at the droplet centers. The dipolar approximation is valid only when the applied electric field is uniform far from the droplets (by applied electric field, we mean the electric field that would be present in the low conductivity phase in the absence of the droplets). Concomitantly, the dipolar approximation requires that the electrodes be situated far from the droplets, a condition that may not always be satisfied in microfluidic applications. When two droplets are nearby in the electric field, their induced dipoles align, leading to an attractive Columbic force between them. The force experienced by the drops is proportional to the square of the electric field, allowing for actuation by both AC and direct current (DC) fields [17].

The distinction between dielectrophoresis, which is often employed in the digital strategies cited in Section 1.2, and electrocoalescence is not entirely straightforward, as both phenomena involve dipolar attraction. Tab. 1 outlines the key differences between what we will refer to as dielectrophoresis and electrocoalescence. Basically, electrocoalescence can occur in uniform applied fields, which is not the case of dielectrophoresis, and at least two droplets are required for electrocoalescence to be active. Note the overlap between the conditions for the two phenomena; if the applied field is nonuniform, the two droplets will simultaneously undergo motion due to dielectrophoresis and electrocoalescence – provided that the droplets and the bulk fluid possess different dielectric constants and conductivities.

For unbounded systems, when the droplets are far apart, the attractive force between them decays like r^{-4}, where r is the distance between droplet centers. When the droplets are close, rearrangement of the surface charges on the drops results in a divergence in the attractive force as they approach one another [14, 18]. Importantly, when the drops are very close, the electric field at the droplet surface is greatly enhanced by polarization charges [14, 18]. The resulting attractive force is much stronger than that might be expected from the applied electric potential. In near contact, the droplet dynamics are affected by van der Waals forces, and the film between them eventually ruptures and leads to coalescence [19]. In the case of electric fields, coalescence may also be driven by the dielectric breakdown of the

Tab. 1 Differences between dielectrophoresis and electrocoalescence

Dielectrophoresis	Electrocoalescence
Nonuniform applied field[a)]	Uniform or nonuniform applied field; nonuniform induced field
Different dielectric constant between the fluid and the particle/drop	Droplet is more conductive than the bulk fluid
Single-body phenomenon	Multibody phenomenon

a) Applied electric field is defined to be the electric field that would be present in the low conductivity phase without the droplets. Induced field is defined to be the electric field in the low conductivity phase in the presence of the droplets.

surrounding fluid [20] or an electrohydrodynamic instability [21]. The last two stages of the process (charge rearrangement and film rupture) are generally much faster than the approach of the two droplets.

Given the technological importance and scientific relevance of electrocoalescence, it is unsurprising that there exists a large body of literature discussing the fundamental aspects of the phenomenon. In the interests of focus, we will briefly discuss two of the most relevant references here. Citations to and within these papers should lead the interested reader to further information in the literature.

Allan and Mason [20] experimentally examined the coalescence of liquid drops in DC electric fields and low velocity gradient shear fields using a Couette-flow geometry. A critical field exists for the onset of coalescence, necessitating fields on the order of 500–1000 V/cm. Coalescence was achieved over a range of voltages and shear rates. For some combinations of shear rate and electric field strength, they observed that, upon approaching, the drops suddenly repelled each other due to an apparent electrical discharge. Allan and Mason's work involved droplets whose radii were small compared to the gap in the Couette device, so wall effects were not significant. This contrasts with typical microfluidic protocols, where the droplets occupy a significant fraction of the channel, often resembling long fluid plugs surrounded by a thin layer of the entraining fluid [22]. The strong wall effects in microfluidic geometries are expected to alter both the results of Allan and Mason, as well as the general model presented by Atten [14].

In an important numerical study, Zhang *et al.* [23] considered the pair-wise dynamics of uncharged, surfactant-free, micron sized drops acted on by gravity and an external electric field in an unbounded, linearly polarizable dielectric fluid or insulator. They showed that the electric field-induced force varies like the square of the droplet size and depends on the relative geometries of the two drops. Moreover, the force is at its maximum when the two droplets are aligned such that they translate in the direction of the field along their line of centers, a result which has since been confirmed experimentally [24]. For a uniform applied field, this implies that the droplet centers are situated on the same field line and experience a "head-on" collision. In a miniaturized system, where the two droplets have undeformed radii equal to or greater than the channel radius (or half-width), and where the channel is aligned with the field, the electrocoalescence force should be maximized, since the droplets must undergo a head-on collision and cannot pass one another without colliding.

13.2
Materials and methods

Fig. 1 depicts our experimental apparatus for the electrocoalescence of droplets in a confined geometry. A 560 μm ID Teflon PTFE capillary (Fisher-Bioblock) was threaded through the annulus of two cylindrical electrodes mounted in a polycarbonate housing with a narrow slit in the center for observation with a binocular microscope (Olympus). The spacing between electrodes can be varied, but was

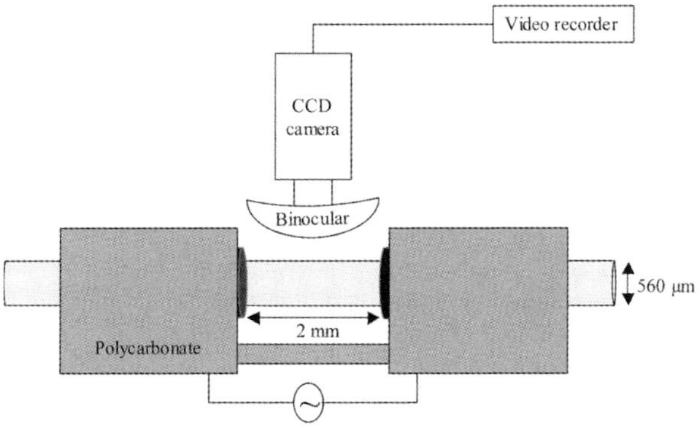

Fig. 1 Schematic of the experimental apparatus.

fixed at 2 mm for all of the experiments described here. The voltage was applied using a Hameg function generator connected to a Trek 677B amplifier, permitting voltages up to 2 kV and frequencies up to 1 kHz. We typically applied a 2 kV sinusoidal voltage to the electrodes to create a 10 kV/cm, 500 Hz AC electric field, although for parametric studies of the critical voltage we applied a variety of field/frequency combinations.

Ideally, the electrode spacing should be as large as possible, so that the droplets are situated well within the gap between the electrodes and the field is uniform far from the droplets. The 2 mm spacing represents the optimal balance between having the strongest possible applied field in the gap between the electrodes, while still fulfilling the requirements for electrocoalescence (*i.e.*, the gap must be wide enough to contain both droplets completely between the electrodes).

The droplets were TBE 5 × buffer (0.45 M Tris-base, 0.45 M boric acid, and 0.01 M EDTA; Sigma, St. Louis, MO, USA) dyed with 0.01 wt% bromophenol blue for observation in a carrier fluid of fluorinated oil (FC-40, 3M) with 0.5 wt% 1H,1H,2H,2H perfluorodecan-1-ol (Fluorochem) added to prevent interactions with the walls. The droplet conductivity was 3 mS/cm and the carrier fluid conductivity was 2.5×10^{-13} mS/cm. We formed the droplets by aspirating at 1 mL/h with a syringe pump (KD Scientific, New Hope, PA, USA) and a Hamilton Gas-Tight 250 μL syringe while oscillating the capillary tip at the water–oil interface of a layered reservoir. The droplet motion was recorded at 25 frames *per* second by a CCD camera (Hitachi) connected to a Sony GVD800 Digital 8 video recorder, and the data were analyzed using custom tracking software and Scion image. Some data were also collected from a 1000 frames *per* second high speed camera (PS 220 Photron Fastcam super 10 K).

Static fluid coalescence experiments were performed by aspirating a pair of droplets into the gap between the electrodes (without applying the electric field) until the droplets were centered in the gap. Once the system equilibrated, we applied the electric field at the desired frequency and strength.

We determined the critical upper field strength for coalescence at a given frequency by starting with a high field strength and gradually decreasing the voltage in 50 V steps until coalescence occurred. Each point is an average over about ten measurements. We attempted to use equal sized droplets with an undeformed diameter of approximately 575 μm, corresponding to a volume of about 100 nL. For these experiments at high field strengths, failed coalescence attempts usually resulted in some exchange of fluid (see Section 3). Thus, if the coalescence failed at a particular voltage, a new droplet pair was aspirated into the gap, the voltage was reduced by 50 V, and the experiment continued. The variance in field threshold between different droplet pairs was less than the 50 V step size, so we use the latter as our error estimate.

The lower bound for coalescence was also determined using two droplets of about 575 μm undeformed diameter with an initial separation of 250 μm. For a given frequency, we increased the applied potential until the droplets began to move toward each other. Once the droplets were animated, they always coalesced. Again, each data point is an average over at least ten measurements. However, the value of the lower bound depends strongly on the droplet size, resulting in uncertainties greater than the 50 V step size between droplet pairs possessing slightly different volumes.

For flowing fluid coalescence experiments, we aspirated the droplet pair into the gap between the electrodes while applying the electric field. For the mixing experiments, we seeded one of the drops with Mica Rheoscopic tracer particles (Kalliroscope, Groton, MA) and followed the procedure for coalescence in static fluids.

13.3
Results and discussion

13.3.1
Calculation of the electric field

The electric field generated by the coaxial cylinder electrode design was calculated using the Poisson equation finite-element solver Quickfield (Version 5.0.3.59, Tera Analysis). The calculation assumes a relative permittivity of $\varepsilon = 2$ for FC-40 and the Teflon tubing and $\varepsilon = 3.3$ for the poly(methyl methacrylate) (PMMA) housing. The electrodes are represented by cylindrical line elements placed inside the Plexiglas housing 1.5 mm from the axis of symmetry, each extending 4 mm from the edge of the computational boundary. A potential of 2000 V between the electrodes generates the field.

The result of the calculation is depicted in Fig. 2. Bold lines represent parts of the device and the thin lines represent equipotentials. The field is relatively uniform in the gap between the electrodes, which should favor electrocoalescence forces over dielectrophoretic forces in our apparatus. Nevertheless, field gradients exist near the electrodes, due to their close proximity to each other and the variable conductivities of the different materials. This may lead to a small dielectrophoretic force opposing the droplet coalescence. Note that the magnitude of the electric field in the FC-40 is significantly less than what might be expected from the applied voltage, owing to the voltage drop at the various interfaces. The field inside the FC-40 is approximately 5000 V/cm for an applied voltage of 2000 V on 2 mm. Finally,

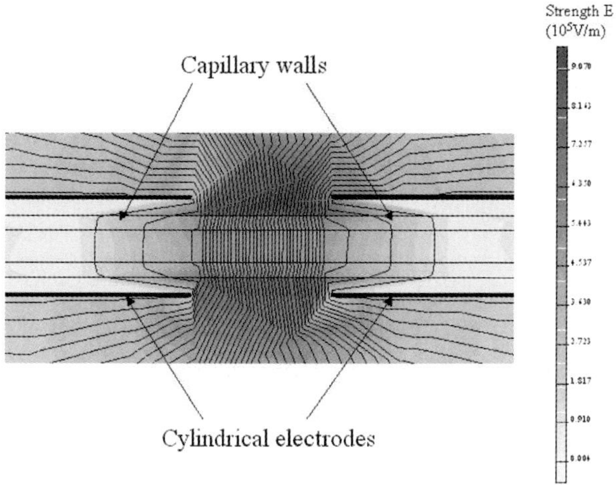

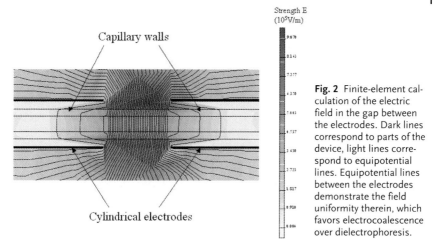

Fig. 2 Finite-element calculation of the electric field in the gap between the electrodes. Dark lines correspond to parts of the device, light lines correspond to equipotential lines. Equipotential lines between the electrodes demonstrate the field uniformity therein, which favors electrocoalescence over dielectrophoresis.

we remark that the asymmetries in the electric field outside the gap are probably artifacts of the calculation procedure arising from imprecise far-field conditions, as the actual device geometry is symmetric. This version of Quickfield only possesses minimal mesh refinement capabilities, so our computation resources did not allow us to eliminate these asymmetries by systematically expanding the computational domain. In any event, the key elements of the calculation (the equipotential lines and field intensity inside the gap) are minimally impacted by the far-field conditions. Moreover, the far-field conditions are not well defined in the experimental apparatus, being a function of the local environment, nearby electrical sources, *etc.*

13.3.2
Droplet coalescence in a quiescent fluid

We first attempted electrocoalescence using a DC field in an otherwise quiescent fluid. However, we never observed coalescence, rather finding that the droplets act as if they have a positive charge. Other studies have reported similar directed motion of presumably neutral droplets toward a particular electrode. For example, Bailes *et al.* [25] observed the motion of the drops toward the negative electrode and attributed the motion to droplet charging during their formation. Allan and Mason [20] also observed motion of neutral droplets toward the negative electrode, claiming that this was due to EOF of silicon oil along the glass plate at the bottom of their Couette apparatus. However, they observe nearly no EOF with another oil (Pale 4 oil) [20].

Given the chemistry of FC-40 and of the tubing, which all are perfluorinated fluorocarbons, the presence of EOF in our system is extremely unlikely. To test for EOF, we aspirated a dilute suspension of Mica Rheoscopic tracer particles into the gap between the electrodes and allowed the system to settle to equilibrium. When we applied the electric field, there was no preferential motion toward either electrode. It is possible, but unlikely, that the EOF is exactly balanced by electrophoretic motion of the particles. However, the particles arrange themselves into a columnar

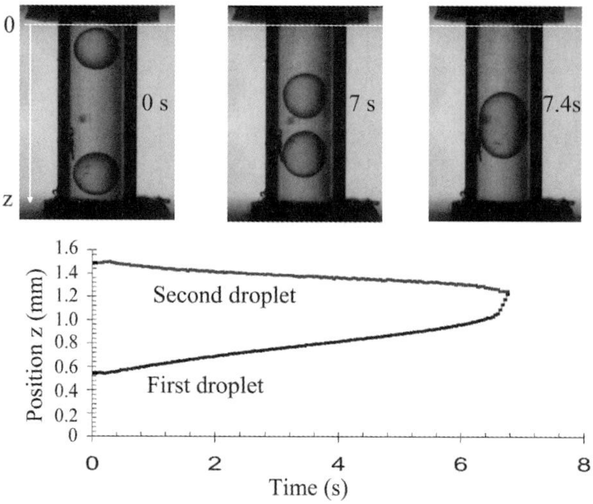

Fig. 3 Electrocoalescence in an otherwise static fluid. First (lower) droplet has an undeformed diameter of 540 μm and the second (upper) droplet has an undeformed diameter of 560 μm. Ordinate of the plot corresponds to the positions of the interfaces which eventually coalesce.

structure in the direction of the field and the assembly exhibits no preferential motion toward an electrode, indicating that the particles have no significant electrophoretic mobility. Moreover, particles that do not align themselves in the columnar structures are immobile. Inasmuch as they have negligible electrophoretic mobility, the EOF must also be negligible. We consequently attribute droplet motion in DC field to their charging.

Since DC fields were insufficient to induce coalescence in our system, we proceeded to investigate AC fields. AC fields suppressed the electrophoresis, and allowed for coalescence in a significant range of field strengths and frequencies. Fig. 3 presents the AC electrocoalescence of two droplets in an otherwise quiescent fluid. Upon applying the field, the initial droplet motion was steady, with the smaller droplet moving at a slightly higher velocity due to its weaker wall effect [26]. Although the dipolar force in an unbounded system scales like r^{-4}, our confined droplets do not accelerate as they approach each other from far away.

We attribute the constant droplet velocity to the breakdown of the dipole assumption invoked in the unbounded model [14]. In our system, the electrodes are not located many droplet radii away from the droplets [14]; rather, all length scales are similar. Consequently, the electrostatic problem cannot be represented solely by a dipolar perturbation of a uniform field. From a modeling standpoint, the full treatment of the electrostatic problem requires replacing the uniform field at infinity with boundary conditions at the electrodes and the droplets. Nevertheless, we would expect that a dipolar-like force would still be a key contributor to the overall force acting on the droplets. Inhomogeneities in the electric field near the edge of the electrodes, as well as the inhomogeneous field in the air outside the electrodes, may give rise to a small dielectrophoretic force opposing coalescence.

In any event, when the droplets become close (*i.e.*, when r is on the order of a droplet radius) the attractive force between them still appears to diverge in the confined system, in agreement with theory [18]. The droplets rapidly accelerate and

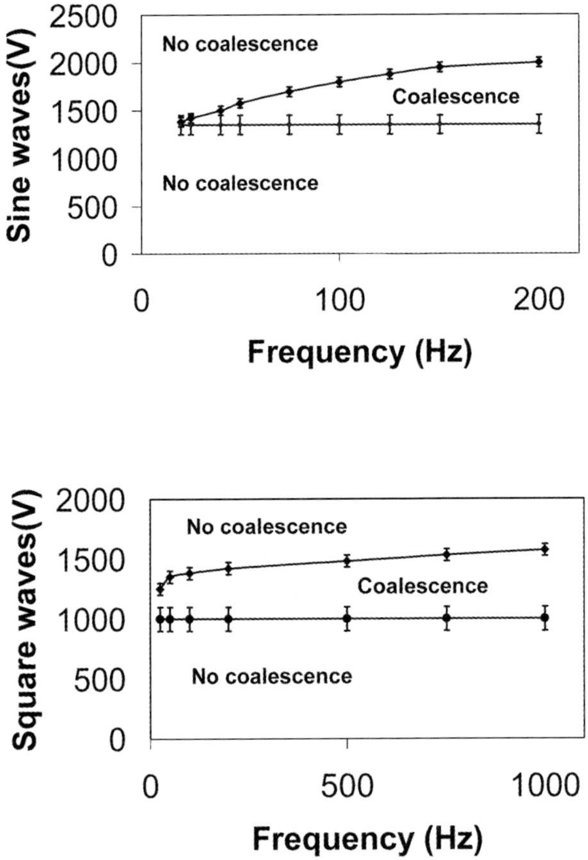

Fig. 4 Phase diagram for the coalescence of two droplets in a static fluid in square and sinusoidal AC electric fields as a function of the amplitude and frequency of the voltage applied to 2-mm spaced electrodes. Droplet undeformed diameters are about 575 μm.

drain the intervening film, despite the large friction with the walls. Once again, the smaller droplet moves more rapidly than its counterpart. The dramatic difference in approach velocities in the two regimes (far apart and close contact) is consistent with the overall phenomena of electrocoalescence [14]. The results here indicate that our system should provide essentially instantaneous coalescence of droplets which are initially close together, such as occurs after two droplets arrive simultaneously at a T-junction.

13.3.3
Phase diagram for coalescence

Fig. 4 presents a phase diagram of electrocoalescence as a function of the amplitude and frequency of the voltage applied to the two 2-mm spaced electrodes. Coalescence occurs at frequencies as low as 10 Hz, but not reproducibly; coalescence is much more reproducible at frequencies around 500 Hz. At a given field strength, the droplet oscillations depend on frequency. Explicitly, the amount of energy absorbed by the droplets depends on whether the field frequency is close to or far from the droplet's "natural" frequency [27]. We estimated the natural frequency by observing the field frequency that yielded the largest oscillations at a given field strength. In our system the droplets appear to have a natural oscillation frequency around 50–100 Hz, corresponding to an applied sinusoidal field frequency of 25–50 Hz. For frequencies above 200 Hz, the drops simply elongate in the direction of the field, and no shape oscillations are observed. The droplets act as if they are in a DC field, albeit without any electrophoretic motion.

The lower bound for coalescence corresponds to the lowest field value for which the droplets move toward one another in less than 1 min. In general, if the droplets fail to move toward each other during the 1 min interval, they will fail to move toward each other over much longer times. In our configuration, the minimum applied field for actuation needs to be higher than the 1 kV/cm typically employed in bulk electrocoalescence devices [17], owing mostly to friction. Nevertheless, our miniaturized system requires a lower applied voltage than bulk systems, due to the small gap between the electrodes, and further miniaturization should increase this advantage of the system.

Surprisingly, there is also an upper bound for electrocoalescence that appears to be correlated with droplet shape oscillations. Typically, droplets start to approach each other, but close to contact they brutally repel one another, forming small satellite droplets between them. This cycle is repetitive, and the procession–recession phenomenon is similar to that observed by Allan and Mason [20].

We propose a simple physical rationale for the coalescence upper bound, which is based upon high speed videomicroscopy observations. As the droplets move close together, the intense field between them results in a Taylor cone [28]. A jet forms at the tip of the Taylor cone, resulting in a liquid bridge between the droplets that disrupts the dipole–dipole interaction. The length of this bridge (*i.e.*, the distance between droplets at the time the jets connect) depends on the strength of the applied field. At lower fields, the droplets are sufficiently close together when the bridge forms so that surface tension-induced coalescence occurs without any repulsion. If the applied field is too high, however, the bridge is longer and unstable. The satellite droplets are then the result of the breakup of a small liquid bridge between the two main droplets. Although our data appear to support these trends, a typical bridging event seems to occur over 2 ms (corresponding to 2 frames at 1000 frames *per* second), which does not give sufficiently detailed imaging to confirm or deny our hypothesis.

Even when the droplets do not coalesce, there appears to be significant exchange of fluid between them during their procession–recession cycle, probably due to unsuccessful bridging events. To test this hypothesis, we prepared one droplet with bromophenol blue and a second without it. After aspirating the two droplets into the gap between the electrodes, we operated the system above the upper critical field strength. After an hour of operation, both droplets were essentially of the same color. Inasmuch as bromophenol blue should be insoluble in FC-40 and the rate of diffusive mass transfer is much too slow to transfer the dye in 1 h, we can conclude that the color change comes from mass transfer by the small discharged droplets from the bridge.

The critical voltages for coalescence as a function of field frequency in both square and sinusoidal AC fields also appear in Fig. 4. In both cases, the lower bound for coalescence only depends on the minimum field strength necessary to animate the droplets. The lower bound is independent of frequency because the applied field is too low to induce significant droplet shape oscillations. Inasmuch as the force is equal to the gradient of the dipolar energy, the critical field amplitude should be higher for sinusoidal waves than for square waves. Indeed, the RMS values of the applied voltages in square and sinusoidal waves are approximately the same.

For both wave types, the upper bound for coalescence is frequency dependent. For square fields, we were able to determine an upper critical field strength for frequencies up to 1 kHz. For sinusoidal fields, the maximum voltage of our equipment always resulted in coalescence at frequencies greater than 200 Hz – we could not reach the upper critical field strength for higher frequencies. Nevertheless, we expect that the upper bound would plateau in a manner similar to its square wave counterpart. The plateau arises because the upper critical field strength is related to droplet shape oscillations and the natural oscillation frequency of the drop. At frequencies near the natural oscillation frequency, the drop gains more energy from the field by mechanically oscillating with it, and the phenomena impeding coalescence (such as drop disintegration) begin to appear at lower field strengths. As noted above, when the applied field frequency greatly exceeds the natural frequency of the drops, shape oscillations are suppressed, so the upper critical voltage should lose its frequency dependence as well as the frequency increases. From these curves, working at high frequency (between 200 and 1 kHz) presents the widest operating range for electrocoalescence.

13.3.4
Droplet coalescence in AC fields with flow

Whereas the static case clearly demonstrates the principle of electrocoalescence on the microscale, most practical circumstances involve fusing droplets that are entrained in a flowing stream. Fig. 5 depicts an example of such an electrocoalescence between two widely separated droplets. This experiment uses a relatively low volumetric flow rate of 50 μL/h, and the electric field is active throughout the duration of the experiment.

The first droplet enters the region between the electrodes and moves at a constant velocity in the absence of the second drop. This indicates that, in absence of a droplet–droplet dipolar force, a single droplet is essentially insensitive to the presence of an AC field. The leading interface of the second droplet appears in the gap between the electrodes after 13 s, but its presence does not initially affect the first droplet as it continues to move at a constant velocity. Only when the trailing droplet is well inside the gap between the electrodes does the dipolar force manifest itself. Thereafter, the coalescence time is essentially the same as in the static case (approximately 8 s) for these widely separated droplets, but the solvent flow slightly alters the dynamics. In this experiment, the dipolar force almost exactly balances the viscous force and arrests the first droplet, while the second droplet moves to-

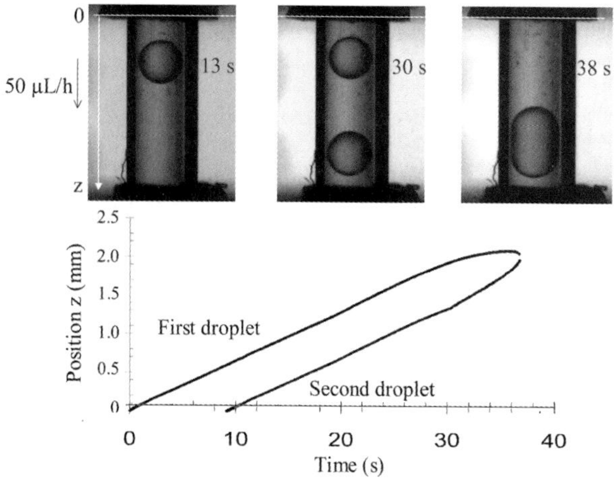

Fig. 5 Electrocoalescence of two water droplets in a 50 μL/h flow. First droplet has an undeformed diameter of 570 μm and the second droplet has an undeformed diameter of 560 μm. Interface tracking is the same as in Fig. 3.

ward it at a constant rate. Once the droplets are close together, the strong dipolar force rapidly drains the intervening fluid and coalescence is achieved. The coalesced drop is simply entrained in the bulk flow, exhibiting no effects of the applied AC field. In both the static and dynamic cases, the droplets coalesce completely without the formation of any satellite drops, indicating that the coalesced droplet is electrohydrodynamically stable [28].

While we have focused here on a relatively slow flow rate to make an accurate measurement of the droplet dynamics, we have performed identical electrocoalescence experiments with stream velocities up to 1 mm/s (data not shown). Inasmuch as the latter corresponds to the velocities used in our PCR system [3], this result demonstrates that the present electrocoalescence system should be easily integrated with continuous flow PCR. The interface trajectories at different speeds are similar to Fig. 5 although the time required for coalescence decreases with the flow rate, as would be expected (data not shown). The essential feature of the coalescence phenomenon is that both droplets must be present between the electrodes. Furthermore, initially close droplets coalesce almost instantaneously, and the upper limit to the initial separation must be decreased as the oil velocity increases. Consequently, the key design feature is the droplet spacing rather than their velocity, which must be determined by an engineering analysis for a given tube radius and electrode spacing.

The system presented here does not require precise droplet synchronization, provided that the spacing between the electrodes is greater than the spacing between droplets and the applied voltage is sufficient. Indeed, the coalescence experiment depicted in Fig. 5 operates close to the minimum electrode spacing; the motion of the leading droplet is arrested very close to the outlet electrode.

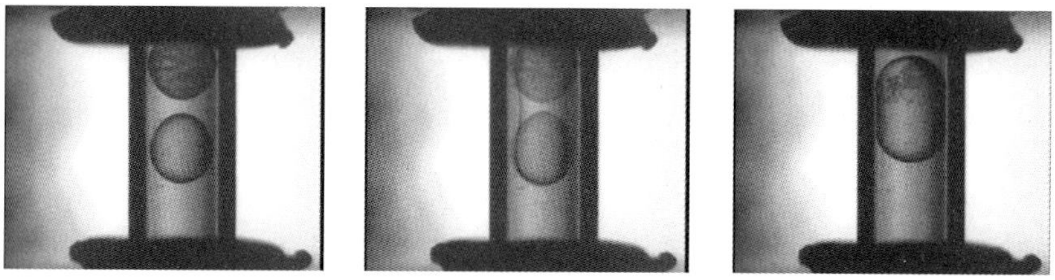

Fig. 6 Electrocoalescence of two water droplets in a static fluid. Top droplet is seeded with small colloidal particles that do not mix upon coalescence.

13.3.5
Droplet mixing upon coalescence

The motion leading to droplet coalescence is relatively violent, so one might expect that this motion also leads to the immediate mixing of the contents of two different droplets when they coalesce. To assess this, we seeded one of the droplets with Mica Rheoscopic tracer particles. As depicted in Fig. 6, the droplet contents do not mix immediately upon coalescence, with the dispersed material being segregated into the top-half of the resultant droplet. As a consequence, it remains necessary to use fluid motion, such as that induced by a curved channel [2], to achieve rapid mixing of the coalesced droplets.

13.4
Concluding remarks

We have presented a system for the electrocoalescence of microfluidic droplets in both quiescent and flowing solvents. Although the study here focuses upon a specific solvent system for brevity, we have achieved similar results for other oil–water emulsions, in surfactant-free systems, and for different electrode spacings. The phase diagram for the coalescence regime in the field-strength/field-frequency domain is more complex than expected, and presents a striking reentrant transition to a regime in which coalescence is prevented at high enough fields. We interpreted this striking behavior as the consequence of frequency dependent dipole disruption due to an unstable bridge formed between the droplets. However, bridge formation is not clear, even at 1000 frames *per* second, and this phenomenon deserves further investigation.

Our coalescence protocol offers several advantages when compared to contact-charging or electrowetting: (i) there is no contact with the electrodes across the thick capillary walls; (ii) the system is easy to implement without the need for sophisticated lithography; and (iii) the system can be applied "on flight" in a flowing stream without distorting the flow far from the droplets. Moreover, by properly spacing the electrodes, the electrocoalescer does not require the exquisite droplet synchronization necessitated by other techniques [6].

We conclude with some comments on further miniaturization of our technique for lab-on-a-chip applications. The length scales employed in our system, namely, a 560 μm channel diameter and 100 nL droplet volumes, do not necessarily commensurate with all microfluidic applications. First, note that capillary-based systems have proven very efficient for microliter droplet volume applications, such as PCR [3], and our system is readily adaptable to virtually any capillary size. Moreover, many of the recent commercial advances in high-throughput analyses were capillary based.

Concerning the relevance to chip-based applications, the *dimensionless* length scales (aspect ratios) characterizing our system are the same as those prevailing in labs-on-a-chip. Explicitly, the undeformed droplet radii are comparable to the size of the channel, which deform into a cylindrical plug surrounded by a thin layer of fluid (see, for example, the chip-based applications in [2, 22]), and the droplets to be coalesced are of similar size. Thus, to the limit that continuum mechanics remains valid in a proposed device, we would expect our device to scale to virtually any size channel. One might argue that further miniaturization is impeded by the fact that electric field-induced force acting on the droplets scales with the square of the undeformed droplet radius [23]. However, in our system, the dipolar force only needs to oppose the viscous force created by draining the oil between the drops. For a given shear rate, the total shear stress on the droplet also scales with the square of the undeformed radius (*i.e.*, with the surface area), so these two effects should cancel each other out. From a fabrication standpoint, the incorporation of two electrodes is a trivial step relative to those necessary to produce a highly integrated chip [29]. Indeed, a PCR chip that utilizes electrocoalescence would already have numerous integrated electronics for heaters and temperature measurements [10]. Finally, reducing the distance between the electrodes and thinning the channel walls will reduce the required voltage, and many microfabrication processes are well suited for producing very thin, insulating walls that can optimize the required voltage.

We are grateful to Michel Petit for constructing the electrode apparatus and Konstantin Zeldovich for the tracking software. M.C. acknowledges a grant from French DGA (Ministère de la Défense). K.D.D. acknowledges a postdoctoral fellowship from the International Human Frontiers Science Program Organization. This work was supported by ARC (Grant No. 3107), the French Ministry of Research ACI "Technologies for Health," and the European VI Framework Project "NABIS," Contract No. NMP4-CT-2003–505311.

13.5 References

[1] Curcio, M., Roeraade, J., *Anal. Chem.* 2003, *75*, 1–7.

[2] Song, H., Ismagilov, R. F., *J. Am. Chem. Soc.* 2003, *125*, 14613–14619.

[3] Dorfman, K. D., Chabert, M., Codarbox, J.-H., Rousseau, G., de Cremoux, P., Viovy, J.-L., *Anal. Chem.* 2005, *77*, 3700–3704.

[4] Olbricht, W. L., Kung, D. M., *J. Colloid Interface Sci.* 1987, *120*, 229–244.

[5] Macaskill, A., Fielden, P. R., Goddard, N. J., Mohr, S., Treves Brown, B. J., *Proceedings of the 8th International Conference on Miniaturized Systems for Chemistry and Life Sci-*

ences (MicroTAS'04), Malmö, Sweden 2004, *Vol. 1*, pp. 207–209.

[6] Tan, Y.-C., Fisher, J. S., Lee, A. I., Cristini, V., Lee, A. P., *Lab Chip* 2004, *4*, 292–298.

[7] Washizu, M., *IEEE Trans. Ind. Appl.* 1998, *34*, 732–737.

[8] Schwartz, J. A., Vykoukal, J. V., Gascoyne, P. R. C., *Lab Chip* 2004, *4*, 11–17.

[9] Ichimura, K., Oh, S.-K., Nakagawa, M., *Science* 2000, *288*, 1624–1626.

[10] Burns, M. A., Mastrangelo, C. H., Sammarco, T. S., Man, F. P., Webster, J. R., Johnson, B. N., Foerster, B., *et al.*, *Proc. Natl. Acad. Sci. USA* 1996, *93*, 5556–5561.

[11] Yoon, J.-Y., Garrell, R. L., *Anal. Chem.* 2003, *75*, 5097–5102.

[12] Velev, O. D., Prevo, B. G., Bhatt, K. H., *Nature* 2003, *426*, 515–516.

[13] Paik, P., Pamula, V. K., Pollack, M. G., Fair, R. B., *Lab Chip* 2003, *3*, 28–33.

[14] Atten, P., *J. Electrostat.* 1993, *30*, 259–270.

[15] Braziers, P. R., Jennings, S. G., Latham, J., *Proc. R. Soc. London Ser. A* 1972, *326*, 393–408.

[16] Eow, J. S., Ghadiri, M., Sharif, A. O., Williams, T. J., *Chem. Eng. J.* 2001, *84*, 173–192.

[17] Eow, J. S., Ghadiri, M., *Chem. Eng. J.* 2002, *85*, 357–368.

[18] Davis, M. H., *Quat. J. Mech. Appl. Math.* 1964, *17*, 499–511.

[19] Vrij, A., *Disc. Faraday Soc.* 1966, *42*, 23–33.

[20] Allan, R. S., Mason, S. G., *J. Colloid Sci.* 1962, *17*, 383–408.

[21] Michael, D. H., O'Neil, M. E., *J. Fluid Mech.* 1970, *41*, 571–580.

[22] Link, D. R., Anna, S. L., Weitz, D. A., Stone, H. A., *Phys. Rev. Lett.* 2004, *92*, 054503.

[23] Zhang, X., Basaran, O. A., Wham, R. M., *AIChE J.* 1995, *41*, 1629–1639.

[24] Eow, J. S., Ghadiri, M., *Colloid Surf. A* 2003, *219*, 253–279.

[25] Bailes, P. J., Lee, J. G. M., Parsons, A. R., *Trans. IChemE* 2000, *78*, 499–505.

[26] Happel, J., Brenner, H., *Low Reynolds Number Hydrodynamics*, Nijhof, Dordrecht, The Netherlands 1983.

[27] Scott, T. C., Basaran, O. A., Byers, C. H., *Ind. Eng. Chem. Res.* 1990, *29*, 901–909.

[28] Taylor, G. I., *Proc. R. Soc. London A* 1964, *280*, 383–397.

[29] Thorsen, T., Maerkl, S. J., Quake, S. R., *Science* 2002, *298*, 580–584.

14
Microfluidic flow focusing: Drop size and scaling in pressure *versus* flow-rate-driven pumping*

Thomas Ward, Magalie Faivre, Manouk Abkarian, Howard A. Stone

We experimentally study the production of micrometer-sized droplets using microfluidic technology and a flow-focusing geometry. Two distinct methods of flow control are compared: (i) control of the flow rates of the two phases and (ii) control of the inlet pressures of the two phases. In each type of experiment, the drop size l, velocity U and production frequency f are measured and compared as either functions of the flow-rate ratio or the inlet pressure ratio. The minimum drop size in each experiment is on the order of the flow focusing contraction width a. The variation in drop size as the flow control parameters are varied is significantly different between the flow-rate and inlet pressure controlled experiments.

14.1
Introduction

Emulsions, foams, aerosols and other dispersions are examples of multiphase fluids. Traditionally these materials find applications and/or arise in processing of home and personal care products, foods, petroleum-based products, mineral flotation, *etc.* In many of these applications the dispersions are polydisperse and poorly controlled, though variants of bulk emulsification methods can achieve control of the drop size and the size distribution. Renewed interest in these fields that involve emulsions has been brought about by the demonstration that microfluidic technology can provide accurate control of droplet size at the micrometer scale, allows precise control of the chemical composition and the thermal environment, and can be utilized so that individual drops act as isolated chemical containers. This control has been demonstrated to be useful in new applications and for developing new technologies, in particular for providing new routes for addressing questions, measurements and products of interest to the biological and chemical communities.

* Originally published in Electrophoresis 2005, 26, 3716–3724

Microfluidic Applications in Biology. Edited by Niels Lion, Joël S. Rossier, and Hubert H. Girault

ISBN-10: 3-527-31761-9

One significant advantage of microfluidic devices is the ability to control the formation of drops at the very scale at which drops are desired. Thus, μm-sized drops can be readily formed using channels and other plumbing that are at the μm scale, and standard visualization procedures can be employed simultaneously. Several different geometries have been investigated for the precise control of drop sizes including liquid-liquid and liquid-gas systems [1–6]. A general theme in all of these studies is that they show the results are reproducible for a wide range of parameters and that there exists the potential for precise control of individual droplets.

It has been recognized that small droplets can act as isolated chemical containers, whose composition and other environmental variables can be carefully controlled and monitored. Thus, for example, biological species may be grown in a precision controlled environment to control population density and specimen yield [7–9]. Microfluidic formation of droplets also enables a variety of encapsulation methods such as producing double emulsions [10, 11], and variants of these strategies have been used for the microfluidic production of vesicles [12] and colloidal crystals [13]. Alternatively, there are many opportunities in biology and chemistry to utilize a microfluidic platform capable of integrating and controlling thousands of channels, which has the potential for rapidly performing many experiments [14] using nanoliters or even picoliters of reagent (note that a spherical droplet with radius 10 microns has a volume of 4 pL and a droplet with radius 100 μm has a volume of 4 nL). Also, single experiments with high precision can be performed where the microfluidic device is used for sensing, detection and separation. For example, it may be possible to rapidly determine if an airborne pathogen is present or detect and sort biological species such as DNA or cells [15]. Recent work has also shown how a microfluidic device containing a viscoelastic liquid can behave much like an electronic transistor as a switch [16]. Further developments are likely to integrate several of the above ideas.

In this paper, we study drop formation in a flow-focusing microfluidic device and compare two methods of supplying the fluids: a syringe-pump methodology that dictates the volume flow rates and a method whereby the inlet fluid pressure is controlled. The flows are at low Reynolds numbers for which the pressure gradient and flow rate are linearly related in a single phase flow. Consequently it might be expected that the two distinct control methods, flow rate *versus* pressure, would lead to the same details of drop formation. Instead, we document distinct differences between these two approaches for the multiphase flows of interest. Although we have not yet succeeded in rationalizing all of these differences, it is clear that the significant influence of surface tension, which is known to introduce nonlinearities into the theoretical description (due to the coupling of the shape and the fluid stresses, such as pressure, in the boundary conditions) is likely to be the major contributor to the significant differences we have identified between volume flow rate and pressure control of drop formation in multiphase flows.

In Section 2 we describe the experimental setup for the formation of water drops in oil streams. We then report in Section 3 the differences between the two methods by varying the ratio of the dispersed to continuous phase flow parameter, either the ratio of flow rates Q_w/Q_o or the ratio of applied pressures P_w/P_o. In order to

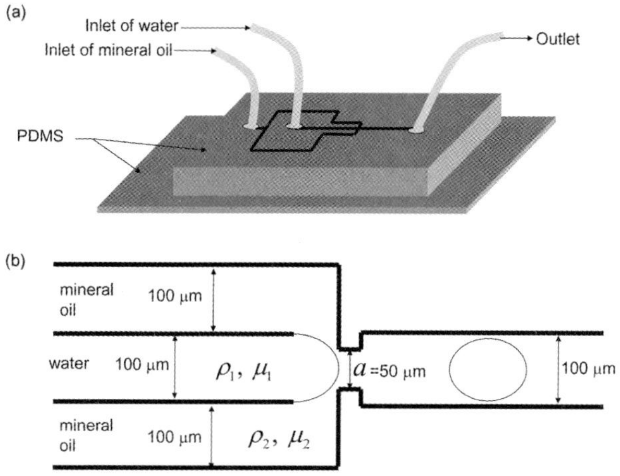

Fig. 1 Microfluidic setup for the two-phase flow in a flow-focusing geometry. The channel height is 75 µm. (a) Side view. (b) Top view.

characterize the drop formation process we report the droplet speed U, frequency of droplet production f, distance between two consecutive drops d, and the drop length l (which is a measure of the drop size). In the Section 4 we discuss surface tension effects as a possible difference between the two experiments.

14.2 Materials and methods

The microfluidic experiments use relief molds that are produced using common soft-lithography techniques [17]. The experiments are performed using a flow-focusing geometry [3, 5] with a single input channel for both fluids. An illustration of the setup is shown in Fig. 1. The channel height is 75 µm, as measured by a profilometer, the inlet channel width is 100 µm, and the outlet channel width is 100 µm with a 50 µm contraction. The channels are made of poly(dimethylsiloxane) (PDMS) and bonded to a thin sheet of PDMS by surface plasma treatment in air.

The fluids are mineral oil (viscosity $\nu = 34.5$ cSt at 40°C and density $\rho = 880$ kg/m^3) and deionized water for the continuous and dispersed phases respectively. The interfacial tension is estimated to be $\gamma = O(10)$ mN/m from published data for surfactant-free mineral oil and water systems [18]. We assume that all of the fluid properties remain constant at their established values at room temperature.

The fluids are driven two ways: by static pressure and by a syringe pump. These methods are distinct since the use of specified pressures sets up corresponding flow rates of the two phases while a syringe pump maintains a specified volume flow rate by appropriate adjustment of the force applied to the syringe, which changes the pressure in the fluid. Although in a single-phase flow these two types of experiment would be equivalent, the same is not true in a multiphase flow since the shape of the fluid-fluid interface, which is determined by the flow itself, complicates the relationship between the pressure drop and the flow rate. In the static

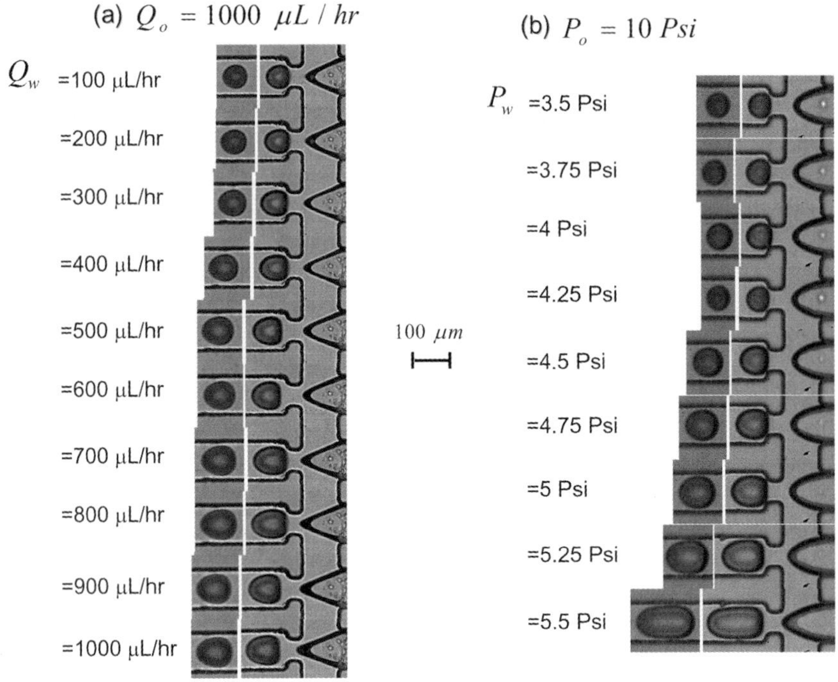

Fig. 2 Drops produced using (a) flow-rate controlled flow-focusing setup where the oil flow-rate is constant and the dispersed phase flow rate varies as indicated; and (b) pressure-controlled setup where the oil inlet pressure is constant and the dispersed phase inlet pressure varies as indicated.

pressure pumping approach the fluids are placed in syringe tubes and the air pressure above the fluid is regulated. The pressure is controlled by precision regulators (Bellofram Type 10) with a sensitivity of 5 mPsi. The pressure is measured using a test gauge (Wika) with a resolution of 150 mPsi. All pressures are reported relative to atmosphere. The syringes are connected to their respective inputs of either the dispersed or the continuous phase channel. The syringe approach for fluid delivery requires the choice of a tube diameter and then uses standard motor-driven pumping to achieve a given flow rate.

The images are obtained using high-speed video at frames rates $O(1-10\ \text{kHz})$ as needed depending on the experiment and drop speeds are determined by tracking the center of mass of individual drops using commercially available software.

14.3 Results

14.3.1 Qualitative differences in flow-rate *versus* pressure-controlled experiments

We first contrast the qualitative features of drop production using flow-rate controlled (syringe pumping) *versus* pressure-controlled flow. Typical results are shown side by side in Fig. 2. A direct qualitative comparison is made by visual inspection of the change in the drop size as either the flow rate of the dispersed water phase Q_w is increased for fixed continuous phase flow rate Q_o of the oil phase (Fig. 2a) or as the inlet pressure of the dispersed phase fluid P_w is increased for fixed inlet pressure of the continuous oil phase P_o. Note: All pressures reported are gauge so that they are relative to atmosphere. The pictures are taken one frame after the liquid thread at the focusing nozzle breaks into a drop. The pictures of drops produced using flow-rate driven pumping show a striking difference to their pressure-driven counterparts. The most notable difference is that as the dispersed phase pressure is increased (by not quite a factor of two) the drop size increases significantly. This response is not true for the flow-rate driven experiments where the drop sizes only change slightly as the water flow rate is increased by a factor of 10.

14.3.2 Quantitative measurements

In a typical experiment drops of a certain size are formed at a given frequency f and move down the channel at a speed U. The frequency f of drop production is measured by direct counting from high-speed video, and the speed U is measured by tracking the drop center of mass for a short distance down the channel. The photos shown in Fig. 2 gave a qualitative indication of the drop size, and we will return to this topic below. Now we consider contrasting the speeds of the drops in the flow-rate driven *versus* the pressure-driven approaches.

The drop speeds *versus* the flow-rate or pressure ratio for different conditions in the continuous phase are shown in Fig. 3. In this figure and all of the other figures below, we use closed symbols to represent the case where the continuous phase flow parameter is held constant, while the dispersed phase flow parameter is increased until a liquid jet is produced and drops no longer form at the orifice. The open symbols are the opposite experiment, where a liquid jet is initially formed at higher values of the operating conditions and then the dispersed phase flow parameter is decreased until the dispersed phase fluid no longer forms drops. In all these experiments we expect a small difference in velocity between the speed of the drops and the average velocity of the continuous phase due to hydrodynamics of the two-phase flow; we have not, however, tried to measure this slip velocity systematically. The error bars are typically the symbol size or smaller. In Fig. 3, for flow-rate driven pumping the drop speeds increase approximately linearly with Q_w. In contrast, for pressure-driven pumping there is a minimum pressure ratio, or pres-

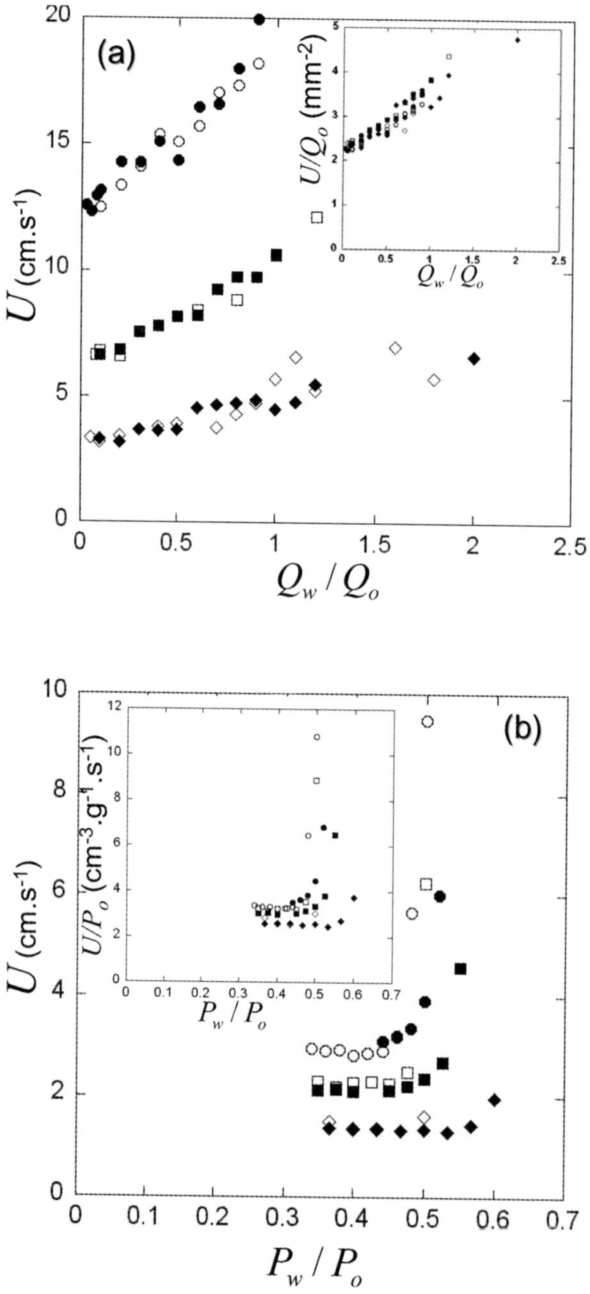

Fig. 3 Measured velocity U *versus* (a) flow-rate where o, ● $Q_o = 2000\ \mu L/h$, □, ■ $Q_o = 1000\ \mu L/h$ and (◇, ◆ $Q_o = 500\ \mu L/h$ and (b) inlet pressure ratio where o, ● P_o = 12.5 Psi, □, ■ P_o = 10 Psi and ◇, ◆ P_o = 7.5 Psi. The open and closed symbols are for experiments where the dispersed phase fluid flow parameter is either increased or decreased respectively.

sure of the water phase, below which no drops are formed, while above this critical pressure the speeds are approximately independent of the pressure until a higher critical value at which the speeds increase rather rapidly.

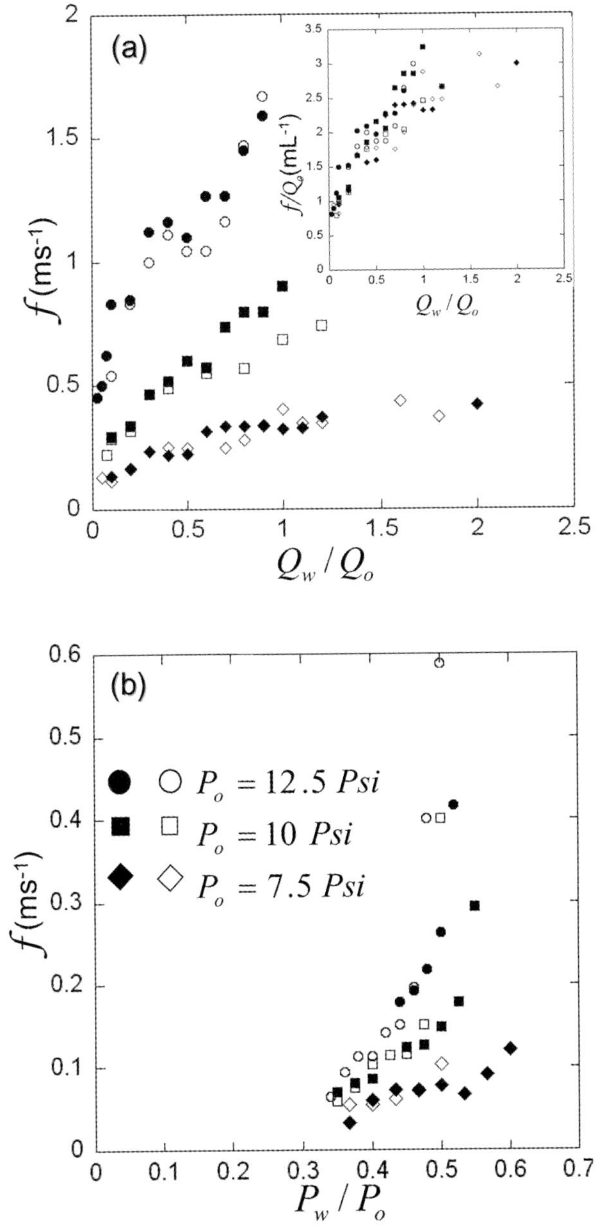

Fig. 4 Frequency f of drop formation *versus* (a) flow rate where o, ● Q_o = 2000 μL/h, □, ■ Q_o = 1000 μL/h and ◇, ◆ Q_o = 500 μL/h and (b) inlet pressure ratio. The open and closed symbols are for experiments where the dispersed phase fluid flow parameter is either increased or decreased, respectively.

The corresponding results for the frequency of drop formation for the two methods of flow control are shown in Fig. 4. The results are reported as the number of drops per millisecond. Again, the error bars are typically the symbol size or smaller. For the conditions used in our experiments there are from zero up to more than 1000 drops formed per second for the flow-rate driven experiments. For the pressure-driven experiments the frequencies are about one-third to one-half as large. In both sets of experiments, the frequencies increase approximately linearly as the independent variable is increased. Also, for all of the experiments shown in Figs. 3 and 4 there is essentially no difference to the results upon either increasing or decreasing the independent variables. For the flow-rate controlled experiments, we have also plotted the frequencies normalized by the external flow rate, and as shown in the inset of Fig. 3 this effectively collapses all of the data.

We next provide data on the relative spacing d between the drops as a function of the speed and frequency of drop formation. If the drop spacing is sufficiently large relative to the channel width we expect the train of drops to be stable, though there is evidence for instabilities in the motion of a train of drops when the drops are too closely spaced (S. Quake, private communication). We expect $fd = U$ *on* geometric grounds and have independently measured all three quantities, d, f, and U. In Fig. 5 we present a plot of fd *versus* U in dimensionless terms by essentially converting to an equivalent Reynolds number (*e.g.*, Ua/ν, where $2a$ is the width of the channel). As expected the plot is nearly linear, independent of the flow control parameter (inlet pressure or flow-rate); we believe the small difference of the slope from unity reflects experimental errors in the independently measured quantities. It is also now clear that the Reynolds numbers Ua/ν are all less than one for all of our experiments so viscous effects should be more important than inertial effects in the dynamical processes.

We next consider in more detail quantitative aspects of the variation of the drop size, l, which, because of the rectangular shape of the channel, is convenient to report as the end-to-end distance. The results are normalized by the width of the orifice a, which is half the width of the channel for all of the experiments reported

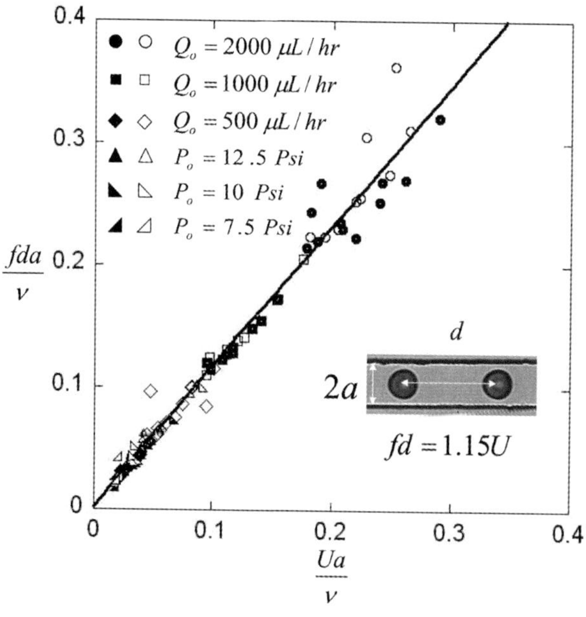

Fig. 5 Frequency f of drop formation times the distance between drops d *versus* measured drop velocity U. Both parameters are made dimensionless by scaling with ν/a.

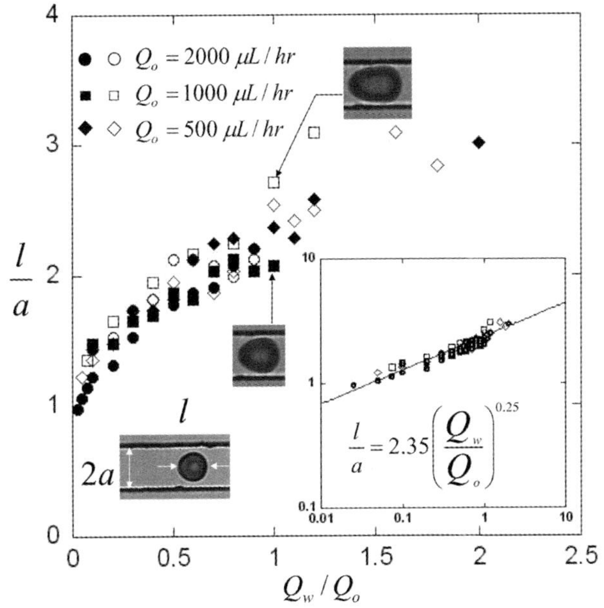

Fig. 6 Dimensionless drop length l/a *versus* flow rate ratio Q_w/Q_o. The inset shows same data using log-log coordinates, where the solid line has slope 0.25.

here. In Fig. 6 the drop sizes are reported as a function of the flow-rate ratio Q_w/Q_o. The corresponding drop sizes as a function of the pressure ratio P_w/P_o are reported in Fig. 8. Comparisons are further shown by plotting the best fit line through log-log plots of the scaled data. This is not meant to suggest a linear relationship but rather to show the difference in trends between the two experiments.

In Fig. 6, the drop size as a function of the flow-rate ratio Q_w/Q_o shows very little variation even when the oil flow rate is varied. At the lowest flow-rate ratios the drop sizes are nearly on the order of the flow-focusing orifice size, $l/a \approx 1$, which indicates the geometric control of drop size common to many microfluidic experiments. The inset shows the same data on a log-log plot with more than a factor of ten variation in Q_w/Q_o, which indicates an approximate power-law relationship, $l/a \propto (Q_w/Q_o)^{0.25}$, obtained by a best fit of the data.

We continue with the data corresponding to the flow-rate driven experiments in Fig. 7a where we plot the distance, d, between the center of mass of two drops, normalized by the orifice size a, *versus* the flow rate ratio Q_w/Q_o. At low flow-rate ratios the distance between drops is large and this distance decreases rapidly to a nearly constant value of about three times the width of the orifice; with this spacing the train of drops is stable. We then combine the above results by showing in Fig. 7b the relationship between the drop size l scaled by the distance between drops d as a function of the flow-rate ratio. When plotted with log-log coordinates, as shown in the inset, the data again appears to be reasonably fit with a power-law relation. Unlike the data scaled by the orifice width, the asymptotic value of $l/d \approx 1$ is reached at $(Q_w/Q_o)_{max} \approx 1.5$.

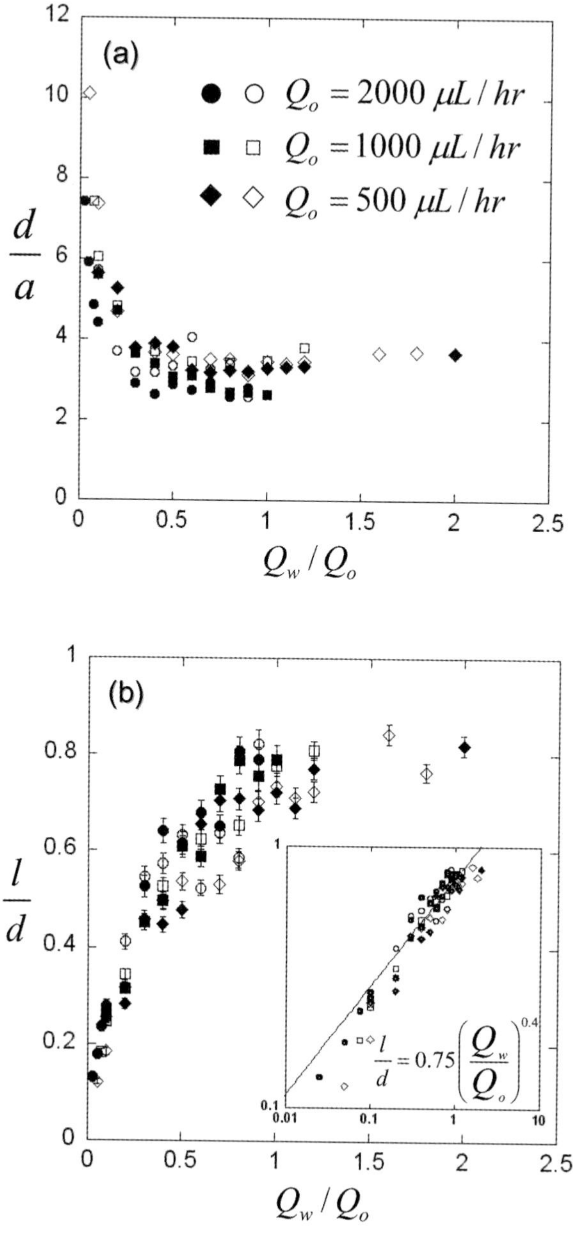

Fig. 7 (a) Distance *d* between the center of mass of two consecutive drops *versus* the flow-rate ratio. (b) Dimensionless drop length *l/a versus* flow-rate ratio. The inset shows the same data plotted in log-log coordinates, where the solid line has slope 0.4, which appears to fit the data over two decades of the flow rate ratio Q_w/Q_o.

The pressure-driven flow rate experiments have some similarities to the flow-rate driven results, though as we shall now see there are both quantitative and qualitative differences. In Fig. 8 we begin with the data for the drop size l/a as a function of the pressure ratio, P_w/P_o. As already indicated above, there is a minimum pressure ratio $(P_w/P_o)_{min} \approx 0.3$ below which the dispersed phase fluid does not penetrate the

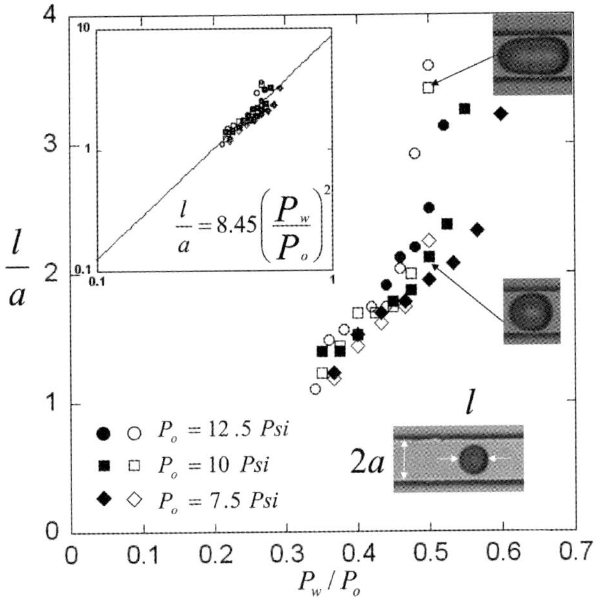

Fig. 8 Dimensionless drop length l/a *versus* inlet pressure ratio P_w/P_o. The inset shows the same data using log-log coordinates, where the solid line has slope 2.

continuous phase. The minimum drop size for each curve is about the orifice size ($\approx$ 50 μm), which indicates the strong geometry dependence in these systems. When plotted in log-log coordinates, as shown in the inset though now there is only a small variation in P_w/P_o, again there appears to be a reasonable fit with a power-law relation, $l/a \propto (P_w/P_o)^2$.

In Fig. 9 we show data corresponding to the inlet pressure-controlled experiments. In Fig. 9a we plot the distance, *d*, between the center of mass of two drops, normalized by the orifice size *a*, *versus* the inlet pressure ratio P_w/P_o. At low inlet pressure ratios $P_w/P_o \rightarrow (P_w/P_o)_{min}$ the distance between drops is large and this distance decreases rapidly to a nearly constant value of a little over three times the orifice width, as observed in the flow-rate control experiments. As with the flow-rate driven experiment we combine the results by showing in Fig. 9b the relationship between the drop size *l* scaled by the distance between drops *d* as a function of the inlet pressure ratio. When plotted in log-log coordinates, as shown in the inset with only a slight variation in P_w/P_o, there again appears to be a reasonable fit with a power-law relation, $l/d \propto (P_w/P_o)^3$. Unlike the data scaled by the orifice width, the asymptotic value of $l/d \approx 1$ is reached at $(P_w/P_o)_{max} \approx 0.55$.

14.4
Discussion

In this section we present some ideas for rationalizing the experimental results reported above. To the best of our knowledge there are no general quantitative theories for two-phase flow in a geometry such as the flow-focusing configuration.

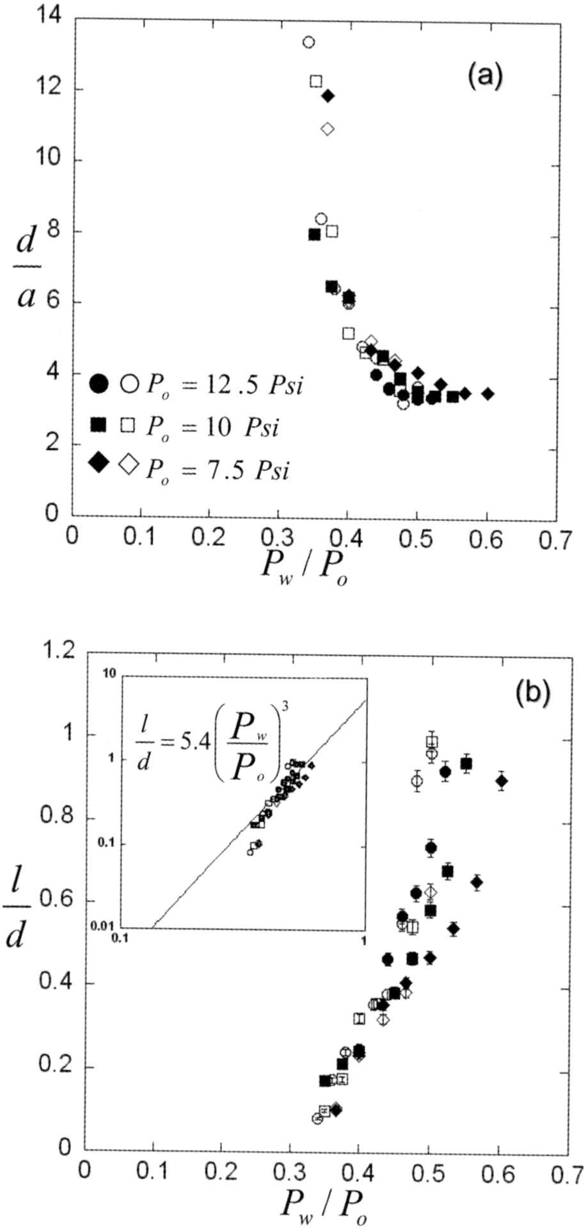

Fig. 9 (a) Distance between the center of mass of two consecutive drops *versus* inlet pressure ratio. (b) Dimensionless drop length *l/a versus* flow-rate ratio. The inset shows the same data plotted in log-log coordinates, where the solid line has slope 3, which appears the fit the data over two decades of the flow rate ratio P_w/P_o.

In the special case that a very narrow thread is formed at and downstream of the orifice, then Ganan-Calvo *et al.* [2] have given quantitative models that are in good agreement with the measurements. All the results presented here are not in this regime. Furthermore, in two-phase flows, the viscosity ratio $\lambda = \mu_w/\mu_o$, where μ_w, μ_o are the dispersed and continuous phase fluid dynamic viscosity, between the drops is a parameter and our experiments have only considered one viscosity ratio, while varying other parameters and considering two methods of flow control.

14.4.1
The role of flow control: flow-rate *versus* pressure

Now we discuss a possible mechanism for the difference in flow-rate and inlet pressure driven droplet production seen in our experiments. To illustrate we qualitatively relate the global transport of momentum for the dispersed phase with the dimensionless expression $P_w \propto g(a,\ P_o, \gamma,\ \mu_w, \mu_o)Q$ where the function $g(a,\ P_o, \gamma,\ \mu_w, \mu_o)$ is the sum of the external resistances that depend on geometry a, external phase fluid pressure P_o, surface tension γ and absolute viscosity of the two fluids μ_w and μ_o.

Most significantly, for the pressure-controlled experiments there is a minimum water pressure below which no drop formation is observed. Just below this pressure the continuous phase fluid pressure is too high for the dispersed phase fluid to penetrate into the main channel after the contraction. But the inlet pressure for the dispersed phase fluid is not zero and there must be another force to create the static or zero flow rate dispersed phase flow. At pressures just below this critical pressure $(P_w/P_o)_{min}$, the external flow exerts shear and normal stresses on the liquid-liquid interface that are greater than the forces exerted by the internal fluid, and a static flow situation is observed at the orifice for the internal fluid where a static spherical cap of fluid is formed, which suggests the significant role of surface tension.

The pressure distribution can have a substantial influence on penetration into the main channel of the spherical cap formed at the orifice. To illustrate this point we use a microscopic force balance to rationalize what happens at the liquid-liquid interface. The boundary conditions for the interface in dimensionless form are written as

$$\Delta p + \tau = Ca^{-1}\bar{\kappa} \qquad (1)$$

Here, Δp is the pressure difference between the dispersed and continuous phase, $\tau = \mu \mathbf{n} \cdot \nabla \mathbf{u} \cdot \mathbf{n}$ is the viscous stress components where $\mathbf{n}$ is a unit vector pointing normal to the interface, $\mathbf{u}$ is the velocity and ∇ is the gradient operator. $\bar{\kappa}$ is the mean curvature along the interface where $Ca = \mu U/\gamma$ is the capillary number.

We estimate the capillary number for our pressure-driven system to be $Ca \approx 0.1 - 1$, based on a drop velocity of $U \sim O(1 - 10\ \text{cm/s})$, indicating the mean curvature term in Eq. (1) is at least of the same order as the fluid pressure and viscous stresses. As a droplet forms near the orifice the volume increases which indicates a decrease in the mean curvature until the outer fluid pressure and viscous forces are high enough to break the drop.

This suggests that for a system where the pressure is specified the dominant contributions to the resistance will come from surface tension, not viscous forces. This may or may not be the case for the flow-rate driven experiments. In either situation, to fully understand this part of the problem we would need to know more about the relationship between viscous stress and surface forces in confined geometries.

Looking further downstream, it is well known in the viscous flow literature that small drops moving in circular or rectangular channels move at nearly the same speed as the external phase fluid. Whether they lag or lead the flow depends on geometry in a very complicated way that has not been completely characterized, but in any event the velocity difference between the two phases is proportional to a small power of the external phase capillary number, *i.e.*, to $Ca^{1/3} = (\mu U/\gamma)^{1/3}$, when the capillary number is small [19]. In our case $\mu U/\gamma \approx 0.1 - 1$, depending on how the flow is driven, so existing theories may yet be useful for providing quantitative insight. Nevertheless the results presented show strong evidence that varying the type of boundary conditions, *i.e.*, fixing fluid flow rate or pressure, when performing experiments and analysis for two-phase fluid systems can produce varying results.

14.5
Conclusions

We have presented results of microfluidic experiments conducted to study micrometer-sized droplet production using a flow-focusing geometry. We study droplet production using either the flow-rate ratio or the inlet pressure ratio as the flow-control parameter. In each experiment the drop size l, velocity U and production frequency f are measured and compared for the different flow-control parameters. Perhaps surprisingly, there are significant differences between these two methods of flow control. The minimum drop size in each experiment is on the order of the flow-focusing contraction width, and illustrates one aspect of the geometric control possible if the results are to be scaled down (or up). The transition in drop size as we vary the flow-control parameter contrasts sharply in the two distinct two-phase flow experiments. It is this distinct difference between the two different manners of flow control used here, which has not been previously noted, that is one of the major conclusions of the present paper.

In each set of experiments the data range for the drop size is qualitatively similar when plotted versus the ratio of the flow parameter, *i.e.*, Q_w/Q_o for the flow-rate driven case and P_w/P_o for the inlet pressure controlled case where w and o denote the dispersed (water) and continuous (mineral oil) phase fluid, respectively. Similarly the maximum value for the onset of a jet which is easily determined using the dynamic length scale $d = U/f$ is the same in both sets of experiments, *i.e.*, the normalized distance d/l between the center of mass of twoo consecutive drops which can never be less than 1. But a power-law fit through the data for the drop size scaled by either a or d *versus* the flow control parameter ratio shows quantitative differences, in spite of the viscous (low-Reynolds number) flow conditions where pressure gradient and average velocity are expected to be proportional. These results suggest a fundamental difference in drop break-up between the two types of experiments, and the influence of surface tension on this free-surface flow is likely to be the origin of the differences.

Droplets rapidly made and manipulated in microfluidic devices have been used in creative ways for chemical and biological applications, such as the measurement of reaction rates and for identifying conditions for protein crystallization [20]. Almost certainly, additional uses of droplets as isolated containers, small reactors, chemical delivery agents, templates for interfacial assembly, *etc.*, will be identified and demonstrated. In such cases, the method for the control of the drop formation process must be chosen and the results reported here contrasting volume flow rate and pressure control should then be of interest.

We thank Unilever Research and the Harvard MRSEC (DMR-0213805) for support of this research. We also thank F. Jousse and colleagues at Unilever and Harvard for helpful conversations. We thank D. Link for helpful feedback on a draft of the paper.

14.6 References

[1] Gañán-Calvo, A. M., *Phys. Rev. Lett.* 1998, *80*, 285–288.

[2] Gañán-Calvo, A. M., Gordillo, J. M., *Phys. Rev. Lett.* 2001, *87*, 274501.

[3] Anna, S. L., Bontoux, N., Stone, H. A., *Appl. Phys. Lett.* 2003, *82*, 364–366.

[4] Link, D. R., Anna, S. L., Weitz, D. A., Stone, H. A., *Phys. Rev. Lett.* 2004, *92*, 054503.

[5] Garstecki, P., Stone, H. A., Whitesides, G. M., *Appl. Phys. Lett.* 2004, *85*, 2649–2651.

[6] Thorsen, T., Roberts, R. W., Arnold, F. H., Quake, S. R., *Phys. Rev. Lett.* 2001, *86*, 4163–4166.

[7] Zangmeister, R. A., Tarlov, M. J., *Anal. Chem.* 2004, *76*, 3655.

[8] Braun, D., Libchaber, A., *Phys. Rev. Lett.* 2002, *89*, 188103.

[9] Burns, M., Mastrangelo, C. H., Sammarco, T. S., Man, F. P., Webster, J. R., *et al.*, *PNAS*, 1996, *93*, 5556–5561.

[10] Okushima, S., Nisisako, T., Torii, T., Higuchi, T., *Langmuir*, 2004, *20*, 9905–9908.

[11] Utada, A., Lorenceau, E., Link, D. R., Kaplan, P. D., Stone, H. A., Weitz, D. A., *Science*, 2005, *308*, 537–541.

[12] Jahn, A., Vreeland, W. N., Gaitan, M., Locascio, L. E., *J. Am. Chem. Soc.* 2004, *126*, 2674–2675.

[13] Shiu, J.-Y., Kuo, C.-W., Chen, P., *J. Am. Chem. Soc.* 2004, *126*, 8096.

[14] Thorsen, T., Maerkl, S. J., Quake, S. R., *Science*, 2002, *298*, 580.

[15] Cho, B. S., Schuster, T. G., Zhu, X., Chang, D., Smith, G. D., Takayama, S., *Anal. Chem.* 2003, *75*, 1671–1675.

[16] Groisman, A., Enzelberger, M., Quake, S. R., *Science* 2003, *300*, 955–958.

[17] McDonald, J. C., Duffy, D. C., Anderson, J. R., Chiu, D. T., Wu, H., Schueller, O. J. A., Whitesides, G. M., *Electrophoresis* 2000, *21*, 27–40.

[18] Cohen, I., Nagel, S. R., *Phys. Rev. Lett.* 2002, *88*, 074501.

[19] Stone, H. A., Stroock, A. D., Ajdari, A., *Annu. Rev. Fluid Mech.* 2004, *36*, 381–411.

[20] Zheng, B., Roach, L. S., Ismagilov, R. F., *J. Am. Chem. Soc.* 2003, *125*, 11170–11171.

15
Aligning fast alternating current electroosmotic flow fields and characteristic frequencies with dielectrophoretic traps to achieve rapid bacteria detection*

Zachary Gagnon, Hsueh-Chia Chang

Tailor-designed alternating current electroosmotic (AC-EO) stagnation flows are used to convect bioparticles globally from a bulk solution to localized dielectrophoretic (DEP) traps that are aligned at the flow stagnation points. The multiscale trap, with a typical trapping time of seconds for a dilute 70 μL volume of 10^3 particles *per* cc sample, is several orders of magnitude faster than conventional DEP traps and earlier AC-EO traps with parallel, castellated, or finger electrodes. A novel serpentine wire capable of sustaining a high voltage, up to 2500 V_{RMS}, without causing excessive heat dissipation or Faradaic reaction in strong electrolytes is fabricated to produce the strong AC-EO flow with two separated stagnation lines, one aligned with the field minimum and one with the field maximum. The continuous wire design allows a large applied voltage without inducing Faradaic electrode reactions. Particles are trapped within seconds at one of the traps depending on whether they suffer negative or positive DEP. The particles can also be rapidly released from their respective traps by varying the frequency of the applied AC field below particle-distinct cross-over frequencies. Zwitterion addition to the buffer allows further geometric and frequency alignments of the AC-EO and DEP motions. The same device hence allows fast trapping, detection, sorting, and characterization on a sample with realistic conductivity, volume, and bacteria count.

15.1
Introduction

15.1.1
General aspects

In the last decade, the advent of microfluidic research has spawned a new array of lab-on-a-chip technologies operating at length scales typically on the order of tens

* Originally published in Electrophoresis 2005, 26, 3725–3737

Microfluidic Applications in Biology. Edited by Niels Lion, Joël S. Rossier, and Hubert H. Girault

ISBN-10: 3-527-31761-9

of microns. Indeed, much time has been spent on developing the essentials, such as micropumps, micromixers, microvalves, and microchannels, for an effective portable microfluidic kit. However, there has been relatively little success in developing portable technologies that can rapidly detect, distinguish, and analyze dilute solutions of bioparticles such as bacteria or viruses.

Current bioparticle analysis techniques typically rely on labor-intensive culturing or PCR amplification to first increase the concentration of a dilute sample. Culturing is then routinely followed by a relatively fast and accurate assay involving fluorescent antibodies, magnetic beads, or fluorescent nanoprobes. While such tests are usually quick, taking several minutes, they are still limited by the time and effort required, usually 1–7 days, to culture or amplify a dilute sample. Typical bacteria counts in medical and environmental samples are on the order of only 100 colony-forming units (CFU) *per* cc, while the present benchmark required for lab analysis is no less than 10 000 CFU/cc [1]. Therefore, significant improvements in detection signal amplification are required before it becomes possible to develop a truly portable and rapid microfluidic bioparticle detection kit.

One possible means of amplifying the bacteria or virus signal is to concentrate the bioparticles at a specific location in order to magnify the fluorescent intensity or electrochemical signal of the sample. Over the past 10 years there has been significant interest in utilizing dielectrophoresis (DEP) forces generated by microelectrodes in order to capture and concentrate charged bioparticles in suspension. DEP has been successfully used to capture a range of bioparticles such as viruses, DNA, and proteins [2–9]; however, there are still many disadvantages in the current DEP trapping techniques. Many of these authors report relatively fast bioparticle trapping times, on the order of a few seconds, however their technique typically involves placing cells in the vicinity of the trap, where upon activation, they are immediately attracted to a local DEP trap and concentrated. If the sample is highly dilute ($<$ 1000 CFU/cc) however, or if one has a large volume of fluid to process, this current technique will usually lead to infeasible processing times.

Limitations in processing times stem from the fact that the DEP velocity of a particle is small, and can be shown to scale quadratically with particle radius and linearly with the applied voltage. The quadratic dependence on the particle radius and practical limitations on the applied voltage render the particle velocity produced by a DEP force miniscule, typically on the order of 10 μm/s for bacteria and 1 μm/s for viruses, and usually leads to a concentration time on the order of hours. Also, the field gradient necessary to drive DEP motion can only be achieved with relatively narrow interdigitated electrodes whose field penetration depth is also limited by the electrode width. As a result, DEP channels are usually less than 50 μm in transverse dimension. Therefore, slow capture and small transverse dimensions are responsible for extremely low throughputs in current continuous flow kits.

Due to the above limitations, there is considerable interest in replacing or augmenting DEP traps with alternating current electroosmotic (AC-EO) flows to increase the volume of fluid that the trap can act on. Such flows are due to field-induced polarization on the AC electrode surface and are not dependent on the particle dimension. They convect the particles by viscous drag and hence can endow the particle with the flow velocity. As such, they are, at least in theory, of longer range and can impart a higher particle velocity than DEP designs [8, 10]. Green *et al.* [11] have investigated in detail microfluidic flows on the surface of a coplanar electrode pair, documented AC-EO flow rates as a function of electrolyte conductivity, signal frequency and potential, and position on the electrode surface.

However, convection by AC-EO alone is not sufficient to trap bioparticles. The particle trajectory would be identical to the AC-EO streamline and due to the volume-conservation property of incompressible flow, none of the stagnation points of the flow field would be attracting – they are either hyperbolic saddle points or elliptic centers. The electrode surface itself is also a slip plane due to the AC-EO slip velocity, although saddle-point stagnation lines or points could appear. The optimal design would be to impose yet another field on the particles at the electrode stagnation points or lines, where the viscous drag is weakest, such that the stagnation points or lines become attracting and can perform as a multiscale bioparticle trap. Such a local field near the stagnation point/line can be a DEP field, as in the case of [2–9], or other short-range local particle forces like gravity or magnetic force.

The purpose of this work is to introduce a new wire trapping design and buffer solution selection to enhance bioparticle concentration, detection, and manipulation of dilute solutions of bioparticles. Bioparticles are not locally trapped on disjoint microelectrodes, but rather they are convected from a bulk solution by a pair of asymmetric vortices across a coplanar serpentine wire, with a large voltage drop and surface polarization, to local DEP traps on the substrate where the local particle force fields can be tangentially attracting. In this manner, both diverging and converging stagnation flows are rendered effective, and rapid traps and the device acts on a much larger volume of fluid than existing DEP technology.

The challenge in rapid global trapping is to align the substrate and electrode stagnation lines of an AC-EO flow with that of a local electric field minima or maxima such that bioparticles are now rapidly convected from the bulk solution to the substrate surface where they are trapped in one of the traps by a local negative or positive DEP (n- or p-DEP) force. Rapid particle sorting of live and dead cells by DEP direction is hence achieved. Zwitterion addition to the buffer also enhances the AC-EO stagnation flow and allows one to achieve optimal trapping conditions by changing the dielectric and surface-conducting properties of the strong electrolyte.

When a dilute aqueous suspension of polymer microspheres, or biological cells such as bacteria, is exposed to an AC electric field, electrical forces can act both on the particles and the surrounding fluid. For AC fields, the governing electrical force acting on a particle is DEP. The electrical force acting on the fluid leads to what is commonly referred to as AC electroosmosis.

15.1.2
DEP

The classical DEP theory [12, 13] produces a DEP velocity of the form

$$u_{DEP} = \frac{1}{3\mu}\varepsilon_m r^2 Re[K(\omega)]\nabla|E|^2 \qquad (1)$$

where for a homogeneous spherical particle, $K(\omega) = (\varepsilon_p^* - \varepsilon_m^*)/(\varepsilon_p^* + 2\varepsilon_m^*)$ is the Clausius–Mossotti (CM) factor, and $\varepsilon^* = \varepsilon - \sigma/\omega$ is the complex permittivity which is dependent on σ, the conductivity, and ω, the applied field frequency. The imaginary part is out of phase with the applied field and, to first order, can be experimentally determined by measuring the torque on a particle in electrorotation experiments [5]. However, the real part of the CM factor is in phase with the applied field and describes the particles polarizability and field-induced dipole moment [6].

A more complex polarization model [14] that includes a conducting Stern layer around the particle surface can be shown to modify the effective particle conductivity and permittivity yielding an effective complex particle permittivity in the form

$$\varepsilon_p* = (\varepsilon_p + \varepsilon_{dl}2\lambda/R) - j(4\pi/\omega)(\sigma_p + \sigma_{dl}2\lambda/R) \qquad (2)$$

where λ is the thickness of the stern layer. From the above expressions the charge relaxation time for the particle can be shown to be [14]

$$\tau = \frac{2\varepsilon_p + \varepsilon_m + \varepsilon_{dl}2\lambda/R}{(2\sigma_m + \sigma_p + \sigma_{dl}2\lambda/R)} \qquad (3)$$

It is well known that u_{DEP} changes sign at the cross-over frequency (COF), $1/2\pi\tau$ [4, 13] where the CM factor $K(\omega)$ vanishes. For most bacteria, $Re[K(\omega)]$ is negative for $\omega >$ COF and the particles move towards the low field region (n-DEP). p-DEP occurs for $\omega <$ COF. It is interesting to note that both the particle size, R, and the double layer thickness, $\lambda = (\varepsilon RT/^2 z2Fc_o)^{1/2}$, appear in the COF only if a conducting Stern layer is included. In fact, when the double layer is thin at high concentrations and low permittivities, it can be shown that COF $\sim 2\pi\sigma_m/\varepsilon_m = 2\pi D/\lambda^2$ for $(\sigma_m/\sigma_{dl}) \gg (\lambda/R)$. Conversely, in the limit of $(\sigma_m/\sigma_{dl}) \ll (\lambda/R)$, COF $\sim 2\pi\sigma_m/\varepsilon_m = 2\pi D/\lambda R$ is particle size-dependent.

15.1.3
AC-EO

AC-EO flow is typically produced with two planar parallel microelectrodes in contact with an electrolyte solution. When an AC voltage is placed between these two electrodes an electric field is produced, which interacts with the electrolyte ions, and in the absence of charge injection, or Faradaic reactions, the electrode surface is polarized by counterions in the electrolyte to form a field-induced electrical double layer. Because the double layer is essentially charged like a capacitor, this particular charging mechanism is typically referred to as capacitive charging.

Based on the work by Gonzalez *et al.* [15], the time-averaged slip velocity on the electrode is

$$u_{AC-EO} = -\frac{\varepsilon_m}{4\mu}\frac{\partial}{\partial x}|\phi - V_0|^2 \qquad (4)$$

where ϕ is the value of the potential at any given location above the electrode surface and V_0 is the potential applied to the electrode. The equation represents the effective slip velocity on the electrode surface. If values of the electric field are known, the bulk fluid velocity can now be solved using the Navier–Stokes equation with Eq. (4) as a boundary condition. The electric field on the electrode surface is frequency-dependent and can be shown to be

$$\sigma\frac{\partial\phi}{\partial y} = i\omega C_{DL}(\phi - V_0) \qquad (5)$$

which is nothing more than a charge balance in the normal direction across the double layer, where $C_{DL} \sim \varepsilon/\lambda$ is the capacitance *per* unit area of the total double layer. The time-averaged electric field can now be solved using the Laplace equation with Eq. (4) as a boundary condition, and using Eq. (5) together with the Navier–Stokes equation the complete hydrodynamic-electrical problem can be solved.

Using typical values of electrolyte conductivity and permittivity ($\sigma = 100$ mS/cm, $\varepsilon_m = 80$), the normalized fluid slip velocity is shown as a function of signal frequency in Fig. 1. It is evident that theory predicts an optimum frequency where a maximum in slip velocity exists. This can be argued from the physical reasoning. Because ion migration to the surface of each electrode requires a finite amount of time, the electrode charging dynamics will have some dependence on the applied AC signal frequency. This dependence can be estimated by simple scaling arguments. The circuit equivalent to the electrode system can be approximated as a double layer capacitor, with charge separation over a length λ, in series with a bulk fluid resistor. The resulting RC charging time for this equivalent resistor-capacitor in series is simply $\lambda L/D$, where L is the electrode separation. Hence, there is an optimum frequency at which one observes a maximum AC-EO flow.

At frequencies below $D/\lambda L$, the half-cycle is long enough such that counterions have enough time to completely saturate the double layer, effectively shielding the electric field from the bulk solution. Additionally, at frequencies above $D/\lambda L$, the counterions do not have enough time to migrate to the electrode surface and form a double layer. Since the time-averaged electrokinetic flow requires both double layer polarization and external field, it must vanish at these two extremes and a maximum AC electrokinetic velocity should occur at a frequency of $D/L\lambda$.

15.1.4
DEP/AC-EO trapping

It is clear from Eq. (4) that a stagnation line exists in the flow field when the tangential electric field vanishes. Ben and Chang [16] have shown that, for capacitive charging on two symmetric planar electrodes, this stagnation line occurs at the

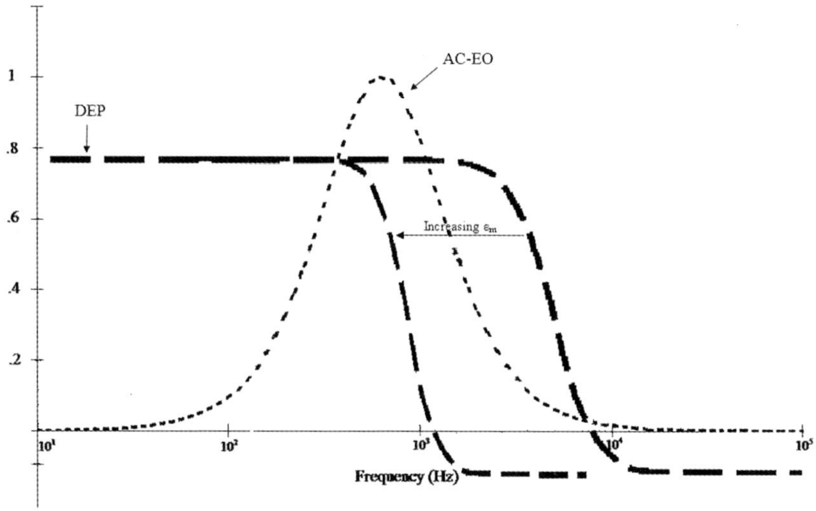

Fig. 1 Real part of the CM factor, representing DEP direction, superimposed on AC-EO slip velocity. Electrolyte permittivity increase shifts Re[$K(\omega)$] to the left into a region of stronger AC-EO flow.

center of the electrode at frequencies lower than $D/L\lambda$. For frequencies near or higher than $D/L\lambda$, the stagnation line shifts toward $1/\sqrt{2}$ of the width as measured from the inner electrode edge.

It is evident that the AC-EO slip velocity is larger than the DEP velocity by a factor of $(L/a)^2$. Since the electric field on any electrode is higher than that of the surrounding medium, and since most particles are of low permittivity compared to a surrounding aqueous solution, they exhibit a p-DEP force at frequencies near $D/L\lambda$, when the capacitive charging electrokinetic flow is most robust. Therefore, because strong electroosmotic convection forces exist under conditions in which particles exhibit p-DEP, and the flow stagnation lines are aligned with that of a high field region, one would expect particles to be convected from the bulk electrolyte and trapped at the converging stagnation line on the high field electrode surface. Additionally, the field between the two electrodes is generally weak and hence represents an n-DEP trap. Hence, if operating at frequencies above the COF of the particle, and assuming strong electroosmotic flow still exists, one would also expect the particles to be trapped in the gap at high frequencies when the particles suffer from n-DEP.

15.1.5
Concentration/separation requirements

It is important to note that DEP and AC-EO typically suffer from widely differing charge relaxation times, and it is for this reason that rapid particle separation or rapid n-DEP trapping, on the order of seconds, has not been feasible. The DEP COF of a particle is directly related to the charge relaxation time of the particle, which is usually much shorter than the charge relaxation time for AC-EO flow. In fact, the typical COFs for common types of particles and bacteria exist where AC-EO flow has tended toward zero [4, 11]. Rapid particle concentration requires both strong AC convection and particles operating under p-DEP or n-DEP. Because most particles exhibit n-DEP at high frequencies where electroosmotic flow is weak, convection-enhanced particle trapping is typically not possible when particles are operating under n-DEP. This detail places limits on current DEP traps. The majority of biological samples used today contain multiple species of bacteria, cells, and waste products. Rapid particle concentration is not practical if the collected sample requires hours of careful preparation, or if the collection device simply collects everything in the sample on an electrode. Eventually, some sort of particle separation would be required either before, during, or after particle trapping had occurred. The current problem is that the rapid separation of a multiparticle sample requires strong electroosmotic convection at frequencies where some particles are operating under p-DEP and other exhibit n-DEP, and as explained before, the difference in charge relaxation times between DEP and electroosmotic phenomena in typical electrolyte solutions prevent this.

The most obvious solution to the separation problem is to shift the CM factor to the left, towards the optimum AC-EO frequency, or the converse, to shift the electroosmotic velocity curve to the right. It has been shown that by increasing the conductivity of the electrolyte solution, one can shift the optimum electroosmotic frequency to the right [11], however this usually leads to a decrease in AC-EO slip velocity.

Another possibility is to modify the CM factor such that the COF of a bioparticle is reduced and exists at a frequency where AC-EO is strong. From Eq. (1), it is clear that if one increases the electrolyte permittivity, the Maxwell–Wagner relaxation frequency will decrease. This can be argued physically in that if the electrolyte permittivity is increased, the number of ions that accumulate on the particle surface for the first half-cycle increases. During the second half-cycle, these ions must migrate around the particle in order for the dipole to change direction. However, because the ion density has increased, it now will take a longer amount of time for the ions to equilibrate and the dipole moment to reverse direction, thus leading to an increased relaxation time or a decreased COF. Therefore, one possible way to reduce bioparticle COF is to increase the permittivity of the electrolyte solution.

The relaxation time for AC-EO is also dependent upon the permittivity of the medium in that $\lambda L/D$ can be shown to scale as ε/σ, where σ is the conductivity of the medium. It is clear that an increase in permittivity will shift the optimum electroosmotic frequency further away from the DEP COF. However, while the magnitude of the DEP force acting on a particle is dependent on electrolyte conductivity, the polarizability of the medium depends largely on a solutions permittivity. Therefore, in order to achieve an optimum balance of both DEP and convection forces, one simply needs to increase the electrolyte permittivity, which will decrease a suspending particles COF, and increase the electrolyte conductivity to counteract the increase in electroosmotic relaxation time. As shown in Fig. 1, a desired separation can most rapidly take place when the COF exists at or near the optimum electroosmotic frequency.

Increasing electrolyte permittivity is quite straightforward. It has been shown that the electrical permittivity of a suspending fluid can be increased by adding ionic molecules, or zwitterions, of high polarizability to an aqueous solution. For example, Arnold and Zimmerman [17] have shown that the addition of 2 mol of a glycine peptide to water increases it relative permittivity by ~252, reducing the Maxwell–Wagner relaxation frequency by ~25%.

15.1.6 Device design

It is clear from the above discussion that, while AC-EO traps should function better than DEP traps, the limitation to low voltages and low-conductivity fluids renders them quite ineffective. The low voltage requirement stems from the fact that, for disjoint electrodes, Faradaic reactions will eventually occur at sufficiently high voltages and low frequencies. This is because the field for disjoint electrodes must cross the electrode/electrolyte interface and, with sufficient voltage drop and sufficient reaction time, will induce an electrode reaction. As thin as the double layer is for strong electrolytes, a sufficiently large electric field should still produce strong AC-EO flow. Hence, limitation to weak electrolytes can be alleviated if Faradaic reaction can be avoided such that a high rms voltage can be applied.

In this work, we remove both limitations by discarding the disjoint electrode design. Instead, a continuous serpentine wire is used such that most of the AC current passes through the wire and not the electrolyte. Large voltage drop exists along the wire to produce enormous capacitive charging on the wire surface but without a significant field across the electrode-electrolyte surface. With this design, capacitive charging up to 2500 V_{RMS} can be achieved to produce AC-EO that are orders of magnitude higher than earlier disjoint electrode designs. Bioparticle concentration, separation, and manipulation can now be achieved in large (cc) volume of strong electrolytes.

As shown from a top view in Fig. 2, the microdevice consists of a thin 50/200 nm Ti/Pt wire on an insulated silicon substrate in contact with an aqueous suspension of bioparticles.

Due to the large difference in electrical conductivity between the wire and electrolyte, the applied current will be largely confined to the wire, thereby eliminating any noticeable electrochemical reactions and pH gradients even at high voltages (~ 2500 V_{RMS} AC). There is still a large voltage drop between different segments of

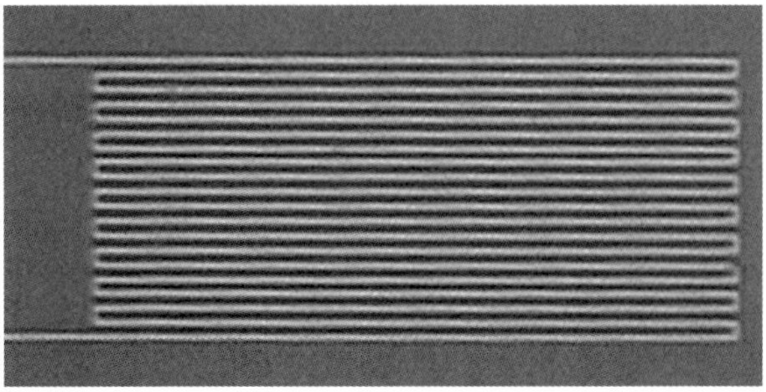

Fig. 2 Fabricated DEP/AC-EO trap.

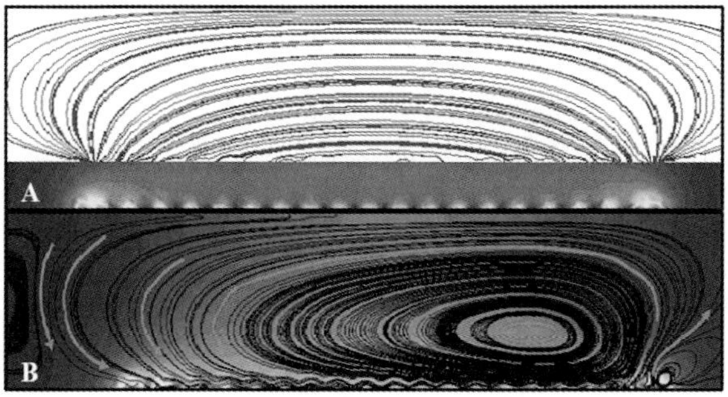

Fig. 3 (A) Side view of the theoretical electric field lines of the device at 5 V_{RMS} and 40 kHz showing the highest values in field is predicted on the outer wires. (B) Theoretical fluid streamlines illustrating two converging stagnation lines, one on right edge of the device and the other on the outer left.

the serpentine wire. However, the Faradaic reaction this intersegment voltage would drive is a high-resistance Faradaic resistor in series with the high-capacitance interfacial Stern layer and double layer capacitors. This Faradaic RC circuit is in parallel with the low-resistance wire. Hence, at the high-frequency employed, much higher than the inverse RC time of the Faradaic resistor, almost all the AC current runs through the wire with little AC Faradaic charge transfer. More importantly, because the serpentine wire can sustain such high voltages, the AC-EO flow is still quite robust for high-conductivity electrolytes.

Although the device still has a large heating power from the current in the wire, Joule heating effects are avoided by fabricating the device atop a heat conducting silicon substrate. It will be shown that heat effects on fluid motion, both through Raleigh–Bernard instability and electrothermal effects, are negligible for this design and that this new configuration allows one to achieve much higher fluid velocities and field strengths than existing electrode geometries.

Following earlier work by Gonzalez *et al.* [15], one can solve the Laplace equation for the electric field using Eq. (5) as a boundary condition. The electroosmotic velocity profile is then solved numerically for a cross-section (cut across wires) of the serpentine wire using Eq. (4) on each wire surface. The magnitude of the electric field and the resulting velocity field are shown in Figs. 3A and B, respectively. Numerical simulations were performed using the FEMLAB finite element software package (COMSOL, Burlington, MA, USA).

From a side view of the device shown in Fig. 3A it is clear that the wires on the outer region of the serpentine have the largest field. Additionally, the velocity profile is anti symmetric with a stagnation point near the two outer wires. The important thing to note, superimposing these two figures, is that a converging stagnation line exists directly above a local field minimum on the right-hand side of the device, as shown in Fig. 4.

The left hand side of the device has a small flow stagnation line, however it is not aligned with a local field minimum. Therefore, if operating under the correct electrolyte and frequency conditions, one would expect bioparticles in the bulk solution to be convected directly to a local n-DEP trap on the right-hand side, collecting much faster than the left hand side where the particles are brought only in the vicinity of a field minimum (~40 μm), but not directly upon it. If operating under p-DEP, the particles should be attracted to the wires with the highest field, as they

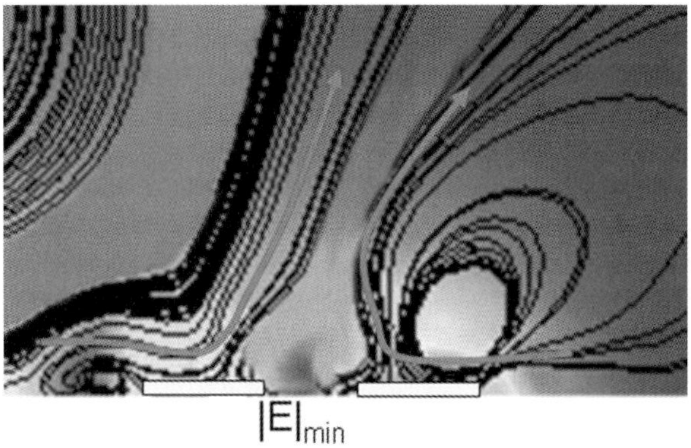

Fig. 4 Fluid streamlines with arrows indicating direction superimposed atop electric field contours for the last two wires on the right side of the device illustrating a stagnation line aligned directly atop a field minimum.

will follow the streamlines and continue to sample the entire device surface. Particle separation can be achieved by noting that the particles suffering n-DEP will be rapidly attracted to the field minimum regions between the wires, while the p-DEP particles will be attracted to the high field wires.

15.2
Materials and methods

Thin-film serpentine platinum wires were fabricated using conventional semiconductor and soft lithography techniques. Briefly, serpentine wire geometries were photopatterned onto a dielectric (SiO_2)-coated silicon wafer. Fifty nanometer titanium/200 nm platinum was then deposited using electron beam evaporation, and the photoresist was then lifted off in an organic solvent to yield thin film wire patterns. The resulting structure consisted of a serpentine structure with wires 35 μm wide and 2500 μm long, arranged in a serpentine structure with a periodicity of 5040 μm, and consisted of ten periods, or 20 parallel wires interconnected at every other end, as shown in Fig. 2.

Experiments were conducted in polymer microchannels which were produced using conventional soft-lithographic techniques. Briefly, microchannel master molds were fabricated using thick SU-8 photoresist (SU8–2075, Microchem, Newton, ME, USA). Uncured polydimethylsiloxane (PDMS) (Sylgard 184, Dow Corning, Midland, MI, USA) was then poured and cured atop the molds, peeled off, and carefully aligned atop the serpentine wire structure. This construction formed a channel that was 500 μm deep, 4000 μm wide, and 3.5 cm long aligned lengthwise above the wire structure. Fluid entrance and exit ports were made in the polymer channel using a 22 gauge syringe needle. Finally, a 300 μm glass capillary was inserted at the entrance port and attached to a syringe, yielding a sample volume of 70 μL. The port and syringe connections were sealed using a quick dry epoxy resin and the syringe placed in a digitally controlled syringe pump. A new polymer channel was used for each experiment. Particle suspensions were then injected into the microchannel *via* a syringe pump. An AC potential was dropped across the

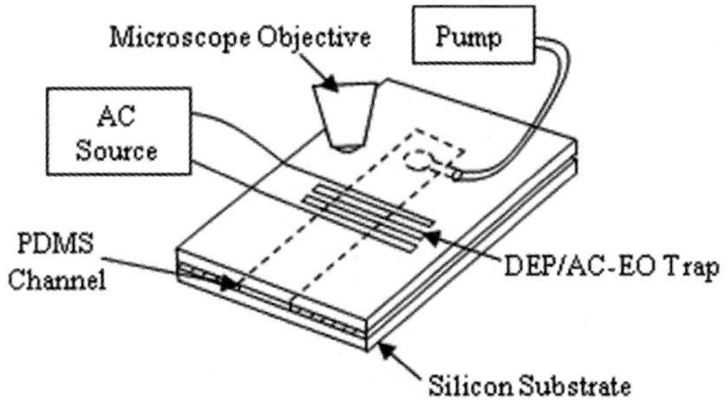

Fig. 5 Experimental setup.

wire structure by a signal generator (Agilent 33220A) connected to an RF amplifier (Powertron Model 250A, Industrial Test Equipment), and the final assembly, as shown in Fig. 5, was then viewed under Fluorescent Microscope for visualization. With reference to the high voltage operation, as mentioned earlier in this work, the power system was further combined with an output transformer (Industrial Test Equipment, P/N 113459–2) capable of outputting up to 2500 VRMS.

Dried bakers yeast cells (Fleishmann Active Dry Yeast) were reconstituted in CO_2-equilibrated deionized (DI) water (12 $M\Omega \times$ cm, pH 6.5), 1 μm fluorescent polystyrene microparticles and 5 m latex microparticles were obtained from Polysciences (Warrington, PA, USA) and diluted to approximately 10^3 particles *per* mL in DI water. Electrolyte permittivity and conductivity was adjusted by adding varying amounts of the zwitterion 6-aminohexanoic acid (AHA) (Sigma-Aldrich, St. Louis, MO, USA) to a prepared polymer suspension. *Escherichia coli* bacteria (F-amp) diluted in DI water to a concentration of ~ 500 CFU/cc was obtained from Scientific Methods.

15.3 Results and discussion

Experiments were conducted in order to verify the enhancing effects that AC-EO has on the trapping of bioparticles. The p-DEP behavior, in the absence of convection forces, of 1 μm polystyrene fluorescent microparticles suspended in pure DI water was first investigated. One micrometer particles were used because by conventional DEP theory, small 1 μm particles exhibit p-DEP behavior at low (< 100 kHz) frequencies. A suspension of particles was injected into the microchannel and an AC potential (5 V_{RMS}) was dropped across the wire at a frequency of 40 kHz. Prior to this experiment, the AC-EO fluid velocity was measured as a function of voltage and applied frequency approximately 15 μm above the wire surface for pure DI water using previously described techniques [11]. As shown in Fig. 6, when 5 V_{RMS} is dropped across the wire at 40 kHz, the fluid velocity is weak at approximately 7 μm/s. It should also be noted that the measured fluid velocity scales quadratically with the applied voltage, while electrothermal flow scales as the applied voltage to the fourth power [18]. Additionally, the channel height is ~ 300 m, giving rise to a Raleigh number of ~ 10, well below the critical value

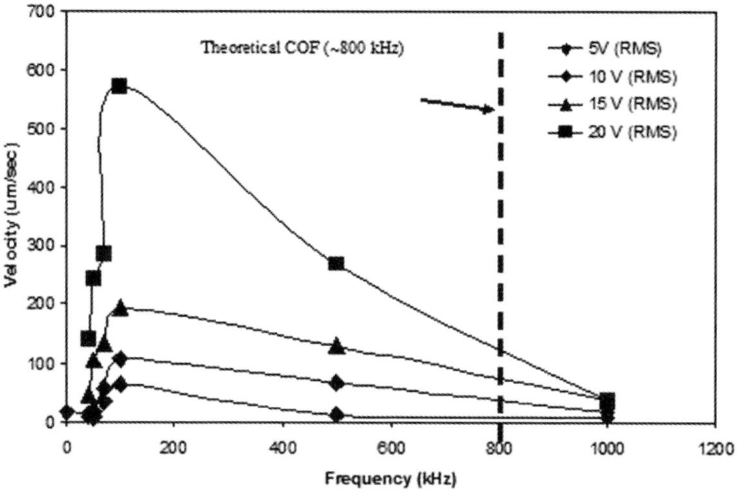

Fig. 6 Measured fluid velocity for DI water as a function of AC frequency for four different applied voltages: 5, 10, 15, and 20 V. Predicted COF for DEP is also indicated. It is clear that the n-DEP trap at $\omega >$ COF would not be aided by the weak AC-EO flow at high frequencies.

required (~ 1707.26) for any flow resulting from a thermal instability [19]. In fact, in order to reach this critical value, one can show that this would require an infeasible temperature gradient on the order of ~15°C/μm. Finally, thermal Joule heating should be uniform along the wire and the vortices they drive are necessarily symmetric in contrast to the highly asymmetric vortex shown in Fig. 3. Therefore, the flow observed experimentally is most likely lateral AC-EO flow.

Fig. 7 shows the optical fluorescent microscopy images of the microparticles patterned by p-DEP. This image took approximately 22 min to form, which is consistent with the trapping time of earlier DEP traps that operate at roughly this voltage. It is obvious that the particles (lighter areas on the device) are experiencing a p-DEP force in that they are attracted to the high field regions of the serpentine wire surface. More importantly, they are highly concentrated on the outer high field wires of the serpentine pattern, thus verifying the electric field calculations as shown earlier in Fig. 3A.

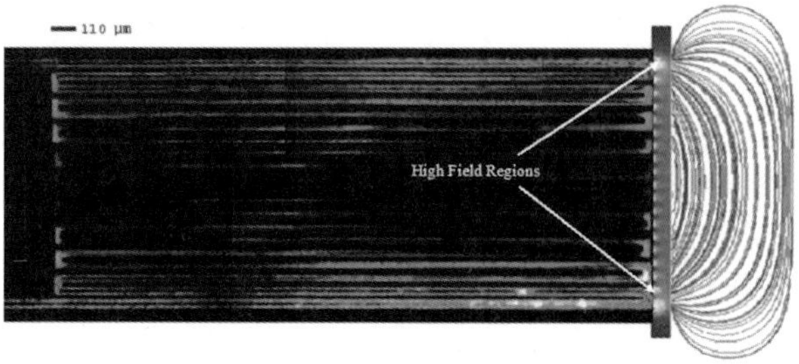

Fig. 7 Image of p-DEP trapping of 1 μm fluorescent particles taken 22 min after device activation at 5 V and 40 kHz, calculated field lines/intensity aligned on far right of device.

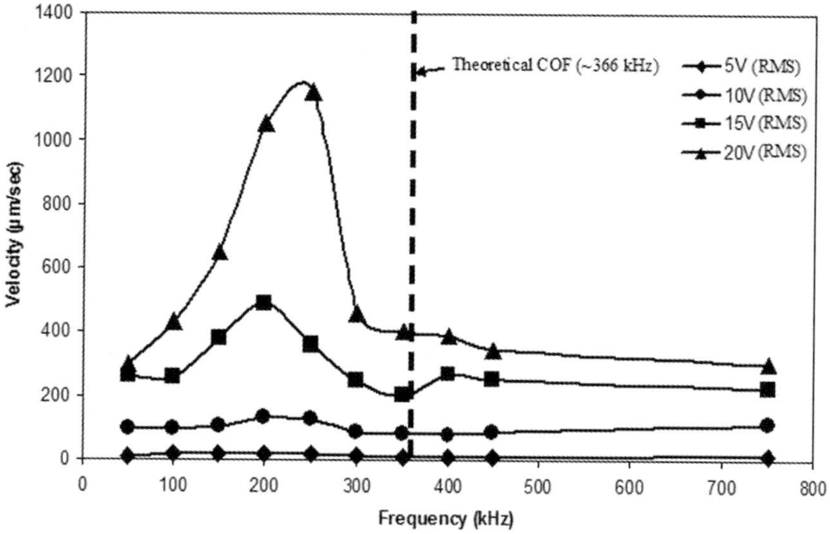

Fig. 8 Measured fluid velocity for 1.5 M AHA solution as a function of AC frequency for four different applied voltages: 5, 10, 15, and 20 V. Zwitterions shift COF to much lower frequencies where the AC-EO velocity is a robust 400 μm/s and can help trap particles at the n-DEP trap.

The n-DEP characteristics of the device in the absence of strong convective forces were also investigated. As mentioned earlier, n-DEP particle collection on the right side of the device should be much faster than conventional n-DEP trapping, without convection, due to the stagnation line that is directly aligned above a local field minimum. A solution of 5 μm latex particles was suspended in DI water and injected into the device. Larger particles were used in this experiment to decrease the particle COF so that n-DEP behavior could be studied. Additionally, a larger voltage was used so that the results of this experiment could be compared to one with zwitterions enhancement. Upon device activation (20 V_{RMS}, 1.2 MHz), particle accumulation in the DEP trap was observed to be slow. There were small amounts of particles collected in n-DEP traps on the device, however these patterns took approximately 30 min to form. From Fig. 6, one can see that at 20 V_{RMS}, the fluid velocity at 1.2 MHz is approximately 40 μm/s, while the calculated DEP velocity can be shown to be approximately 10 μm/s. It is obvious that in order to decrease the time required for collecting particles in the n-DEP traps, one needs to increase the electroosmotic convection fluid velocity or the local field at the stagnation line. We shall do both below by the addition of zwitterions.

A new suspension of 5 μm particles was created using a 1.5 M solution of AHA. The solutions relative permittivity was measured in previous work and shown to be ~ 200 [17], approximately a factor of 2 higher than that of DI water (~ 80). The AC-EO velocity was measured as a function of applied frequency prior to the experiment and is shown in Fig. 8.

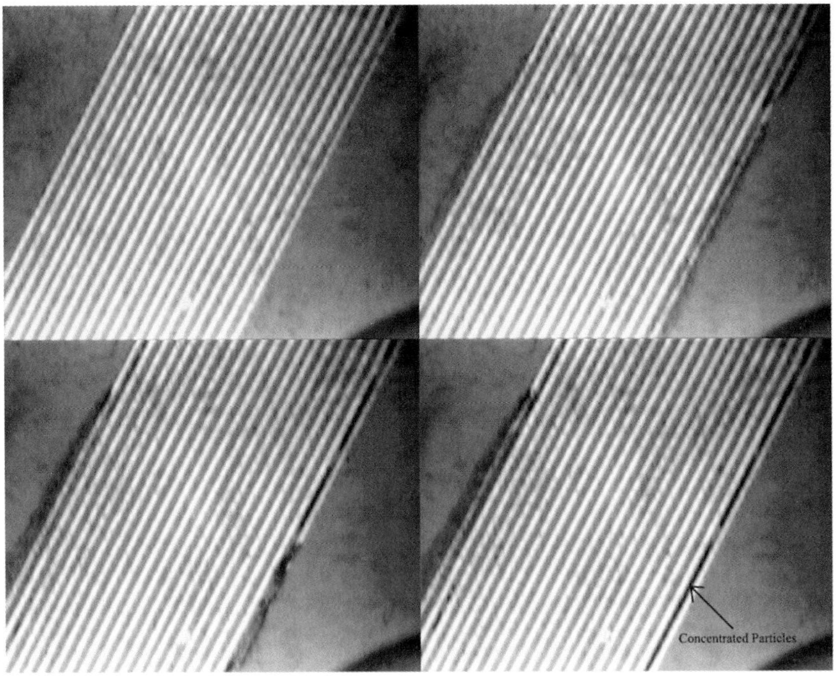

Fig. 9 Images of 5 μm particle collection over a period of 35 s for 20 V_{RMS} and 400 kHz: top left image, 0 s; top right, 10 s; bottom left, 20 s; bottom right, 32 s.

As shown in Fig. 8, the optimal frequency is approximately 200 kHz, a factor of 2 higher than that of DI water, while that of the measured electroosmotic velocity is also a factor of 2 greater. More importantly, n-DEP can now exist at much lower frequencies where significant AC-EO flow (~ 400 μm/s) can be utilized to accelerate particle trapping. The solution was injected into the device, an AC voltage (20 V_{RMS}) was applied, and the frequency was slowly increased until an n-DEP particle collection was observed. At a frequency of 400 kHz, unlike the slow particle trapping in DI water, rapid n-DEP particle trapping in the stagnation aligned negative field region on the right-hand side of the device was observed. As shown in Fig. 9, trapping time was reduced by two orders of magnitude, taking only 32 s for 5 μm particles to arrange in a highly ordered cubic array of approximately 2500 μm in length. The left-hand side of the device also trapped particles, however the time required to form the complete 2500 μm line took approximately 4 min. The 20 V_{RMS} is beyond most of the earlier AC-EO traps with disjoint electrodes due to the appearance of Faradaic reactions. With the addition of zwitterion, the convection flow is further enhanced as the optimal frequency for AC-EO can now be employed and still allow n-DEP trapping. Moreover, with any significant AC-EO convection, the rate-limiting step is probably the trapping speed at the local field and the addition of zwitterion also amplifies the local n-DEP force.

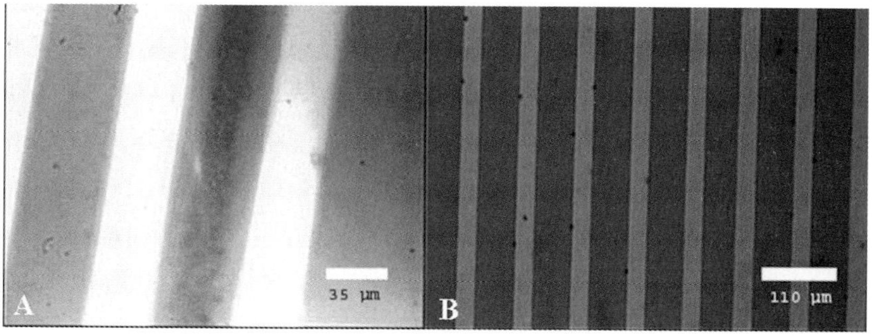

Fig. 10. n-DEP behavior of yeast cells with the addition of zwitterions (A) and no zwitterions added (B) at 20 V_{RMS} and 400 kHz.

The observable decrease in particle collection time and increase in fluid velocity is also affected by the permittivity adjustment of the electrolyte. By adding zwitterion to the solution, the electrolyte permittivity was increased and therefore led to a decrease in the particle COF and placed the n-DEP frequency range of the particle in a range where strong electroosmotic flow existed. Additionally, as shown in Eq. (4), the electroosmotic slip velocity is also proportional to the electrolyte permittivity, and the addition of the zwitterion also led to a factor of 2 increase in the observable electroosmotic velocity. Therefore, by simply increasing the permittivity of the electrolyte solution, the frequency at which a particle experienced an n-DEP forced was reduced, driving the particle into an operating range where increased convection forces existed which in turn convected the particle across the channel gap and directly into an n-DEP trap, reducing the required collection time. This concept is further supported by the fact that the aligned stagnation trap on the right side of the device required only seconds to trap particles, while that of the trap on the left side, 100 µm away from a stagnation line, required a much longer time, taking on the order of several minutes.

It is clear from the above experiments that trapping time can be enhanced significantly with well-aligned AC-EO stagnation lines. Based on the above results, bakers yeast cells were also used to test the trapping concept. As shown in Fig. 10B, cells in pure DI water exhibited a very weak response to the applied voltage (20 V_{RMS}, 400 kHz), and after 15 min, there was a negligible collection. However, as illustrated in Fig. 10A, the addition of a 1.5 M solution of zwitterion to the suspension under identical conditions led to a collection of yeast cells in the stagnation aligned field minimum in approximately 12 s, which is very rapid for the large sample volume of 70 µL.

E. coli bacteria (F-amp) at a concentration of approximately 500 CFU/mL, suspended in an aqueous solution of 1.5 M AHA, were also tested. The device was operated at 20 V_{RMS} and 400 kHz, and as shown in Fig. 11A, after approximately 12 s, the bacteria were concentrated on the high field region at the edge of the rightmost wire. More importantly, they appear to preferentially collect on the outside wire edge, as opposed to the inside edge. Referring to Fig. 4, it is clear that

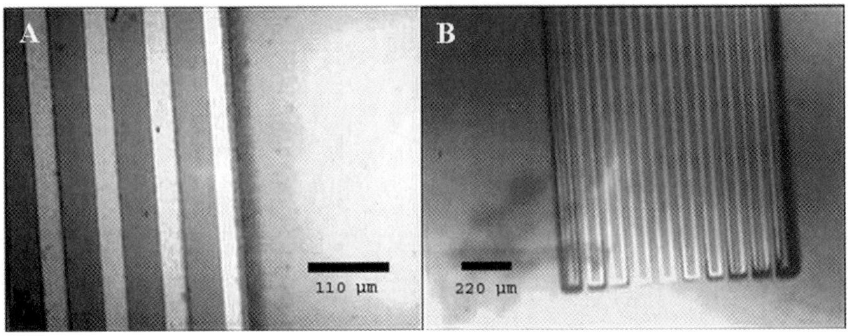

Fig. 11. (A) Convection-enhanced bacteria accumulation on high field region of wire after 12 s at 20 V_{RMS} and 400 kHz. (B) Full view of device after 30 s indicates highly concentrated regions of bacteria at the high field edges of the device.

there exists strong tangential EO flow over the wire surface and any bacteria in the bulk solution can be convected across the substrate surface to a wire edge from either the left or the right, where they can then be trapped at a high field region. From the calculated fluid velocity profile, any bacteria approaching from the right will be first exposed to a region of high field at the outer wire edge of the rightmost wire, while any bacteria approaching from left will be convected towards the wire to the left of the rightmost wire and into a region of field minimum. Because the bacteria are repelled from regions of low field, and cannot oppose fluid convection, they will be introduced into the fluid streamlines above the rightmost wire, where they will be convected, and trapped at the high field wire edge. Fig. 11B shows a full view of the device after 30 s and clearly shows the accumulation of bacteria on the high field regions of the wire.

Because the device has well-defined field minimums and maximums in the vicinity of strong AC-EO flow, an experiment was carried out in order to observe the separation abilities of the device. A dilute suspension of 1 and 5 μm microspheres in a 1.5 M AHA solution was injected into the microchannel. Shown in Fig. 12A, when the device was operated at 20 V_{RMS} and 400 kHz, the 1 μm polystyrene particles were attracted to the high field wire corners, while the 5 μm latex particles were attracted to the low field gap between the two wires. It is clear from this picture that the COF is particle size-dependent. The fact that the 5 μm particles have a lower COF than the 1 μm particles at least qualitatively verifies the COF scaling arguments described earlier.

The experiment was repeated under identical voltage frequency conditions with a suspension of 1 and 5 μm microspheres in DI water. Shown in Fig. 12A, the 1 μm particles collect in the same manner, however the 5 μm particles are convected to the stagnation region, where they are locally repelled by the n-DEP trap. This is seen in Fig. 12B as a concentrated region of 5 μm particles are levitating above the trap, most likely due to a balance between convective and DEP forces.

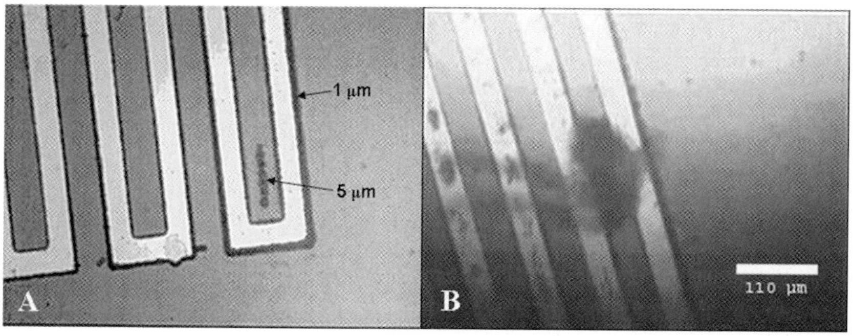

Fig. 12. (A) 1.5 M AHA, 1 μm particles experiencing p-DEP at 20 V_{RMS} and 400 kHz and attracted to wire edges, 5 μm particles experience n-DEP, attracted to wire gap. Image taken after 15 s. (B) DI water, 1 μm particles experience p-DEP, 5 μm particles repelled from field minimum at the same voltage and frequency.

The device was also used for particle manipulation. A dilute solution of 5 μm latex particles in 1.5 M AHA was placed in the microchannel and a voltage was applied (20 V_{RMS}, 500 kHz) for 4 min, which provided ample time for the microparticles to collect on both sides of the device. Following the collection, the applied signal frequency was adjusted in a step-change manner from 500 to 50 kHz and the particles were observed to lift off from the substrate and become reintroduced into the velocity field. As shown in Fig. 13, the particles are seen to lift off and two plugs of particles appear to approach each other. The plug on the left then overshoots the plug on the right, which gets trapped on the serpentine wires.

The step change in frequency from 500 to 50 kHz causes the trapped particles to cross-over from n-DEP to p-DEP. At 50 kHz, the once highly attracting field minimum is a highly repulsive region, which forces the particles outward where they are reintroduced into the velocity field. In looking at the calculated velocity profile, shown in Fig. 3, it is clear that if trapped particles are repelled from the field minimum, they will follow the fluid streamlines until they reach a more favorable location (p-DEP trap). The particles on the right side of the device will be forced upward and to the left, while the particles on the right will be convected across the surface of the device. In looking down on the device, it would appear that the particles move towards each other, however careful imaging, shown in Fig. 13, shows that the particles originally trapped on the right-hand side of the device clearly flow above the particle slug arriving from the left. Additionally, because the particles on the left are convected across the device surface, and are acting under p-DEP, they are immediately trapped on the high field wires, while that of the right-hand slug continues to flow over the device. These images clearly support the calculated velocity profile over the device and that particles can be manipulated to different field regions by exploiting their COFs.

Finally, experiments were performed to study the device in the high voltage regime. A solution of 1.5 M AHA was injected into the microchannel and a potential of 2500 V_{RMS} at a frequency of 500 kHz was dropped across the wire. Within

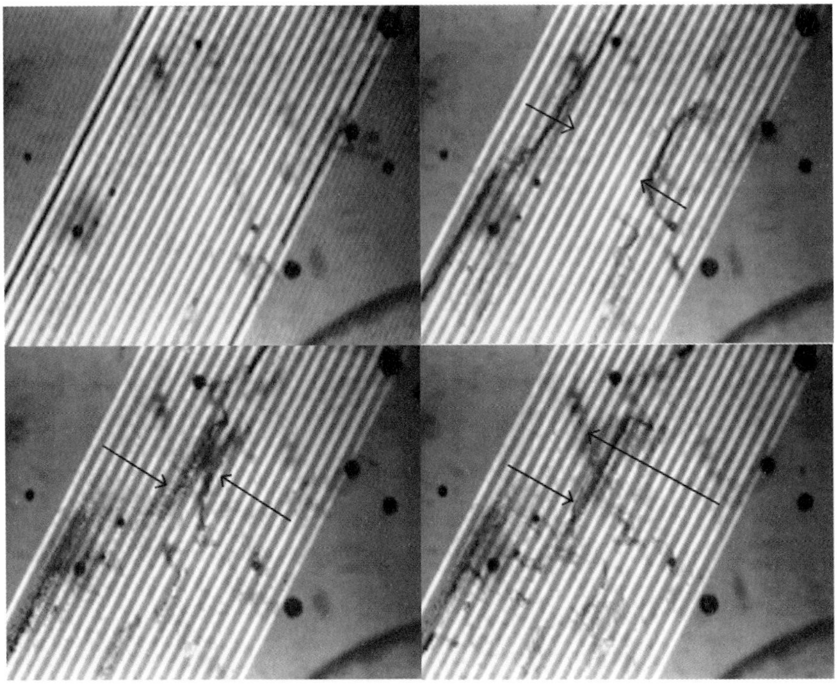

Fig. 13. Rapid particle release: Top left, 0 s; top right, 0.25 s; bottom left, 0.50 s; bottom right, 1 s. Careful look at bottom right image shows the right particle slug overshooting the left particle slug. Applied voltage is 20 V_{RMS} and the applied frequency is 500 kHz.

seconds two large counterrotating vortices were observed on each end of the serpentine wire device. This type of fluid motion is expected and is consistent with all the previous work involving AC-EO generated fluid flows. Prior to applying the potential to the wire, the backside of the silicon substrate temperature was measured to be 24.45°C. While no vapor formation was observed in the microchannel during operation, the substrate temperature increased, and in approximately 10 s after the voltage was applied, achieved an equilibrium value of 35.4°C. A high voltage probe was placed across the device and the voltage drop across the serpentine wire was measured to be ~2455 V_{RMS}. The device was continually operated for a period of 2 h and during this time the substrate temperature increased to 37.2°C, however no Faradaic reactions were observed and the device demonstrated stable operation.

A second experiment was performed in order to determine under what conditions Faradaic reactions are produced by the wire. A solution of 1.5 M AHA was subjected to 2500 V_{RMS} at varying frequencies and was observed for any evidence of Faradaic reactions. Initially, when operating at 500 kHz, no reactions were observed. The frequency was slowly decreased from 500 to 132 kHz, where significant bubble generation was observed. Increasing the frequency from 132 to 200 kHz immediately stopped bubble formation.

The fact that no significant heat, bubble, and vapor generation was observed at high frequencies indicates that there is no significant Faradaic reactions being generated by the wire device. This is surprising given the high applied voltage. However, based on the experiments we believe reactions are negligible because of the combination of high AC frequency and the cross-section geometry of the wire.

15.4 Concluding remarks

This paper proposes an improved method over conventional DEP techniques and earlier AC-EO traps for collecting, separating, and manipulating various types of micron size particles in suspension. By designing a serpentine device that allows a high voltage to be applied and by tuning the permittivity of a particle suspension using ionic molecules, optimum conditions can be achieved where there exists strong electroosmotic convection (~1 mm/s) over a large volume of electrolyte (70 μL) within a frequency range where particles can exhibit both n-DEP and p-DEP. By aligning fluid stagnation lines directly on local field minima and maxima and by aligning their characteristic frequencies with zwitterions, the required time to trap particles was reduced by two orders of magnitude. Additionally, the rapid concentration of bakers yeast cells and *E. coli* bacteria was also established. The concept of an enhanced separation technique was demonstrated. The critical factor for aligning the frequencies of AC-EO and DEP for convection-enhanced trapping is the electrolyte permittivity and double layer conductivity, and this work has produced several unanswered questions that need future investigation. For example, the permittivity of the electrolyte is assumed to be affected by AHA, however it is unknown as to how this ionic molecule affects the permittivity of the particle surface, or the membrane of a cell or bacterium. The observed decrease in particle COF could have been caused by a modification of both electrolyte and particle permittivity brought on by the zwitterions. These ions could also change the double layer conductivity of the particles and hence produce a tangential conduction effect not related to the permittivity and not included in the classical DEP theory discussed in this work. Such detailed considerations will be analyzed with a parallel theoretical study. Finally, the serpentine device appears to generate high electric fields without observable Faradaic reactions. While this is counterintuitive, experiments indicate that high-frequency AC wire operation appears to be responsible for reaction elimination. However, a more detailed study needs to be performed in order to validate why a continuous wire structure can sustain voltages on the order of kilovolts without boiling or electrolyzing the fluid sample.

15.5 References

[1] Ivnitski, D., Abdel-Hamid, I., Atanasov, P., Wilkins, E., *Biosens. Bioelectr.* 1999, *14*, 599–624.

[2] Auersward, J., Knapp, H. F., *Microelectron. Eng.* 2003, *67*, 879–886.

[3] Gomez, R., Bashir, R., Sarikaya, A., Ladisch, M. R., Sturgis, J., Robinson, J. P., Geng, T., Bhunia, A. K., Apple, H. L., Wereley, S., *Biomed. Microdevices* 2003, *3*, 201–209.

[4] Morgan, H., Hughes, M. P., Green, N. G., *Biophys. J.* 1999, *77*, 516–525.

[5] Pethig, R., Hughes, M. P., *Anal. Chem.* 1999, *71*, 3441–3445.

[6] Arnold, W. M., Schwan, H. P., Zimmermann, U., *J. Phys. Chem.* 1987, *91*, 5093–5098.

[7] Suehiro, J., Hamada, R., Noutomi, D., Shutou, M., Hara, M., *J. Electrostat.* 2003, *57*, 157–168.

[8] Hoettges, K. F., McDonnell, M. B., Hughes, M. P., *J. Phys. D* 2003, *36*, L101–L104.

[9] Wu, J., Ben, Y., Battigelli, D., Chang, H.-C., *Ind. Eng. Chem. Res.* 2005, *44*, 2815–2822.

[10] Wong, P. K., Wang, T.-H., Deval, J. H., Ho, C.-M., *IEEE/ASME Trans. Mechantronic* 2004, *9*, 366–376.

[11] Green, N. G., Ramos, A., Gonzalez, A., Morgan, H., Castellanos, A., *Phys. Rev. E* 2000, *61*, 4011–4018.

[12] Jones, T. B., *Electromechanics of Particles*, Cambridge University Press, Cambridge 1995.

[13] Pohl, H. A., *Dielectrophoresis*, Cambridge University Press, Cambridge 1978.

[14] O'Konski, C. T., *J. Phys. Chem.* 1960, *64*, 605–619.

[15] Gonzalez, A., Ramos, A., Green, N. G., Castellanos, A., Morgan, H., *Phys. Rev. E* 2000, *61*, 4019–4028.

[16] Ben, Y., Chang, H.-C., to appear in *CRC Handbook of MEMS*, 2005.

[17] Arnold, W. M., Zimmermann, U., *Biochem. Soc. Trans.* 1993, *21*, 475S.

[18] Morgan, H., Green, N., *AC Electrokinetics: Colloids and Nanoparticles*, Research Studies Press, Hertfordshire, England 2003.

[19] Drazin, P. G., Reid, W. H., *Hydrodynamic Stability*, Cambridge University Press, Cambridge 1981.

16
Dielectrophoresis induced clustering regimes of viable yeast cells*

John Kadaksham, Pushpendra Singh, Nadine Aubry

We experimentally study the transient clustering behavior of viable yeast cells in a dilute suspension suddenly subjected to a nonuniform alternating current (AC) electric field of a microelectrode device. The frequency of the applied electric field is varied to identify two distinct regimes of positive dielectrophoresis. In both regimes, the yeast cells eventually cluster at electrodes' edges, but their transient behavior as well as their final arrangement is quite different. Specifically, when the frequency is much smaller than the cross-over frequency, the nearby yeast cells quickly rearrange in well-defined chains which then move toward the electrodes' edges and remain aligned as elongated chains at their final location. However, when the frequency is close to the cross-over frequency, cells move individually toward the regions of collection and simply agglomerate along the electrodes' edges. Our analysis shows that in the first regime both the dielectrophoretic (DEP) force and the mutual DEP force, which arises due to the electrostatic particle–particle interactions, are important. In the second regime, on the other hand, the DEP force dominates.

16.1
Introduction

Dielectrophoresis is now well established as a way to produce a deterministic force for the manipulation of microscale and nanoscale particles, including biological particles. Such manipulations are crucial for molecular and cellular analyses, as needed in many applications ranging from agriculture, food industry, homeland security, environmental monitoring to biotechnology. For example, the fields of genomics and proteomics have recently made significant progress toward the identification of genes and proteins that can influence disease states. This molecular understanding of diseases, however, requires further advances in our ability to

* Originally published in Electrophoresis 2005, 26, 3738–3744

Microfluidic Applications in Biology. Edited by Niels Lion, Joël S. Rossier, and Hubert H. Girault

ISBN-10: 3-527-31761-9

manipulate and separate in a precise and controlled way target particles and cells within their suspending medium. Dielectrophoresis has been shown to be well adapted to such manipulations [1].

Pohl [2] was the first to perform extensive experimental research on the phenomenon of dielectrophoresis, showing that polarizable particles could be moved to the regions of high or low electric fields depending on their polarizability, and therefore be manipulated and separated. Pohl, however, had doubts that the method could be applied to small particles, particularly submicron sized particles, as the dielectrophoretic (DEP) force acting on the particle, which scales as the volume of the particle, would decrease significantly while the Brownian force on the particles would become relatively stronger. He noted that "a minimum particle size of about 4000–6000 Å is required under reasonable physical conditions of dielectrophoresis, if a "victory" over Brownian scrambling is demanded."

Several recent studies [3–5] performed using microdevices have shown that it is indeed possible to move submicron and nanosized particles deterministically, provided the DEP force applied to them is sufficiently strong. For example, it can be shown that the generation of the required DEP forces for moving micron sized particles can be achieved with an electric field gradient of the order of 10^9–10^{10} V/m^2. Interestingly, in microdevices, such strong electric field gradients can be generated by applying a voltage of just a few volts, as the distance between the electrodes is of the order of a few microns. The electric field strength in these devices is of the order of 10^5 V/m.

Another important issue one must consider is that the presence of strong electric fields can cause motion in the suspending liquid, which under certain conditions can be quite strong and can drag particles along with it. For example, as most real liquids are conducting or mildly conducting, the presence of high electric fields can cause significant Joule heating, resulting in electrothermal motion [5–7]. In addition, another frequency-dependent force called the alternating current (AC) electroosmotic force [8] can also act on the liquid at low frequencies. Green and Morgan [8] have shown that such motion of the suspending liquid can be either used in conjunction with the DEP force for particle separation and collection, or eliminated by choosing appropriate frequencies.

In an AC electric field, the time-averaged DEP force, assuming that the particle is isolated and the dipole limit is valid [9], is given by

$$\mathbf{F}_{DEP} = 2\pi a^3 \varepsilon_0 \varepsilon_c \beta \, \nabla \mathbf{E}^2 \qquad (1)$$

where a is the particle radius, ε_c is the permittivity of the fluid, $\varepsilon_0 = 8.8542 \times 10^{-12}$ F/m is the permittivity of free space, and **E** is the RMS value of the electric field. Expression (1) is also valid for a DC electric field where **E** is simply the electric field intensity. The coefficient $\beta(\omega)$ is the real part of the frequency-dependent Clausius–Mossotti factor given by

$$\beta(\omega) = Re\left(\frac{\varepsilon_p^* - \varepsilon_c^*}{\varepsilon_p^* + 2\varepsilon_c^*}\right)$$

where ε^*_p and ε^*_c are the frequency-dependent complex permittivity of the particle and fluid, respectively. The complex permittivity, $\varepsilon^* = \varepsilon - j\sigma / \omega$, where ε is the permittivity, σ is the conductivity, and $j = \sqrt{-1}$.

By definition, β has values between -0.5 and 1.0. If it is negative, the DEP force is in the opposite direction of the gradient of the electric field magnitude, and in this case particles collect in the regions of low electric field. In contrast, if it is positive, particles collect in the regions of high electric field. The cross-over frequency is the frequency at which the real part of the Clausius–Mossotti factor changes sign.

In addition to the DEP force due to the external electric field, as particles get close to each other they modify the electric field and exert forces on each other, which we refer to as particle–particle interactions or mutual dielectrophoresis [2]. It is well known that such particle–particle interactions in uniform electric fields result in long chains and columns of particles spanning the domain in the direction of the applied electric field. This is referred to as the pearl chaining phenomenon [2, 10], and has been investigated also for binary mixtures of particles in uniform electric fields [11].

The particle–particle interaction force on the *i*-th particle due to the *j*-th particle in a nonuniform DC electric field, based on the point dipole approximation, is

$$\mathbf{F}_{D,ij} = \frac{12\pi\varepsilon_0\varepsilon_c\beta^2 a^6}{r^4}\begin{pmatrix} \mathbf{r}_{ij}\left(\mathbf{E}_i \cdot \mathbf{E}_j\right) + \left(\mathbf{r}_{ij} \cdot \mathbf{E}_i\right)\mathbf{E}_j + \left(\mathbf{r}_{ij} \cdot \mathbf{E}_j\right)\mathbf{E}_i \\ -5\mathbf{r}_{ij}\left(\mathbf{E}_i \cdot \mathbf{r}_{ij}\right)\left(\mathbf{E}_j \cdot \mathbf{r}_{ij}\right) \end{pmatrix} \tag{2}$$

where $\mathbf{r}_{ij}$ is the unit vector in the direction from the center of *i*-th particle to the center of *j*-th particle. This force is present even if the applied electric field is uniform.

It is interesting to note from expressions (1) and (2) that the force induced by the particle–particle interactions is proportional to β^2 while the DEP force itself is proportional to β. Here we wish to recall that for AC fields β varies with frequency and that at the cross-over frequency $\beta = 0$. For frequencies near the latter, β is thus small. As β approaches zero, assuming (2) is approximately valid for AC fields (strictly speaking, in this case β should be replaced by the magnitude of the Clausius-Mossotti factor) the point dipole approximation predicts that the particle–particle interaction force vanishes faster than the DEP force (see also Addendum). It follows that, for frequencies near the cross-over frequency, it may be possible to manipulate particles individually by means of dielectrophoresis alone, that is, without the formation of chains. This can be beneficial in certain applications, *e.g.*, biological, where it may be desirable to keep particles isolated while they are being manipulated.

In this paper, we experimentally investigate the clustering behavior of viable yeast cells. Experiments were conducted using yeast cells suspended in water using castellated microelectrodes with an electrode gap of 50 μm (also see our direct numerical simulations of this problem [12, 13]).

16.2 Materials and methods

16.2.1 Microfluidic device

The device used in this paper is the microfluidic platform for manipulating microscale and nanoscale particles previously reported [14, 15]. The device is an integrated dual-microelectrode array chamber which was designed and fabricated using standard microfabrication processing techniques, and has geometry similar to the microelectrode device used for DEP experiments described in [16, 17]. Starting with a silicon wafer, 50 nm of dry thermal oxide was grown onto the wafer followed by 150 nm of stoichiometric nitride for insulation purposes. Planar gold electrodes were then realized by evaporating a 20 nm chromium layer followed by a 250 nm gold thin film layer onto the insulating oxide layer of the wafer. Photoresist was spun on the wafer and then lithographic patterning of the electrode geometry definition was performed using a photomask plate (Compugraphics, CA). An acetone lift-off process was used to define the microelectrodes and remove excess gold and photoresist from the field area. Electrical connection pads were provided on the wafer in order to make external electrical connections while electrical connections to the pads were made by hand soldering 0.25 mm diameter wire to the 1 mm electrical connection busses on the chip. The device contains a total of 16 different sets of microelectrodes, including four different geometrical designs with four different sizes of electrodes and spacing. The chip device can be enclosed in a chamber with inlet and outlet tubing to provide a fluid flow condition, or it can be operated in the static-flow condition. The open static-flow condition was used for the study carried out in this paper.

The experiments reported below were performed using the periodic microelectrode structure shown in Fig. 1, with the electrode width of 50 μm and an interelectrode gap of 50 μm as well. This microelectrode arrangement has well-defined elec-

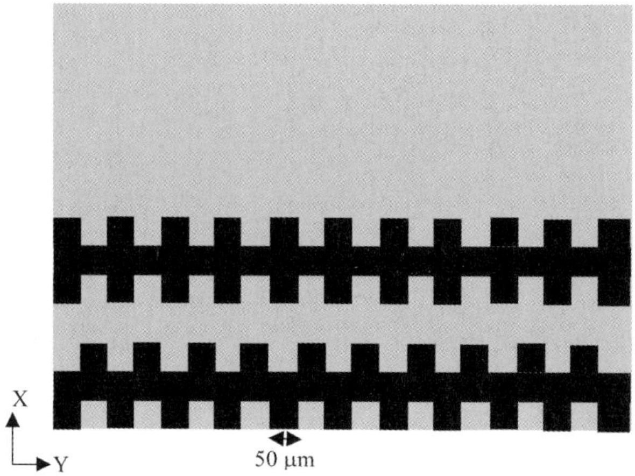

Fig. 1 Schematic representation of the bottom surface of the device containing the microelectrodes. Black geometry represents the electrode while the gray area corresponds to the silicon substrate. Yeast suspension is pipetted onto this surface.

a

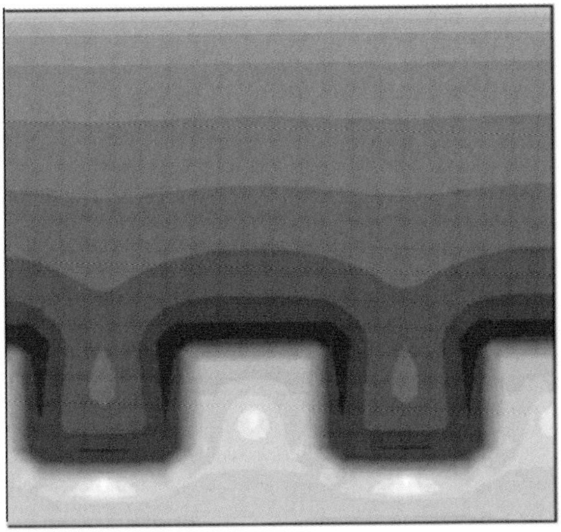

b

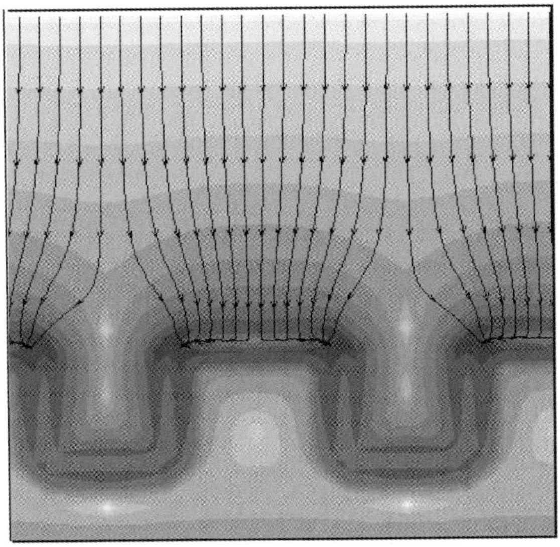

Fig. 2 Electric field and $\mathbf{E} \cdot \nabla \mathbf{E}$ distributions above the top electrode in Fig. 1. (a) Contour plot of the log of the magnitude of the electric field on the bottom surface containing the electrode; black corresponds to the regions where the electric field is the highest and white corresponds to the regions where the electric field is the lowest. (b) Contour plot of log $(\mathbf{E} \cdot \nabla \mathbf{E})$, showing the DEP force lines on the plane of the electrodes in the microdevice. Black and white areas correspond to regions of high and low intensity of the DEP force, respectively.

tric field maxima at the edges of the electrodes and electric field minima in the bay regions between the edges and on top of the electrodes [17]. In Fig. 2, the electric field distribution is shown in the region above the end electrode. Notice that the lines of DEP force at a distance of $\sim$100 μm from the electrode are approximately vertical and the magnitude of $\mathbf{E} \cdot \nabla \mathbf{E}$ is approximately constant on horizontal planes. The electric field was computed using the numerical code described in [12, 13, 18].

16.2.2
Preparation of viable yeast cells suspension

Hereafter, we use a suspension of live yeast cells as in [2, 5, 16], and concentrate our analysis on the phenomenon of dielectrophoresis only although other physical phenomena may be present as well [5]. Yeast cells, or *Saccharomyces cerevisiae*, were grown overnight at 32°C in adextrose solution made with deionized water. The yeast cells were then harvested from the growing solution using a sterile pipette and resuspended in Millipore water. The suspension was then centrifuged and the supernatant liquid decanted. The cells were again resuspended in fresh Millipore water and centrifuged to isolate the cells. The process was repeated until the suspension conductivity was of the desired value of about 1.0 mS/m at a frequency of 7 MHz. The conductivity of the suspension before and after the experiments was measured using a broadband dielectric spectrometer (BDS-80; Novocontrol, GmBH, Hundsangen, Germany) and was found to be approximately constant for the time duration of 30 min over which the experiments were conducted. The dielectric sample cell used for the measurements was a cylindrical one having a diameter of 11 mm and height of 5.5 mm, filled with the yeast suspension. The viability of the cells was checked by introducing methylene blue in the suspension. Approximately 97% of the cells repelled methylene blue, indicating that they were viable. The suspended live yeast cells had an average radius of 3.9 μm.

16.2.3
Dependence of the DEP velocity on the frequency of the electric field

We next describe the approach used for measuring the DEP force induced velocity of viable yeast cells from the video recordings made at a fixed location of the microdevice. The velocity was obtained for a set of frequencies between 500 Hz and 10 MHz. The time taken by an isolated cell to move over a fixed distance (between two fixed points) was measured and the velocity was deduced in a straightforward manner by dividing the distance by the time taken. Cells close to others were excluded from this calculation so that electrostatic particle–particle interactions were negligible for the cells selected. It was noticed that for all frequencies investigated here, particles move toward the electrode edges and therefore experience positive dielectrophoresis. Fig. 3 reports velocity measurements averaged over a few cells, showing that the velocity of individual cells is about 6 μm/s at 1 kHz, and approximately 0 at 10 MHz.

If we assume that the particle acceleration is negligible, equating the Stokes drag with the DEP force gives

$$6\pi a U = 2\pi a^3 \varepsilon_0 \varepsilon_c \beta \nabla E^2 \qquad (3)$$

where U is the cell velocity.

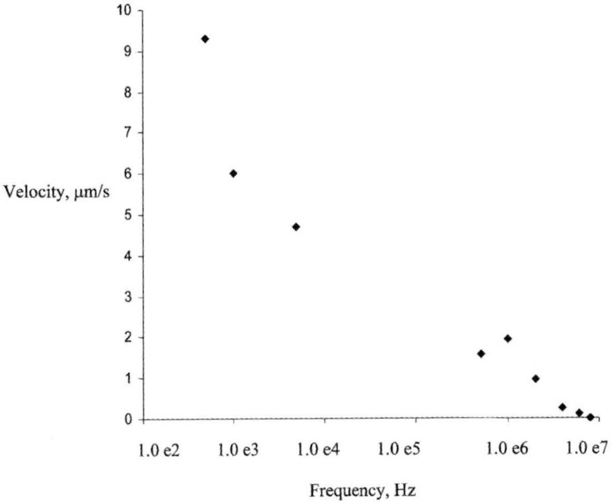

Fig. 3 Measured average velocity of yeast cells as a function of the frequency of the applied electric field, showing high velocities at low frequencies and nearly zero velocities around 10 MHz.

Solving the above equation for β, we obtain

$$\beta = C \cdot U \qquad (4)$$

where $C = \dfrac{3\eta}{a^2 \varepsilon_0 \varepsilon_c \nabla \mathbf{E}^2}$. Thus, β is proportional to the velocity of the cell, provided C is constant. Therefore, if the velocity of an isolated cell is zero, the above equation implies that β is also zero, which, in our system, occurs at about 10 MHz. This frequency is therefore the cross-over frequency for our suspension. The quantitative deduction of β from the cell (particle) velocities presented in Fig. 3 is not needed for our purpose and goes beyond the scope of our paper.

Fig. 3 shows that at low frequencies, the cell velocity is the highest, indicating that the value of β is relatively large. This, in turn, implies that the magnitude of the DEP force is large and the electrostatic particle–particle interactions are strong. Notice that Fig. 3 shows the DEP velocity of the cells only for those frequencies for which the dominant force is the DEP force. In the frequency range of 5–500 kHz the particle motion is influenced by the fluid flow due to AC electroosmosis, and therefore it is not possible to accurately determine the velocity induced by the DEP force alone (Eq. 3). Nevertheless, it is clear from Fig. 4, which shows dielectrophoresis at 10 kHz, that even in this frequency range the cells undergo positive dielectrophoresis, as the cells that are initially sufficiently close to the electrode are attracted and held by the electrodes. The cells that are not sufficiently close to the electrode are swept away by the fluid flow due to AC electroosmosis.

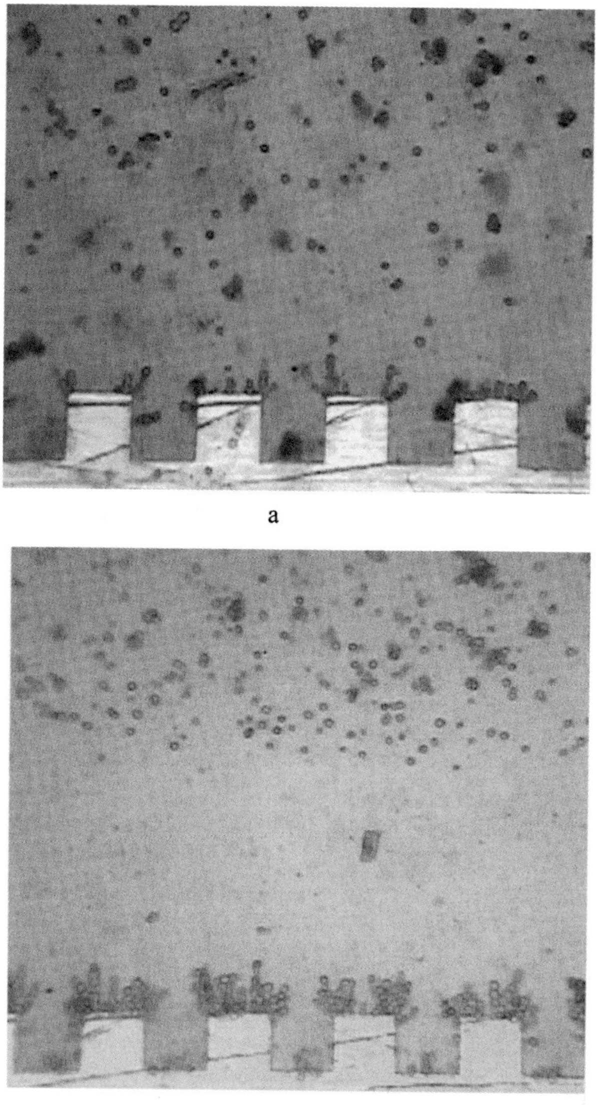

Fig. 4 Visualization of the suspension for an applied electric field frequency of 10 kHz, (a) immediately after applying the electric field and (b) at time $t = 60$ s after applying the electric field. Cells sufficiently close to the electrode get collected at the electrode edge due to positive dielectrophoresis, whereas cells away from the electrode are carried away by the fluid flow due to AC electro-osmosis.

16.3 Results

A small volume of the yeast suspension prepared in the manner described in Section 2.2 was pipetted and placed on the microelectrode structure. The suspension was then sealed with a cover slip and allowed to settle for a few seconds, and then the electrodes were energized. A variable frequency AC signal generator (BK Precision Model 4010A) was used to apply a voltage to the electrodes over a fre-

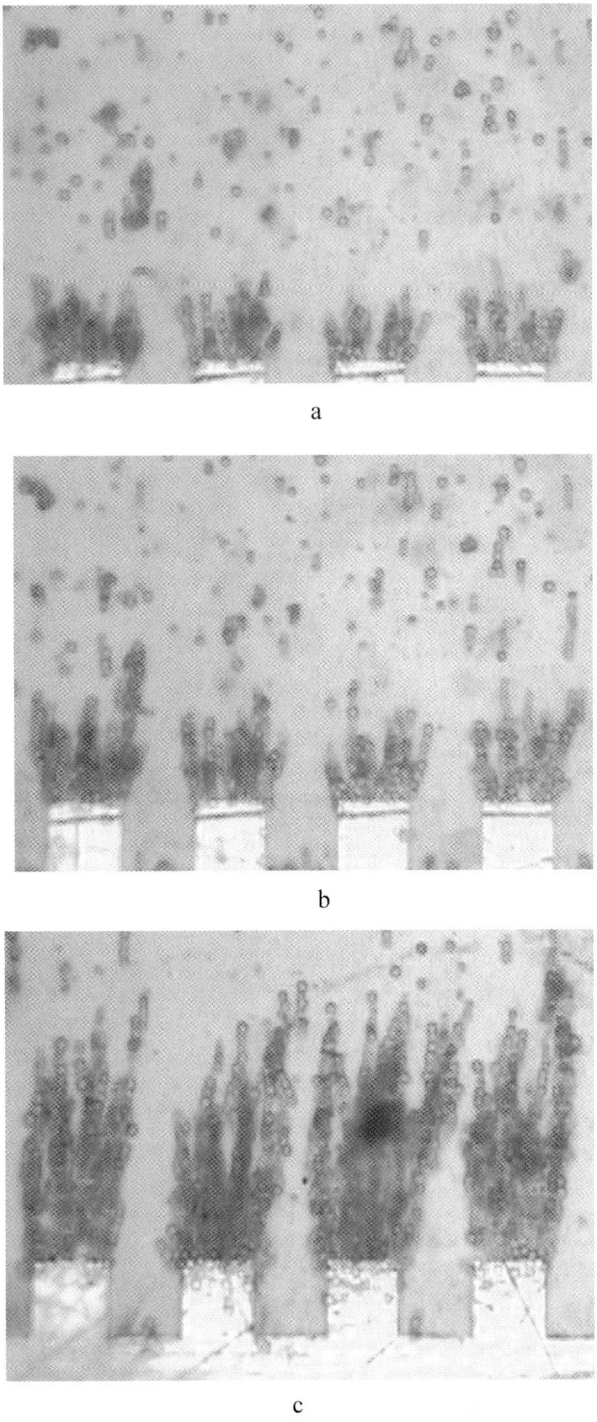

Fig. 5 Visualization of the suspension for an applied electric field frequency of 1 kHz at time (a) $t = 5$ s, (b) $t = 10$ s, and (c) $t = 90$ s. Particles close to each other form chains, which then move to the electrode edges.

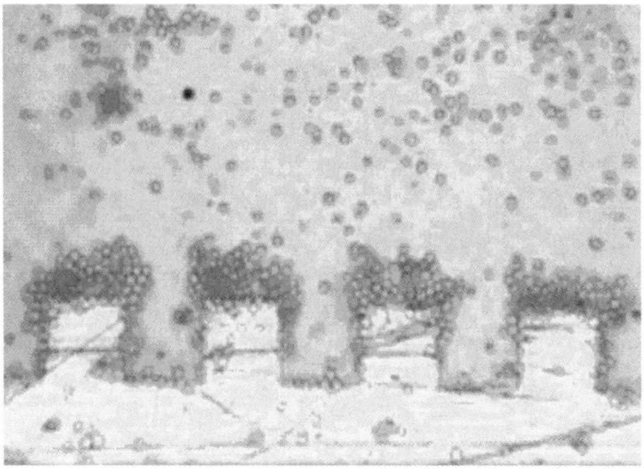

Fig. 6 Same as Fig. 5 for an electric field of frequency 2 MHz. Photograph was taken at time $t = 90$ s after applying the electric field. While collection of particles is observed, particle chaining has mainly vanished, indicating that the electrostatic particle–particle interactions (mutual dielectrophoresis) are negligible.

quency range of 200 Hz to 10 MHz. The applied voltage was 4 V P/P (peak-to-peak). The applied voltage, frequency, and resulting current were monitored with an oscilloscope and digital voltmeter connected in parallel with the electrodes. Observations of the DEP motion for the electrode at the edge of the device were made and recorded using a Digital Color CCD camera connected to a Nikon Metallurgical MEC600 microscope.

We first describe our results at the frequency of 1 kHz. Once the electric field is switched on, as expected, the cells near the electrode are immediately collected at the electrode's edges (Figs. 5a–c). The cells collect at the edge of the electrode in the form of chains along the DEP force lines (Fig. 5a), and these chains are visible immediately after the electric field is applied. The cells away from the electrode start moving toward the electrode, while simultaneously forming chains by joining with neighboring cells, as seen in Figs. 5a and b. The cells and chains formed move toward the electrode's edges and eventually collect there. Some of the chains are also observed to regroup as columns. The average length of the chains at the electrode's edges at time $t = 90$ s is observed to be about 150 μm (Fig. 5c).

We now consider the case where the applied electric field frequency is 2 MHz. Once the electric field is switched on, the cells close to the electrode get collected at the electrode's edges, but their velocities are much smaller than in the previous cases. In addition, the collected cells at the edges of the electrodes, instead of forming distinct chains as in the previous case, are seen to agglomerate at the edges. The cells located far away from the electrode also move progressively toward the electrode, but they do not join together to form chains. Instead, they move individually, even when they are quite close to each other. The cells collected at the edges of the electrode at time $t = 90$ s are shown in Fig. 6. From this figure it is clear that the cells collected at the electrode's edge do not regroup into chains. Instead, they are observed to agglomerate at the edge of the electrode with no clear structure, which demonstrates that the electrostatic particle–particle interactions are negligible.

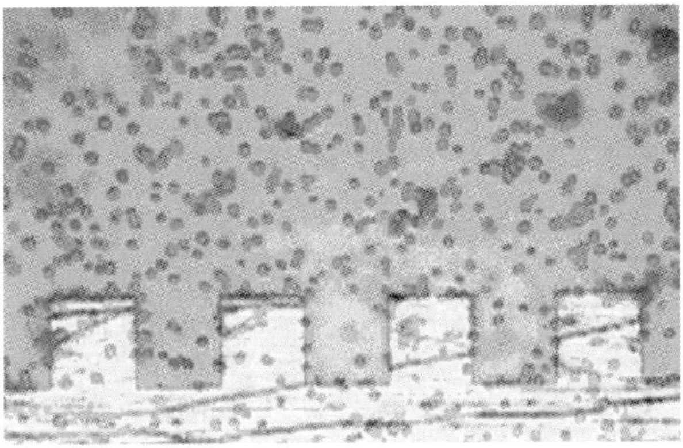

Fig. 7 Same as Fig. 5 for an electric field of frequency 10 MHz. Photograph was taken at time $t = 90$ s after applying the electric field. Particles are hardly moving, and therefore not being collected on the electrode edge and not regrouping, indicating that both the DEP force and the electrostatic particle–particle interactions are approximately zero in this case.

As the frequency of the electric field is increased further, the DEP force acting on the cells becomes very small and thus only a few cells are collected at the electrode's edges at $t = 90$ s. Finally, since the DEP force becomes approximately zero at 10 MHz, no particle motion could be identified (see Fig. 7) as both the DEP and particle–particle interaction forces become approximately zero.

16.4 Discussion

Our experiments show that when the frequency of the applied AC electric field is smaller than approximately 10 MHz the yeast cells are subjected to positive dielectrophoresis, and hence they cluster at the edges of the electrode. The nature of clusters, however, depends on the frequency of the electric field. When the frequency is around 1 kHz, the cells close to each other, including those away from the electrode, first organize in chains aligned parallel to the direction of the local electric field. Then these chains move toward the edges of the electrode and join with the chains already located at the edges of the electrode to form longer chains, which may contain as many as 35 cells.

On the other hand, when the frequency is around 2 MHz, which is close to the cross-over frequency of 10 MHz, cells move toward the edges of the electrodes individually; the tendency to form chains is not as noticeable. This shows that electrostatic particle–particle interactions do become negligible when the frequency is close to the cross-over frequency and β is much smaller than 1.

These results are in agreement with the model presented in Section 1 and in the Addendum, which predicts that the particle–particle interaction force depends on the second power of β, and therefore becomes negligible compared to the DEP force, which itself depends linearly on β, when β is small. We may therefore conclude that for frequencies close to the cross-over frequency it is possible to remove the particle–particle interaction force and manipulate particles individually entirely by means of the dielectrophoresis force.

We gratefully acknowledge the support of the New Jersey Commission on Science and Technology through the New-Jersey Center for Micro-Flow Control under Award number 01–2042–007–25, and the W.M. Keck Foundation for providing support for the establishment of the NJIT Keck laboratory for Electrohydrodynamics of Suspensions.

16.5 References

[1] Gascoyne, P. R. C., Vykoukal, J. V., *Proc. IEEE* 2004, *92*, 22–42.

[2] Pohl, H. A., *Dielectrophoresis*, Cambridge University Press, Cambridge 1978.

[3] Hughes, M. P., Morgan, H., *J. Phys. D* 1998, *31*, 2338–2353.

[4] Hughes, M. P., *Nanotechnology* 2000, *11*, 124–132.

[5] Ramos, A., Morgan, H., Green, N. G., Castellanos, A., *J. Phys. D* 1998, *31*, 2338–2353.

[6] Castellanos, A., Ramos, A., Gonzalez, A., Green, N. G., Morgan, H., *J. Phys. D* 2003, *36*, 2584–2597.

[7] Ramos, A., Morgan, H., Green, N. G., Castellanos, A., *J. Colloid Interf. Sci.* 1999, *217*, 420–422.

[8] Green, N. G., Morgan, H., *J. Phys. D* 1998, *31*, L25–L30.

[9] Jones, T. B., *IEEE Eng. Med. Biol. Mag.* 2003, *22*, 33–42.

[10] Hao, T., *Adv. Colloid Interf. Sci.* 2002, *97*, 1–35.

[11] Giner, V., Sancho, M., Lee, S., Martinez, G., Pethig, R., *J. Phys. D* 1999, *32*, 1182–1186.

[12] Kadaksham, J., Singh, P., Aubry, N., *J. Fluids Eng.* 2004, *126*, 170–179.

[13] Kadaksham, J., Singh, P., Aubry, N., *Mech. Res. Commun.* 2005 in press.

[14] Kadaksham, J., Batton, J., Singh, P., Aubry, N., *Proceedings of ASME International Mechanical Engineering Congress and RD&D Expo*, Washington, DC 2003.

[15] Kadaksham, J., Batton, J., Singh, P., Golubovic-Liakopoulos, N., Aubry, N., Dielectrophoretic manipulation of micro- and nano scale particles in microchannels, *Nanotechnology World Forum*, 2003, Marlborough, Massachusetts, June 23–25, 2003.

[16] Pethig, R., Huang, Y., Wang, X.-B., Burt, J. P. H., *J. Phys. D* 1992, *25*, 881–888.

[17] Wang, X.-B., Huang, Y., Burt, J. P. H., Markx, G. H., Pethig, R., *J. Phys. D* 1993, *26*, 1278–1285.

[18] Kadaksham, J., Singh, P., Aubry, N., *Electrophoresis* 2004, *25*, 3625–3632.

16.6 Addendum

The importance of the particle–particle interaction force can be seen by computing the ratio of $\mathbf{F}_D$ and $\mathbf{F}_{DEP}$, *i.e.*

$$\frac{\mathbf{F}_D}{\mathbf{F}_{DEP}} = \frac{6\beta a^3}{r^4}\left(\begin{array}{l}\mathbf{r}_{ij}(\mathbf{E}_i \cdot \mathbf{E}_j) + (\mathbf{r}_{ij} \cdot \mathbf{E}_i)\mathbf{E}_j + (\mathbf{r}_{ij} \cdot \mathrm{tf} = \text{"PS}_S\text{CSLFBD"}\mathrm{E}_j)\mathbf{E}_i \\ -5\mathbf{r}_{ij}(\mathbf{E}_i \cdot \mathbf{r}_{ij})(\mathbf{E}_j \cdot \mathbf{r}_{ij})\end{array}\right) / \nabla \mathbf{E}^2$$

The above expression can be nondimensionalized by assuming that the gradient of the square of the electric field scales as E_0^2/L and the electric field as E_0, and that the distance between the cells and the interelectrode gap is R and L, respectively. This gives

$$\frac{\mathbf{F}_{\mathrm{D}}}{\mathbf{F}_{\mathrm{DEP}}} = \frac{6\beta a^3 L}{R^4}\mathbf{R}$$

where $\mathbf{R} = \begin{pmatrix} \mathbf{r}_{ij}\left(\mathbf{E}_i \cdot \mathbf{E}_j\right) + \left(\mathbf{r}_{ij} \cdot \mathbf{E}_i\right)\mathbf{E}_j + \left(\mathbf{r}_{ij} \cdot \mathbf{E}_j\right)\mathbf{E}_i \\ -5\mathbf{r}_{ij}\left(\mathbf{E}_i \cdot \mathbf{r}_{ij}\right)\left(\mathbf{E}_j \cdot \mathbf{r}_{ij}\right) \end{pmatrix} / \nabla' \mathbf{E}^2$

which is $O(1)$ and dimensionless. Therefore, $\frac{\mathbf{F}_{\mathrm{D}}}{\mathbf{F}_{\mathrm{DEP}}} \sim \frac{6\beta a^3 L}{R^4}$

If we further assume that the concentration of particles in the suspension is ϕ, then the average distance between the particles is

$R \approx \left(\frac{4\pi}{3\phi}\right)^{1/3} a$. This gives

$$\frac{\mathbf{F}_{\mathrm{D}}}{\mathbf{F}_{\mathrm{DEP}}} \sim \frac{6\beta L}{\left(\frac{4\pi}{3\phi}\right)^{4/3} a} =, P_4$$

The above parameter is similar to the parameter obtained in [11, 12] for a moderately concentrated suspension where the distance between the particles R was taken to be $2a$.

Clearly, if P_4 is of order 1, the particle–particle interaction and DEP forces are of comparable magnitude. In this case, we expect cells to collect at the electrodes' edges while simultaneously forming chains due to strong particle–particle interactions. However, if P_4 is less than order 1, particle–particle interactions are negligible and we expect cells to move individually and collect without forming chains. This explanation obviously assumes that the point dipole approximation is valid and that the total electrostatic force can be divided into two parts given by expressions (1) and (2).

17
3-D electrode designs for flow-through dielectrophoretic systems*

Benjamin Y. Park, Marc J. Madou

Traditional methods of dielectrophoretic separation using planar microelectrodes have a common problem: the dielectrophoretic force, which is proportional to $\nabla|E|^2$, rapidly decays as the distance from the electrodes increases. Recent advances in carbon microelectromechanical systems have allowed researchers to create carbon 3-D structures with relative ease. These developments have opened up new possibilities in the fabrication of complex 3-D shapes. In this paper, the use of 3-D electrode designs for high-throughput dielectrophoretic separation/concentration/filtration systems is investigated. 3-D electrode designs are beneficial because (i) they provide a method of extending the electric field within the fluid. (ii) The 3-D electrodes can be designed so that the velocity field coincides with the electric field distribution. (iii) Novel electrode designs, not based on planar electrodes designs, can be developed and used. The electric field distribution and velocity fields of 3-D electrode designs that are simple extensions of 2-D designs are presented, and two novel electrode designs that are not based on 2-D electrode designs are introduced. Finally, a proof-of-concept experimental device for extraction of nanofibrous carbon from canola oil is demonstrated.

17.1
Introduction

17.1.1
Dielectrophoretic separation/concentration/filtration systems

The extraction of specific types of particles from a fluid is desirable for applications that involve concentration of specific particles [5], purification (filtration), and separation [6]. Although many methods for concentration, purification, or separa-

* Originally published in Electrophoresis 2005, 26, 3745–3757

Microfluidic Applications in Biology. Edited by Niels Lion, Joël S. Rossier, and Hubert H. Girault

ISBN-10: 3-527-31761-9

tion of particles that have a diameter greater than 1 μm have been devised, manipulation of submicron particles has proven difficult. This is because fabricating physical filters that effectively filter submicron particles is difficult and because the relative magnitude of Brownian forces increases while the influence of deterministic forces such as gravity is diminished as a particle is scaled down [1, 2].

There are many methods available for directly or indirectly (by moving the fluid) actuating a particle in a fluid. Electroosmosis, electrophoresis, dielectrophoresis, optical tweezers (optical trapping), magnetic manipulation, pressure-based movement, heat convective flows, and acoustic actuation are all techniques that allow movement of a particle suspended in a fluid without actually touching the particle [1]. Of these techniques, the electrokinetic methods (electroosmosis, electrophoresis, and dielectrophoresis) scale favorably when scaled down and are easily controlled through simple application of voltages on surface electrodes [1]. All forms of electrokinetic manipulation use direct current (DC) or alternating current (AC) electric fields to manipulate species within the fluid or the fluid itself. In electrophoresis and electroosmosis, electric fields are used to apply Coulombic forces onto charged species within the fluid. In dielectrophoresis, the difference of polarizability between particle and solution in a nonuniform electric field gives rise to a net force acting on the particle [4]. In positive dielectrophoresis, particles are more polarizable than the solution and tend to move toward high-field regions, while in negative dielectrophoresis, particles are less polarizable compared to the solution and migrate toward low electrical field regions. Unlike electrophoresis and electroosmosis, using dielectrophoresis, uncharged particles can be moved without significant movement of the fluid. Three example applications out of the many potential applications for dielectrophoretic separators/concentrators/filters are listed in the following paragraphs. Other applications that are not detailed include fluid-based self-assembly of electronic components [20], particle-based nanolithography [1], and field flow fractionation [2].

Dielectrophoresis is being considered as a method of separating metallic and semiconducting carbon nanotubes. Although carbon nanotubes are being touted as one of the most promising materials of the future, all current methods for producing carbon nanotubes create a mixture of semiconducting and metallic nanotubes [3, 14]. There is great interest in separation of these two types of nanotubes so that there can be further advances in devices that utilize the electrical properties of carbon nanotubes. Dielectrophoretic separation is one of the methods being actively researched to separate different types of carbon nanotubes [3, 14].

Another field of interest where dielectrophoresis was first demonstrated for use in separation is cell sorting [2, 6]. Separation of living and dead cells [6, 9], human breast cancer cells from blood [7, 8], leukocytes from blood [8, 10], heterogeneous mammalian cell cultures [11], and different bacteria [12, 13] has been demonstrated. Dielectrophoretic separation and concentration techniques may lead to advances in diagnostics and pathogen detection by increasing signal magnitude, S/N, detection speed, and decreasing the lower LOD.

One more field where removal of particulates is important is tribology ("The science of the mechanisms of friction, lubrication, and wear of interacting surfaces that are in relative motion." [15]). Researchers estimate that 80% of all machine failures are due to wear [16] and that 1.3–1.6% of an industrialized nation's Gross National Product (GNP) is lost to degradation of machinery due to friction-related wear [17]. The abnormal abrasive wear due to lubricant contamination in marine diesel engines eclipses that of normal wear (wear that occurs even without abrasion) and the gap becomes wider with time [18]. It was found that although oil filters used in automotive engines are designed to filter particles in the 15–30 μm range, particles with diameters below 10 μm caused 44% of the wear to engine cylinders [19]. We believe that dielectrophoretic forces can be used to remove small (<10 μm) particles while providing online monitoring of oil contamination (see Sections 3.4, 4).

Because the dielectrophoretic force is inversely proportional to the cube of the distance between electrodes, one order of magnitude decrease in distance can increase the force by three orders of magnitude [1, 2]. The favorable scaling of dielectrophoresis has led to much progress in using microelectrodes to apply voltages within a solution. Planar metallic microelectrodes fabricated using conventional integrated circuit (IC) and microelectromechanical system (MEMS) techniques have typically been used in past dielectrophoresis experiments. Although there has been great success in separation of various particles using these planar electrodes [2, 3, 6–14], most of the designs for separator/concentrator/filter systems are plagued by a common problem: low-throughput. The problem with traditional methods using planar microelectrodes is that the dielectrophoretic force, which is proportional to $\nabla|E|^2$, rapidly decays as the distance from the planar arrays increases. This is one of the stumbling blocks that have prevented dielectrophoresis from being widely used in high volume applications. There have been attempts in the past of using screens, conducting plates, and microfabricated filters for effective flow-through particle separators (a good review of these methods is given in [2]), but most designs require either application of high voltages (due to the distance between the electrodes) or involve complex fabrication (such as requiring multiple substrates involving transparent indium tin oxide electrodes for visual feedback or requiring bulk micromachining).

17.1.2 Carbon microelectromechanical systems (C-MEMS)

Carbon MEMS, or C-MEMS [21–26], describes a manufacturing technique in which carbon devices are made by treating a prepatterned organic structure to high temperatures (typically above 900°C) in an inert or reducing environment. It has been shown recently that 3-D high-aspect-ratio carbon structures can be made from patterned thick SU-8 negative photoresist (Microchem, Newton, MA, USA) layers (see Fig. 2) [23]. SU-8 negative photoresist is a high-transparency UV photoresist that enables creation of "LIGA [33]-like" (LIGA—Lithographie (lithography), Galvanoformung (electroplating), Abformung (moulding)) structures using traditional UV photolithography. The geometry is largely preserved during the carbonization process although some shrinkage occurs. The shrinkage is isometric; as a result,

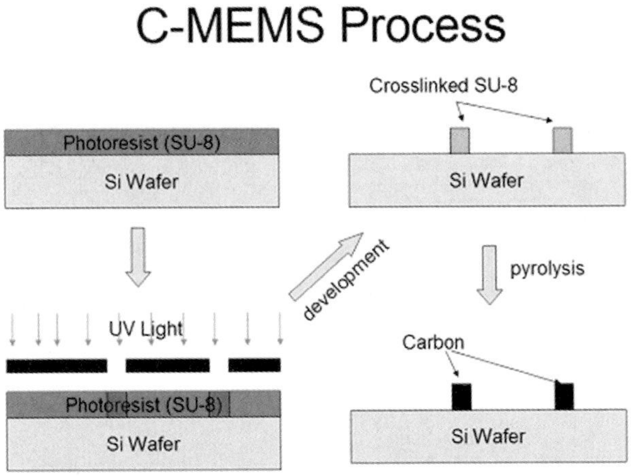

Fig. 1 Basic process flow chart for a typical C-MEMS process using SU-8 negative photoresist.

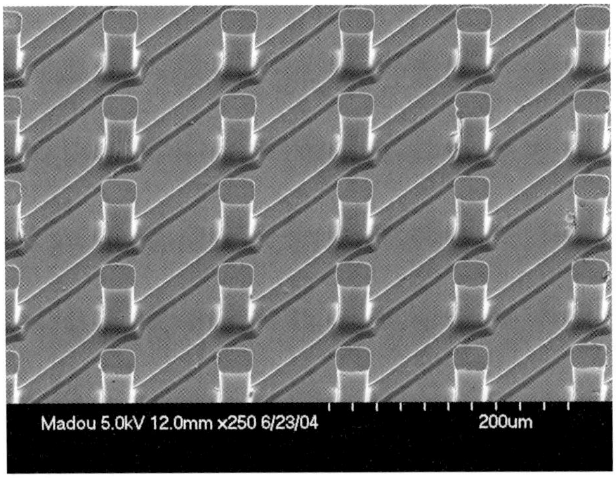

Fig. 2 SEM image of a dielectrophoretic trap array fabricated by carbonizing SU-8 photoresist.

the shape of the original photoresist structures is preserved. Fig. 1 illustrates the basic steps in a typical C-MEMS process. The details of the fabrication process are beyond the scope of this paper, and were detailed in a separate paper [23]. The capability of creating microscale and nanoscale 3-D carbon structures has opened up a wide range of new applications for pyrolytic carbon including microbatteries [26], glucose sensors, super capacitors, biofuel cells, dielectrophoretic electrode arrays for micromanipulation of microparticles and nanoparticles [1], electrodes for biorecording/stimulation, and electrode arrays for DNA hybridization and detection. The electrical properties of the C-MEMS carbon derived from SU-8 and AZ4620 photoresists has been quantitated [27], and the dielectrophoretic trapping of polystyrene beads has been demonstrated using 3-D C-MEMS electrodes [1]. Fig. 2 is a scanning electron microscope (SEM) image of a C-MEMS dielectrophoretic trap array fabricated by carbonizing SU-8 photoresist, and Fig. 3 is an optical microscope image of polystyrene beads trapped in a dielectrophoretic trap

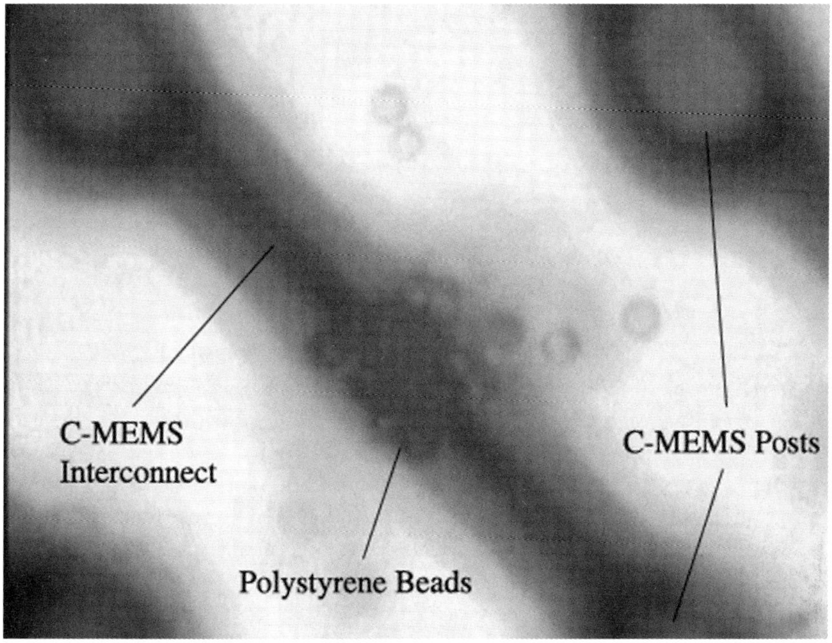

Fig. 3 A cluster of 9.62 μm polystyrene beads trapped in a 3-D dielectrophoretic trap located at the center of four high-aspect-ratio posts. Carbon interconnects were patterned before the carbon posts were patterned and the carbon interconnect layer provided the electrical connection to the posts. Particles are suspended above the posts because of the repelling force of the carbon interconnect structure underneath the trap. DI water was used as the fluid medium. Note: An array design slightly different from the one shown in Fig. 2 was used for this experiment.

created using 3-D C-MEMS electrodes. Our group has developed self-aligning insulation methods for insulating the interconnect layer connecting the C-MEMS structures. The exact fabrication details will be described in a separate paper. We believe that using C-MEMS technology, high-throughput 3-D dielectrophoretic separation systems can be made with relative ease. Although 3-D microelectrodes for dielectrophoresis have been created in the past [28], the past designs involve time-consuming fabrication techniques. High-throughput is achieved because the electric field is extended into the solution when 3-D electrodes are used. The advantages of using C-MEMS electrodes over other techniques of creating high-throughput systems are: (i) complex microscale 3-D electrodes with high-aspect-ratios can easily be shaped and patterned using conventional lithography, (ii) carbon has a high window of stability, thus allowing application of higher voltages, (iii) there is no need for bulk micromachining or patterning electrodes on multiple planes (top and bottom), and (iv) the distance between electrodes is precisely controlled through the lithography process. In this paper, 3-D electrode designs based on existing 2-D designs are evaluated with respect to velocity field and electric field distribution. A novel design rule of correlating the velocity field and electric field for enhanced dielectrophoretic separation efficiency is introduced, and two 3-D electrode designs that are not based on any existing 2-D design are presented. Finally a proof-of-concept system is introduced.

17.2
Materials and methods

17.2.1
Finite-element modeling of velocity and electric fields

FEMLAB 3.0a Multiphysics Modeling software (COMSOL, Stockholm, Sweden) was used for the modeling of electric fields as well as velocity fields. FEMLAB is a finite-element analysis tool that performs partial differential equation (PDE)-based multiphysics modeling. The Navier–Stokes application mode was used for the flow-field simulations, and the conductive media DC application mode was used for the electric field simulations (the DC mode can be used because we are dealing with magnitudes). The specifics of these modes (equations, assumptions, *etc.*) are detailed in the FEMLAB documentation (especially the Modeling Guide) and will not be detailed in this paper. The reader is referred to the FEMLAB documentation for more information. Because the specific simulation parameters differ for each system (depending on the fluid and particles used), general parameters were used for all the simulations. The fluid was assumed to be deionized (DI) water (conductivity σ of 10^{-4} S/m), and the electrodes were assumed to be perfectly conductive. Postprocessing was done in MATLAB version 6.5 (Mathworks, Natick, MA, USA) after exporting relevant data from the 3-D model into MATLAB to obtain some of the relevant plots. Although the magnitude of the electric fields will change with different fluid conductivities, the field distribution will be similar. Voltage drop within the 3-D electrodes may have a significant effect on the electric field distribution if less conductive electrode materials are used, but this effect is ignored in the simulations. The effects of double layer formation and charging cannot be ignored in fluids with high concentrations of ions, but these effects are ignored as well. Finally, electrohydrodynamic forces, heating effects, and other effects other than pressure-driven fluid flow and dielectrophoresis are not taken into account. For all flow simulations, a pressure difference of 2 kPa is assumed in the horizontal or vertical direction. The velocity field distribution will be similar, but different in absolute magnitude if different flow pressures are applied as long as the flow is kept in the laminar regime. The electric and velocity field simulations in this paper simulate the electric and velocity fields solely with regard to the shape of the electrodes. Other effects such as the particle's effect on the field are ignored. More sophisticated techniques for numerical simulation are given in [30, 31].

17.2.2
Filtration of carbon nanofibers in canola oil

Two fluidic chambers were made using planar 2947 glass microslides (Corning Glass Works, Corning, NY, USA), GE Silicone II (GE Sealants and Adhesives), poly(dimethylsiloxane) (PDMS, created using the Sylgard 184 Silicone Kit from Dow Corning, Midland, MI, USA), Scotch magic tape 810 (3M, St. Paul, MN, USA), and stainless-steel wires (1.58 mm diameter extruded wires). First the wires were polished using 1500 401Q wet/dry microfine imperial sand paper sheets (3M). Thin strips of Scotch tape were wrapped around the ends of the wires. Three layers of Scotch tape (approximately 250 μm) were used as a spacer between the two wires. The two wires were then taped together to form the main separation electrodes. The ends of the wires which were taped together were sealed with Silicone II sealant to prevent leakage to the sides during operation of the device. PDMS sheets were cut into 2 mm thick strips or patterns according to the design. The PDMS sheets and electrode array were assembled between two cleaved glass slides with the help of Silicone II sealant. The two designs used are shown in Figs. 4, 5. In the first design (Fig. 4), a PDMS strip was used for the front and back walls of the fluidic channel, and an outlet orifice was drilled into the back wall. Silicone II was used to create the side walls. In the second design (Fig. 5), U-shaped patterns were cut out of a 2 mm thick PDMS sheet. These patterns as well as the electrode array were then assembled in a manner similar to the first device between two cleaved glass slides. Canola oil (Ventura Foods, LLC, City of Industry, CA, USA) was used as the fluid suspension because of its low conductivity and relative robustness (compared to aqueous solutions). Carbon nanofibers were obtained from W.M. Boughton, Hawaii Industrial Laboratory, Wailuki, HI. The nanofibers have a density of 0.033 g/cc and an iron content of 2.36 wt%. SEM photos of the fibers revealed an approximate diameter of 200 nm for most of the fibers. An HP 8111A pulse/function generator (Hewlett-Packard, Palo Alto, CA, USA) was fed into an AMS-1B30 high-voltage bipolar amplifier (Matsusada Precision, Shiga, Japan), and the

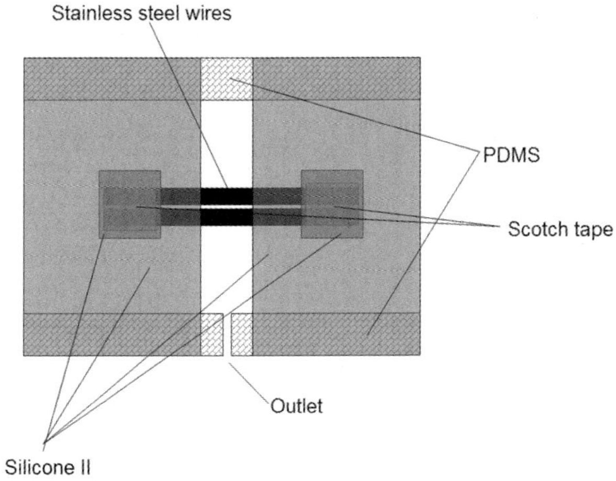

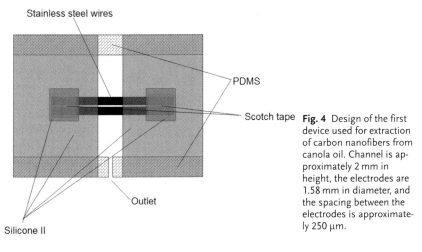

Fig. 4 Design of the first device used for extraction of carbon nanofibers from canola oil. Channel is approximately 2 mm in height, the electrodes are 1.58 mm in diameter, and the spacing between the electrodes is approximately 250 μm.

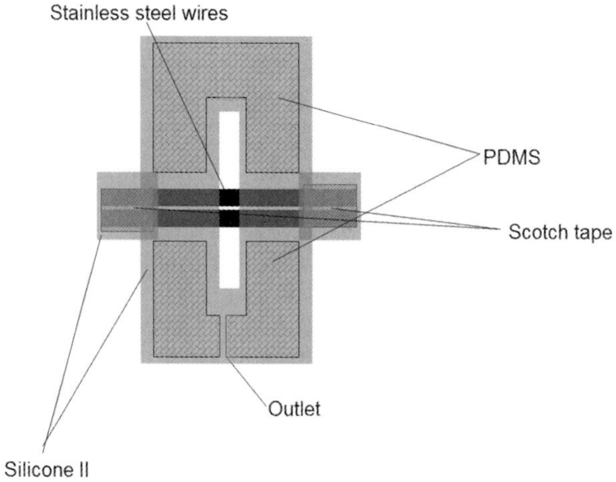

Fig. 5 Design of the second device used for extraction of carbon nanofibers from canola oil. Channel is approximately 2 mm in height, the electrodes are 1.58 mm in diameter, and the spacing between the electrodes is approximately 250 μm.

voltage was applied to the metal wire electrodes through alligator clip contacts. Although the function/pulse generator was capable of generating frequencies up to 20 MHz, the high-voltage amplifier full-scale bandwidth was limited to only 0–10 kHz. After carbon nanofibers were suspended in the canola oil solution, a syringe was filled with the solution. The syringe needle was carefully inserted into the PDMS front wall of the channel, and the plunger was manually pressed to pump the fluid through the fabricated channel.

17.3 Results and discussion

17.3.1 Electric field simulation of 3-D *versus* planar electrodes

Fig. 6 depicts surface plots of $|E|^2$ for 20 μm diameter planar electrodes and 50 μm high 3-D electrodes. The dielectrophoretic force can be inferred from the slope of the surface plots. The figure clearly illustrates the difference in field penetration between planar and 3-D electrodes. Conceptually, the field can be extended by however high the posts can be made. Fig. 7 shows how the dielectrophoretic trapping force ($\propto \nabla|E|^2$) for a four-electrode negative trap differs as a function of distance from the bottom of a microfluidic channel (1–5 μm) for planar and 50 μm high 3-D electrodes. The dielectrophoretic force corresponds to the slope of these plots. The planar trap shows a sharp decay in force as the distance increases while the 3-D trap remains strong. If we assume the height where appreciable dielectrophoretic force is present for planar electrodes is 5 μm and assume a channel height of 100 μm, only 5% of the particles within the fluid will experience significant dielectrophoretic force (assuming uniform distribution of particles). If 100 μm high 3-D electrodes (for example, C-MEMS electrodes) that extend to the ceiling of

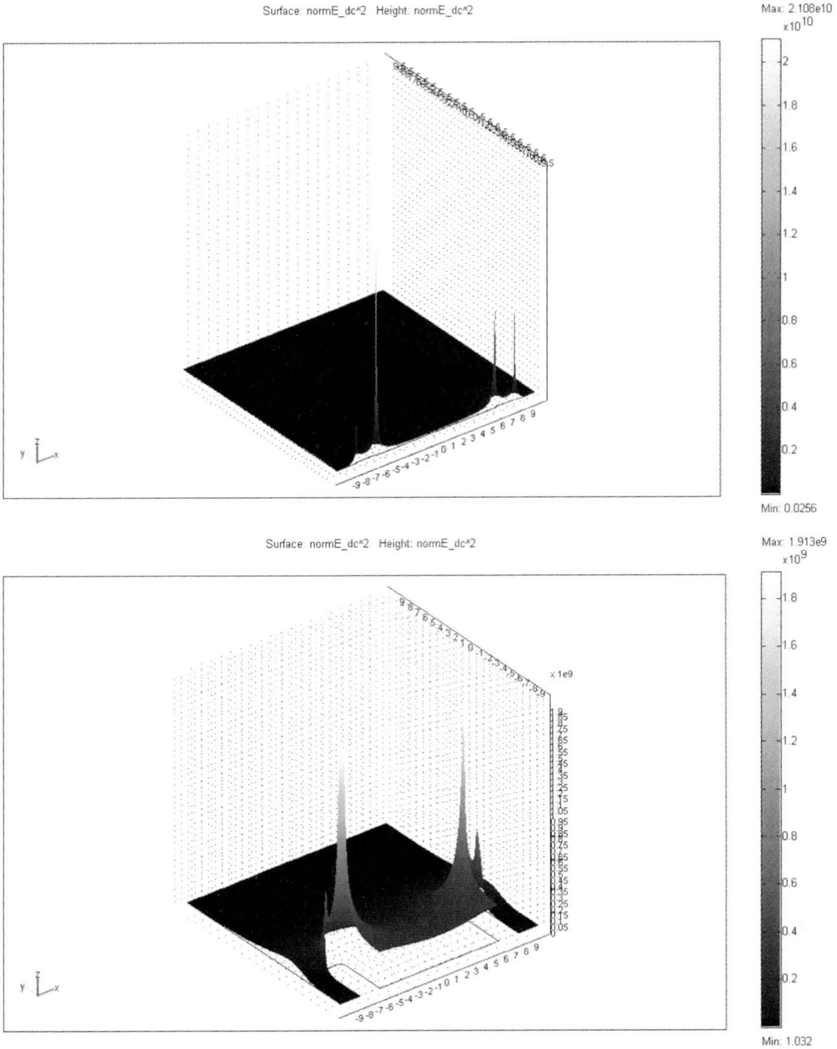

Fig. 6 Simulation results showing the magnitude of the square of the electrical field (V^2/m^2) between pairs of planar (1 μm high) electrodes (top) and 50 μm high 3-D electrodes (bottom). Magnitude of $|E|^2$ is depicted both with grayscale and height. Dielectrophoretic force, which is proportional to the gradient of the magnitude of the squared electric field, can be inferred from the 3-D surface. Plane shown is the vertical plane. Both types of electrodes have a diameter of 20 μm and a center-to-center distance of 140 μm, but the planar electrodes have a height of 1 μm while the 3-D electrodes have a height of 50 μm. Fluid is assumed to be DI water, and a voltage of ±1 V is applied. Electrode edges were slightly chamfered to reduce the edge effects and to accurately represent real world conditions.

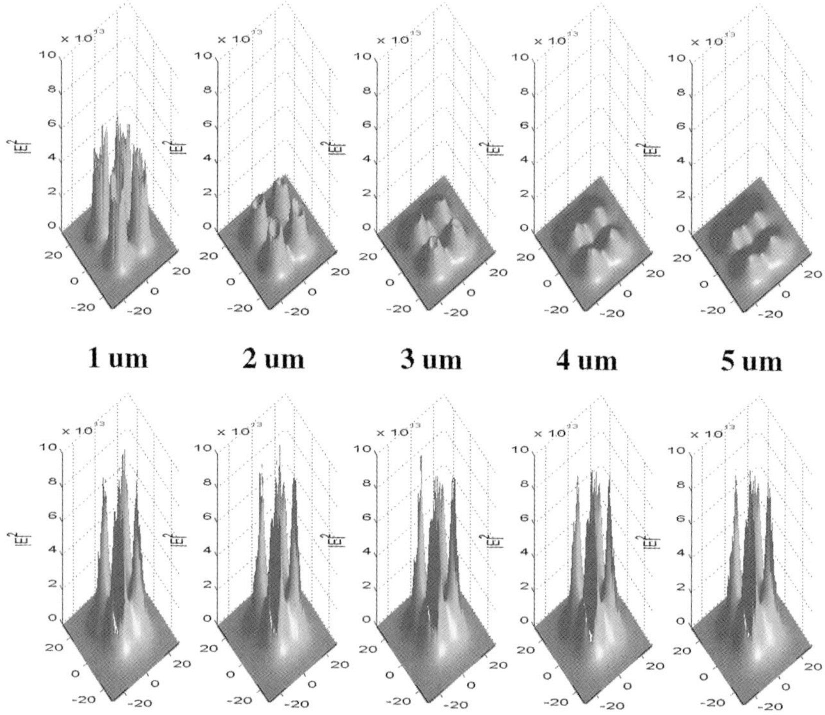

Fig. 7 3-D surface plots depicting the magnitude of $|E|^2$ (V^2/m^2) in various horizontal cross sections at different heights above the channel floor for a planar quadrupole trap (top) and a 50 μm high 3-D quadrupole trap (bottom). Dielectrophoretic force is proportional to the gradient of $|E|^2$, therefore a particle that exhibits positive dielectrophoresis at the given frequency will move up the slope, and the opposite would be true for a particle exhibiting negative dielectrophoresis. Top plot depicts $|E|^2$ for a planar array of electrodes as the distance from the bottom of the channel increases from 1 to 5 μm, and the bottom plot shows $|E|^2$ for 50 μm high 3-D electrodes. Planar trap shows a sharp decay in force while the 3-D trap remains strong. Postdiameters are 10 μm, the center-to-center distance between adjacent posts is 20 μm, and the voltage applied is ±30 V. Voltage on each electrode is opposite in polarity to the adjacent electrodes.

the microchannel are used, all the particles will experience appreciable dielectrophoretic force. A 3-D design used for concentration/filtration/separation can thus easily achieve 2000% the efficiency of a similar design fabricated using planar electrodes. This ratio increases further if the channel height as well as the 3-D electrode height is increased.

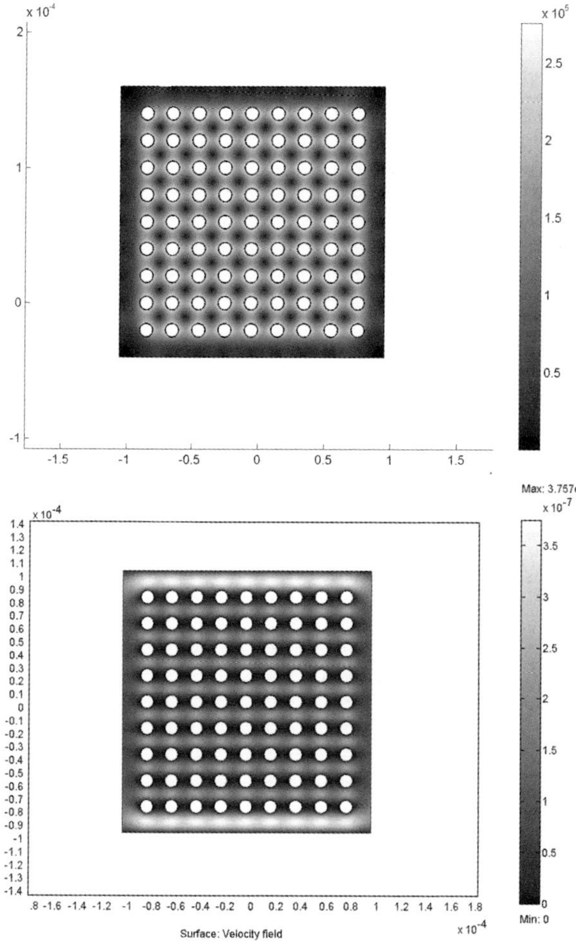

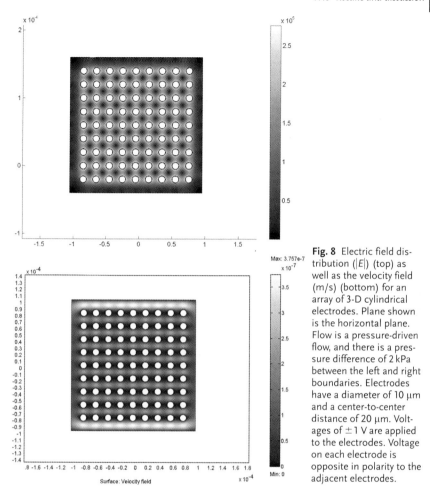

Fig. 8 Electric field distribution ($|E|$) (top) as well as the velocity field (m/s) (bottom) for an array of 3-D cylindrical electrodes. Plane shown is the horizontal plane. Flow is a pressure-driven flow, and there is a pressure difference of 2 kPa between the left and right boundaries. Electrodes have a diameter of 10 μm and a center-to-center distance of 20 μm. Voltages of ±1 V are applied to the electrodes. Voltage on each electrode is opposite in polarity to the adjacent electrodes.

17.3.2
Electric field and velocity field simulations of 3-D electrode designs based on existing 2-D designs

Unlike the conventional planar microelectrodes used in flow-through separator designs, 3-D electrodes will affect the flow velocity field as well as the electric field. The electric field distribution ($|E|$) as well as the velocity field in the horizontal plane for an array of cylindrical electrodes are shown in Fig. 8, and the electric and velocity field distributions in the horizontal plane for an array of 3-D castellated electrodes are depicted in Fig. 9. We are assuming that the electrodes extend from the bottom of the flow channels to the ceilings, and we are ignoring the flow profile in the vertical direction for simplicity. Particles exhibiting positive dielectrophoresis will be attracted to the high-field regions (white), and particles exhibiting negative dielectrophoresis will be attracted to the low-field regions (black). For effective separation, it would be

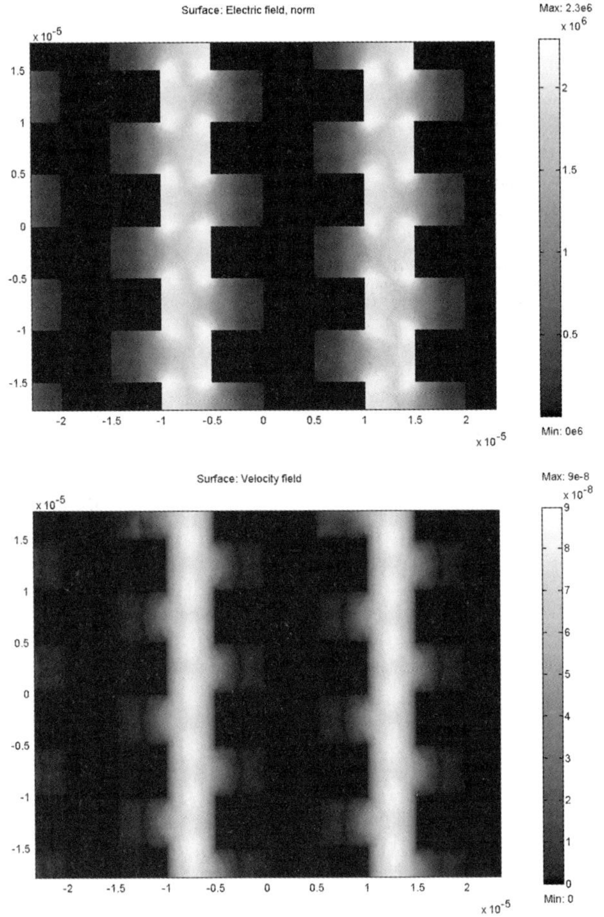

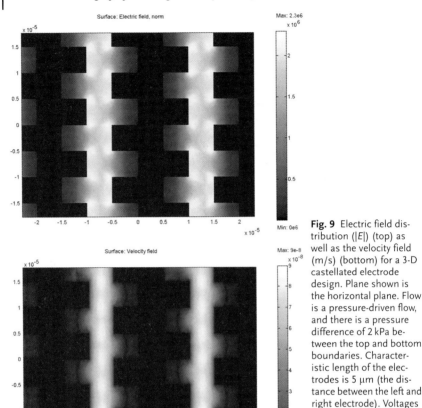

Fig. 9 Electric field distribution (|*E*|) (top) as well as the velocity field (m/s) (bottom) for a 3-D castellated electrode design. Plane shown is the horizontal plane. Flow is a pressure-driven flow, and there is a pressure difference of 2 kPa between the top and bottom boundaries. Characteristic length of the electrodes is 5 μm (the distance between the left and right electrode). Voltages of ±5 V are applied to the electrodes. Voltage on each electrode is opposite in polarity to the adjacent electrode.

beneficial to have high flow velocities present in the high-field regions, and low flow velocities present in the low-field regions if particles exhibiting positive dielectrophoresis are to be retained in the flow-through separation device. High flow velocities in low electric field regions and low flow velocities in high electric field regions are beneficial for a device where particles exhibiting negative dielectrophoresis are to be retained. In other words, we believe that more efficient devices can be designed if the velocity field overlaps significantly with the electric field distribution. Although the electric fields and velocity fields correspond somewhat in the case of the 3-D castellated electrode design (Fig. 9), the electric and velocity fields do not coincide at all in the case of the cylindrical electrode design shown in Fig. 8. If we rotate an array of electrodes similar to that shown in Fig. 8 by 45°, we can obtain a flow velocity field that corresponds much more closely to the electric field. A computer simulation result showing the flow velocity field for such a configuration is shown in Fig. 10. Note that, in this configuration, the flow velocities are lowest near the electrodes (the high electric field regions) and that the velocity field coincides somewhat with the electric field.

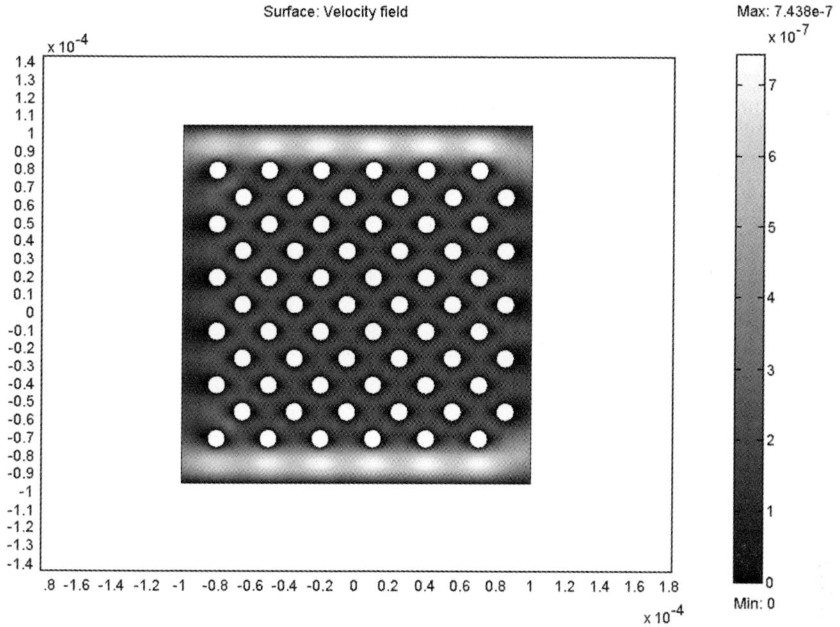

Fig. 10. Velocity field of an array of electrodes similar to that shown in Fig. 8 rotated by 45° in the horizontal plane. Flow is a pressure-driven flow, and there is a pressure difference of 2 kPa between the left and right boundaries.

17.3.3
Correlation of velocity field and electric field to enhance dielectrophoretic separation in flow-through systems

Although there has been previous work using 3-D structures to change the electric field within a fluid [28, 29], to the best of our knowledge, there has been no attempts at designing 3-D electrodes so that the velocity field coincides with the electric field. Additionally, most 3-D electrode designs have been simple extensions of 2-D designs. We believe that there is merit to 3-D electrode designs that are not based on previous 2-D architectures and are designed to maximize the correlation between electric and velocity fields. An example of such a design is shown in Fig. 11. The electric and velocity fields of a novel 3-D electrode design for use in dielectrophoretic separation/concentration/filtration devices are presented in this figure. The design consists of long semicylindrical electrodes that are placed in close proximity of each other. Field strengths will be lowest near smooth surfaces and highest near sharp edges as well as locations where the electrodes are closer apart. A particle that exhibits positive dielectrophoresis at a given frequency will be attracted to the narrow spaces between the semicircular electrodes. To better illustrate the dielectrophoretic force a particle "feels," a 3-D surface plot of $|E|^2$ is given in Fig. 12. The dielectrophoretic force is proportional to the gradient of $|E|^2$, there-

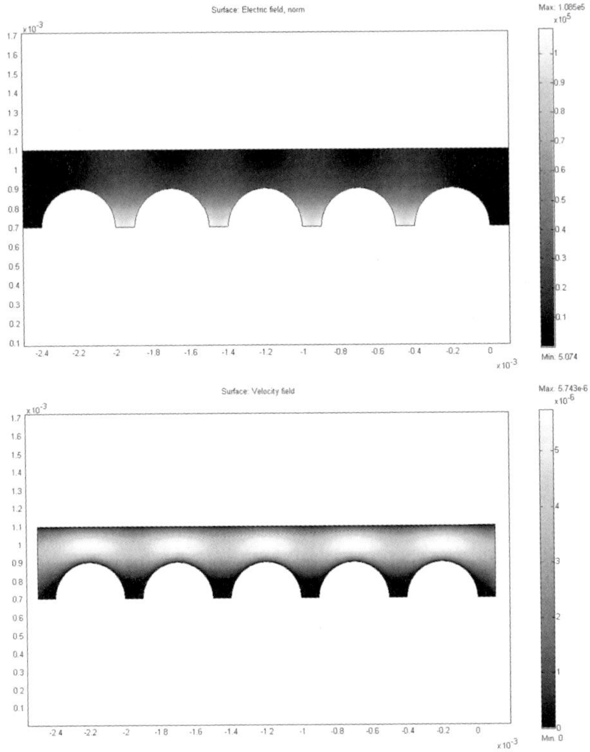

Fig. 11. Electric field distribution (|*E*|) (top) as well as the velocity field (m/s) (bottom) for a novel 3-D semicircular electrode design for dielectrophoretic separation. Plane shown is the vertical plane. Flow is a pressure-driven flow, and there is a pressure difference of 2 kPa between the left and right boundaries. Diameter of the electrodes is 400 μm and the distance between the electrodes is 100 μm. Voltages of ±5 V are applied to the electrodes. Voltage on each electrode is opposite in polarity to each adjacent electrode.

fore a particle that exhibits positive dielectrophoresis at the given frequency would move "up" the slope, and the opposite would be true for a particle exhibiting negative dielectrophoresis. The velocity of the fluid passing through the channel will be greatest in the solution above the electrodes (the low-field region), and the lowest where the particles experiencing positive dielectrophoresis are trapped (the high-field regions). This design has the following advantages: (i) The velocity field coincides very well with the electric field distribution, causing particles that exhibit positive dielectrophoresis to be trapped in the crevices while particles that exhibit negative dielectrophoresis are effectively washed away. (ii) The bottom surface of the flow-through channel between the electrodes can be functionalized. If the particulate we wish to concentrate exhibits positive dielectrophoresis, an assay can be done in parallel with the concentration/separation step. For example, for a DNA assay, a monolayer of DNA can be immobilized on the surface between the electrodes. The 3-D electrodes would locally increase the DNA concentration between the electrodes increasing the signal intensity as well as decreasing the response time for the assay.

Multiple methods are envisioned to shape the electrodes described in the previous paragraph. One method is to carbonize a shaped polymer using the C-MEMS process. The polymer can be shaped using many methods: (i) Photoresist can be patterned into high-aspect-ratio structures using a high-transparency resist such as

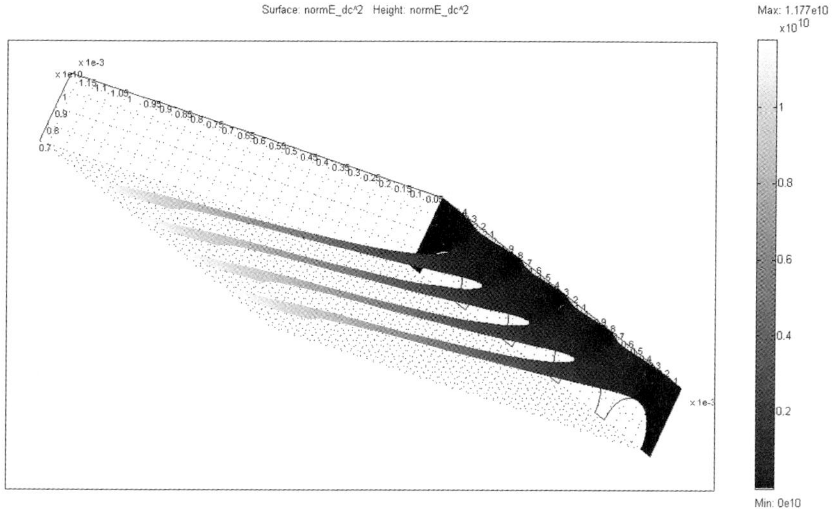

Fig. 12. 3-D surface plot depicting the magnitude of $|E|^2$ (V^2/m^2) for the electrode configuration depicted in Fig. 11. Dielectrophoretic force is proportional to the gradient of $|E|^2$, therefore a particle that exhibits positive dielectrophoresis at the given frequency will move up the slope, and the opposite would be true for a particle exhibiting negative dielectrophoresis.

SU-8. The resist can then be flowed by sharply increasing the temperature above the glass transition temperature (T_g). (ii) A polymer can be molded using a variety of molding techniques. (iii) A liquid polymer can be silk-screened onto the substrate. (iv) The polymer can be expelled through multiple nozzles onto a substrate to create all the electrode traces simultaneously. Methods 1–3 are illustrated in Fig. 13. Other methods of creating long semicircular electrodes include molding a molten metal and extruding a molten metal through multiple nozzles.

Another new electrode design that has most of the advantages of the previous design consists of closely spaced wire electrodes. Two possible flow-through designs are shown in Figs. 4, 5. Computer simulated results showing the electric field distribution and velocity field of two arrays of closely spaced wire electrodes are shown in Fig. 14, and a 3-D surface plot of $|E|^2$ is given in Fig. 15. This design has the following advantages: (i) Multiple arrays of electrodes can be stacked for increased throughput. (ii) This design is much easier to fabricate because the spacing between the electrodes can be controlled using thin spacer materials. This is illustrated in Fig. 16. The thickness of the spacer will determine the magnitude of the electric field as well as the decay of the field gradient for a certain voltage. Thinner gaps between the wires are advantageous for creating high-field gradients, but there is a possibility of shorting or arcing due to the close proximity of the wires. It may be advantageous to make the wires short so that flexing of the wires is minimized. Also, the diameter of the wires will have a direct impact on the field created by the device.

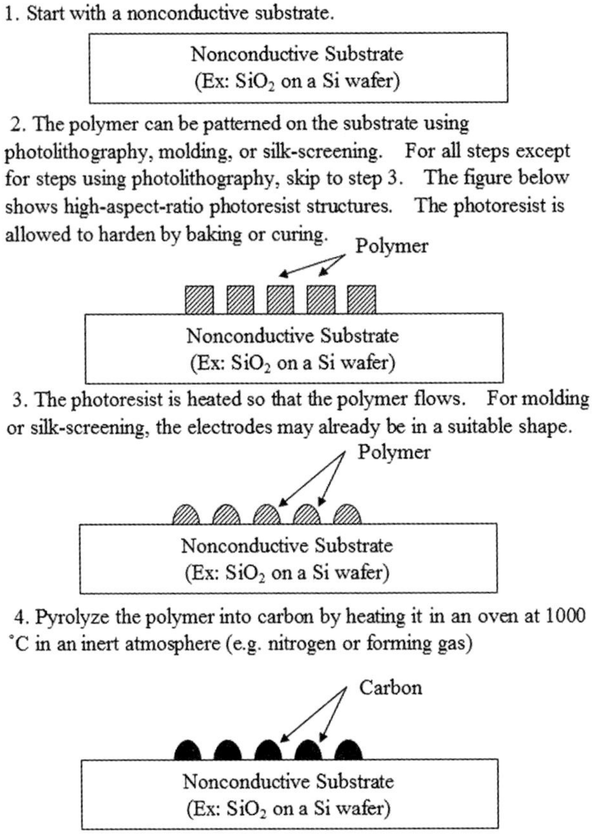
1. Start with a nonconductive substrate.
Nonconductive Substrate
(Ex: SiO_2 on a Si wafer)
2. The polymer can be patterned on the substrate using photolithography, molding, or silk-screening. For all steps except for steps using photolithography, skip to step 3. The figure below shows high-aspect-ratio photoresist structures. The photoresist is allowed to harden by baking or curing.
Polymer
Nonconductive Substrate
(Ex: SiO_2 on a Si wafer)
3. The photoresist is heated so that the polymer flows. For molding or silk-screening, the electrodes may already be in a suitable shape.
Polymer
Nonconductive Substrate
(Ex: SiO_2 on a Si wafer)
4. Pyrolyze the polymer into carbon by heating it in an oven at 1000 °C in an inert atmosphere (e.g. nitrogen or forming gas)
Carbon
Nonconductive Substrate
(Ex: SiO_2 on a Si wafer)

Fig. 13. Methods of creating long semicircular electrodes using photolithography, molding, or silk-screening is depicted. Methods involve a carbonization step that converts the polymers into carbon.

17.3.4
Experimental validation of design: purification of canola oil with carbon nanofiber contaminants

The designs shown in Figs. 4, 5 were fabricated and the experiments were performed using the designs and equipment detailed in Section 2.2. Fig. 17 is a photo of an actual fabricated device based on the design shown in Fig. 5. Before experimenting with the flow-through device, the reaction of the carbon nanoparticles to applied electric fields was determined. The carbon nanofibers exhibited positive dielectrophoresis at frequencies between 0 and 10 kHz. We believe that most of the polarizability difference between the carbon nanofibers and the oil is due to the difference in conductivity. The voltage amplifier was not able to fully amplify signals at higher frequencies. A 5.09 kHz sine wave was used for all of the experiments. The dielectrophoretic strength increased (evident by observing the movement of nanoparticles) when higher voltages were applied, but arcing occurred within a short time when voltages in excess of 250 $V_{p\text{-}p}$ were applied due to carbon particles providing a conduction path between the electrodes. To perform the flow-

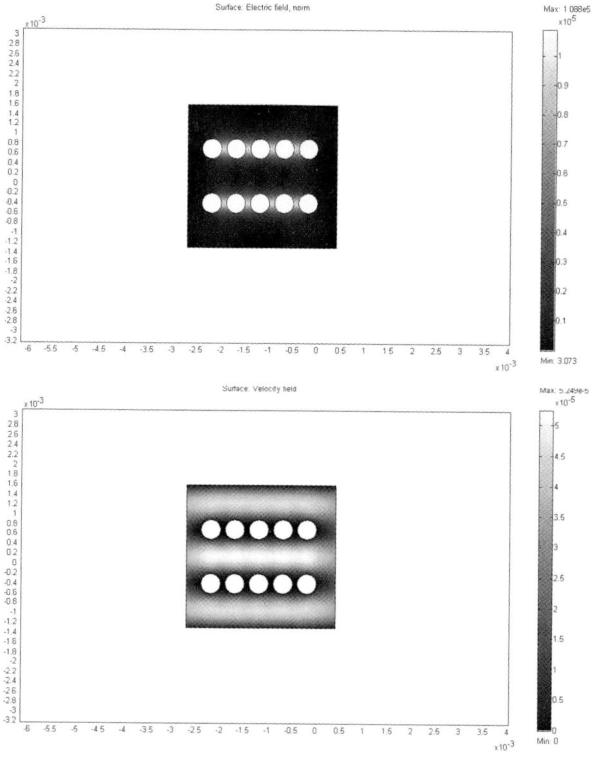

Fig. 14. Electric field distribution (|*E*|) (top) as well as the velocity field (m/s) (bottom) for a novel 3-D wire electrode design for dielectrophoretic separation. Wires are kept at a constant distance by the use of thin spacers. Plane shown is the vertical plane. Flow is a pressure-driven flow, and there is a pressure difference of 2 kPa between the left and right boundaries. Diameter of the electrodes is 400 μm and the distance between the electrodes is 100 μm. Voltages of ±5 V are applied to the electrodes. Voltage on each electrode is opposite in polarity to each adjacent electrode.

through experiments, the device was first filled with clean canola oil using a clean syringe, and then the canola oil/carbon nanofiber mixture was slowly injected into the channel. When a voltage of 250 $V_{p\text{-}p}$ was applied to the electrodes, the carbon was quickly trapped between the electrodes. A photo taken during flow-through experimentation showing carbon particle trapping is shown for the two devices in Fig. 18. After a quick initial deposition of carbon between the electrodes, arcing occurred through the carbon. Continued arcing caused bubble formation within the channel. Arcing was delayed when a voltage of 152 $V_{p\text{-}p}$ was applied, but the carbon nanofibers accumulated at a slower rate. We observed that once a small amount of nanofibers is trapped between the electrodes, the trapping effect is noticeably reduced (particles were passing the two electrodes without being pulled between the electrodes). Closer inspection revealed that the carbon particles were forming "pearl chains" [2] between the two electrodes. We believe that these carbon particles are changing the electric field distribution, affecting the collection efficiency of the device. The aggregation behavior of the carbon particles can be reduced by applying several frequencies simultaneously, or by switching the applied frequencies [11], but arcing/shorting is unavoidable when large amounts of conductive particles are collected between the two electrodes. Although detrimental in many cases, we believe that this arcing/shorting behavior can be exploited as a particulate sensing mechanism. The device can double as an online monitoring

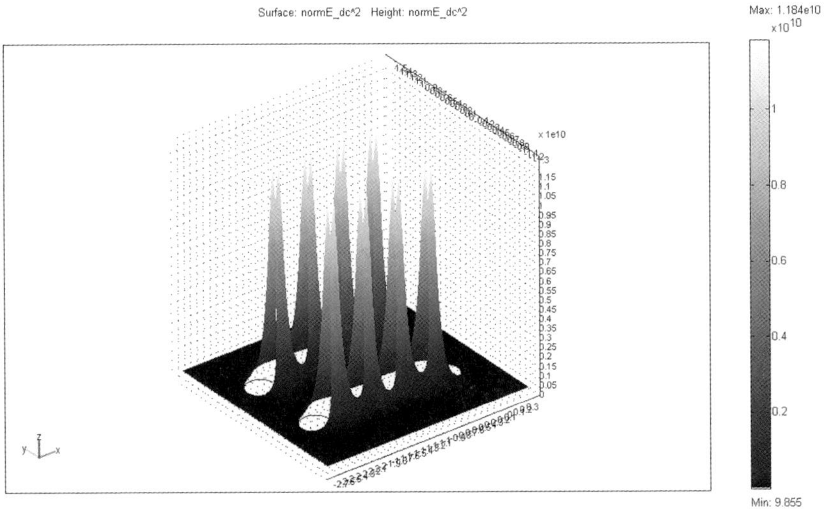

Fig. 15. 3-D surface plot depicting the magnitude of $|E|^2$ (V^2/m^2) for the electrode configuration depicted in Fig. 14. Dielectrophoretic force is proportional to the gradient of $|E|^2$, therefore a particle that exhibits positive dielectrophoresis at the given frequency will move up the slope, and the opposite would be true for a particle exhibiting negative dielectrophoresis.

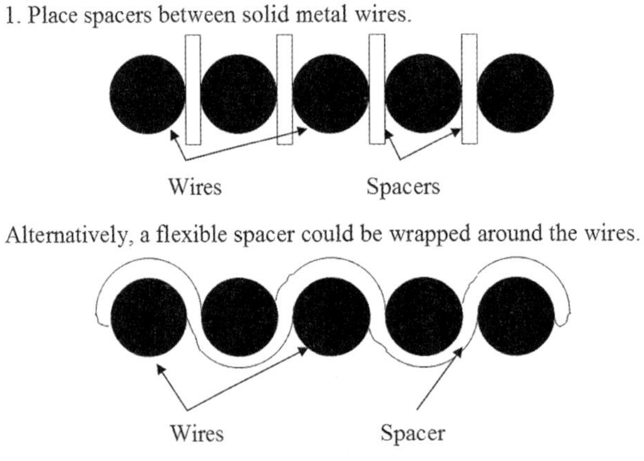

Fig. 16. Use of thin spacer materials to control the distance between wire electrodes is illustrated.

tool that senses the amount of conductive particulates trapped between the electrodes by sensing the impedance or current between the two electrodes. Sustained arcing can be prevented by using a current-limiting voltage source. Flow-through devices using similar designs can be used to filter conductive contaminants suspended in an electrically insulating medium. This is especially important in applications where the contaminant size is small (<15 μm), making it difficult to filter the fluid using physical filters. Filtering efficiency can be improved by increasing the number of closely spaced electrodes, and filtering throughput improved by stacking many arrays of closely spaced electrodes. A photo of the prototype for an electrode

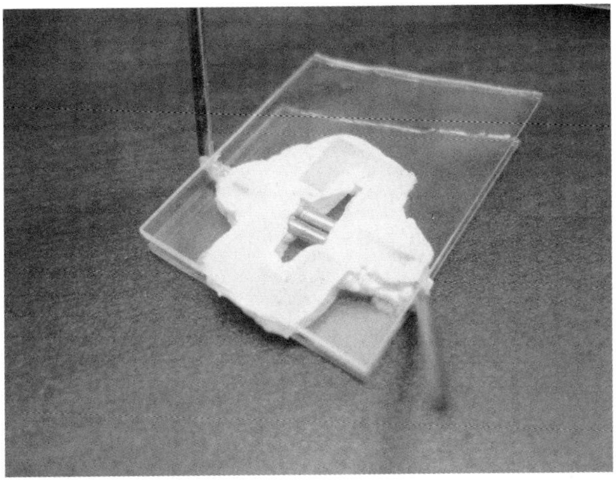

Fig. 17. Photo of the device based on the design shown in Fig. 5 after fabrication.

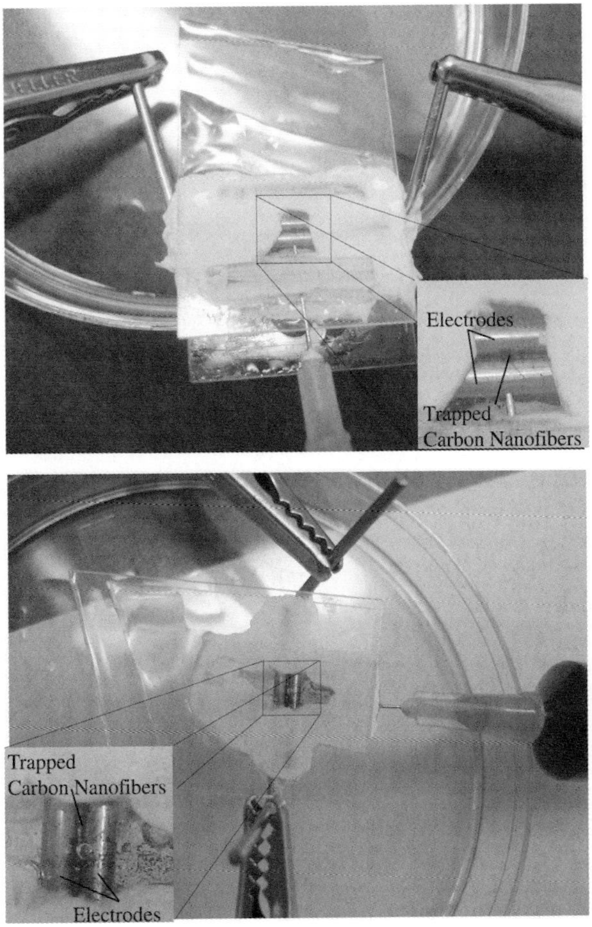

Fig. 18. Photo taken during flow-through experimentation showing carbon particle trapping. Two insets show close-up views near the electrodes where the trapping occurs. The top photo is a photo of a device fabricated according to the design shown in Fig. 4. The bottom photo is a photo of a device fabricated according to the design shown in Fig. 5.

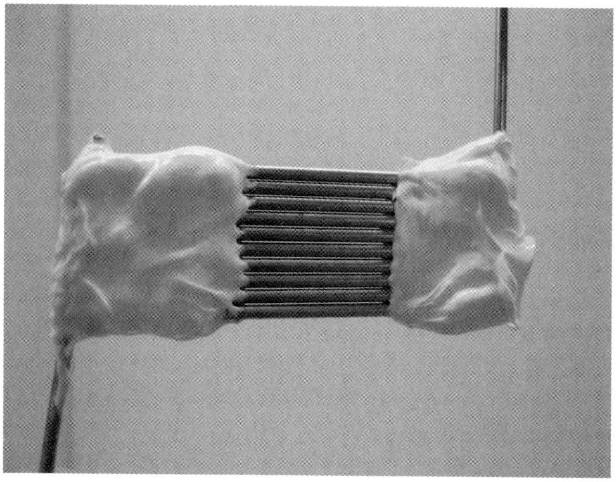

Fig. 19. Photo of the prototype for an electrode array consisting of ten electrodes. Every other wire electrode was soldered and connected to a common wire so that the whole device only has two contact points. Paper was used as the spacer material between the electrodes (~100 m). Separators and soldered connections were encased in Silicone II.

array consisting of ten electrodes is shown in Fig. 19. Every other wire electrode was soldered and connected to a common wire so that the whole device only has two contact points. Paper was used as the spacer material between the electrodes (~100 μm). The separators and soldered connections were encased in Silicone II.

17.4
Concluding remarks

The use of 3-D electrode designs for high-throughput dielectrophoretic separation/concentration/filtration systems is beneficial because: (i) They provide a method of extending the electric field within the fluid. (ii) The 3-D electrodes can be designed so that the velocity field coincides with the electric field distribution. (iii) Novel electrode designs, not based on planar electrodes designs, can be developed and used. The electric field distribution and velocity fields of 3-D electrodes based on 2-D designs as well as two novel electrode designs have been presented, and a proof-of-concept experimental device based on one of the novel designs was successfully used to extract nanofibrous carbon from canola oil.

We believe that a device similar to the device tested can be used to filter conductive contaminants (*e.g.*, metal particles from an engine, carbon soot [32]) suspended in an electrically insulating medium (*e.g.*, a lubricant). For example, such a device could supplement a physical oil filter in an automobile and trap conductive particles that are too small to be physically filtered. The device could double as an online monitoring system measuring the level of contaminants in the lubricant and provide feedback to the owner of the vehicle so that timely oil change operations can be performed.

The authors would like to thank Alia Marafie, Rabih Zaouk, Mihri Ozkan, Jim Zoval, and Chunlei Wang for the many extremely useful discussions on FEMLAB, C-MEMS, and fabrication techniques. The authors would also like to thank the UCI Microsystems laboratory for allowing access to the high-voltage amplifier. All the C-MEMS devices shown were fabricated at the Integrated Nanosystems Research Facility at the University of California, Irvine. This work was supported in part by National Science Foundation under contract No. DMI-0428958 through Mechanical and Aerospace Engineering, University of California, Irvine.

17.5 References

[1] Park, B. Y., Zaouk, R., Madou, M., *SPIE Microlithography, Emerging Lithographic Technologies VIII*, SPIE, Bellingham, WA 2004, pp. 566–578.

[2] Morgan, H., Green, N. G., *AC Electrokinetics: Colloids and Nanoparticles*, Research Studies Press, Baldock, Hertfordshire, England 2003.

[3] Weisman, R. B., *Nat. Mater.* 2003, *2*, 569–570.

[4] Pohl, H. A., *Dielectrophoresis*, Cambridge University Press, Cambridge, UK 1978.

[5] Pohl, H. A., Schwar, J. P., *J. Appl. Phys.* 1959, *30*, 69–73.

[6] Pohl, H. A., Hawk, I., *Science* 1966, *152*, 647–649.

[7] Becker, F. F., Wang, X.-B., Huang, Y., Pethig, R., Vykoukal, J., Gascoyne, P. R. C., *Proc. Natl. Acad. Sci. USA* 1995, *92*, 860–864.

[8] Wang, X.-B., Yang, J., Huang, Y., Vykoukal, J., Becker, F. F., Gascoyne, P. R. C., *Anal. Chem.* 2000, *72*, 832–839.

[9] Docoslis, A., Kalogerakis, N., Behie, L. A., Kaler, K. V. I. S., *Biotechnol. Bioeng.* 1997, *54, 239–250.*

[10] Becker, F. F., Wang, X.-B, Huang, Y., Pethig, R., Vykoukal, J., *J. Phys. D* 1994, *27*, 2659–2662.

[11] Gascoyne, P. R. C., Huang, Y., Pethig, R., Vykoukal, J., Becker, F. F., *Meas. Sci. Technol.* 1992, *3*, 439–445.

[12] Markx, G. H., Dyda, P. A., Pethig, R., *J. Biotechnol.* 1996, *51*, 175–180.

[13] Lapizco-Encinas, B. H., Simmons, B. A., Cummings, E. B., Fintschenko, Y., *Electrophoresis* 2004, *25*, 1695–1704.

[14] Krupke, R., Hennrich, F., v. Lohneysen, H., Kappes, M., *Science* 2003, *301*, 344–347.

[15] Pickett, J. P. (Ed.), *The American Heritage Dictionary of the English Language*, 4th Edn., Houghton Mifflin Company, Boston 2000.

[16] Fodor, J., *Tribal Int.* 1979, *12*, 127–129.

[17] Jost, H. P., *Wear* 1990, *136*, 1–17.

[18] Wakuri, Y., Ono, S., *Experimental Studies on the Abnormal Wear of Cylinder Liners and Piston Rings in a Marine Diesel Engine*, International Symposium on Marine Engineering, Tokyo, November 12–15, 1973, pp. 47–58.

[19] Khorshid, E. A., Nawwar, A. M., *Wear* 1991, *141*, 349–371.

[20] Lee, S. W., Bashir, R., *Appl. Phys. Lett.* 2003, *83*, 3833–3835.

[21] Madou, M. J., Lal, A., Schmidt, G., Song, X., Kinoshita, K., Fendorf, M., Zettl, A., White, R., in: *Electrochemical Society Proceedings, Vol. 19*, 1997, p. 61.

[22] Ranganathan, S., McCreery, R., Majji, S. M., Madou, M. J., *J. Electrochem. Soc.* 2000, *147*, 277–282.

[23] Wang, C., Jia, G., Taherabadi, L., Madou, M. J., IEEE JMEMS 2005, *14*, 348–358.

[24] Singh, A., Jayaram, J., Madou, M., Akbar, S., *J. Electrochem. Soc.* 2002, *149*, E78–E83.

[25] Kim, J., Song, X., Kinoshita, K., Madou, M. J., White, R., *J. Electrochem. Soc.* 1998, *145*, 2314–2319.

[26] Wang, C., Taherabadi, L., Jia, G., Madou, M. J., *Electrochem. Solid-State Lett.* 2004, *7*, A435–A438.

[27] Park, B. Y., Taherabadi, L., Wang, C., Zoval, J., Madou, M. J., *J. Electrochem. Soc.* 2005, in press.

[28] Voldman, J., Toner, M., Gray, M. L., Schmidt, M. A., *J. Electrostat.* 2003, *57*, 69–90.

[29] Lapizco-Encinas, B. H., Simmons, B. A., Cummings, E. B., Fintschenko, Y., *Electrophoresis* 2004, *25*, 1695–1704.

[30] Kadaksham, J., Singh, P., Aubry, N., *Electrophoresis* 2004, *25*, 21–22, 3625–3632.

[31] Kadaksham, J., Singh, P., Aubry, N., *J. Fluids Eng.* 2004, *126(2)*, 170–179.

[32] Kawamura, M., Ishiguro, T., Fujita, K., Morimoto, H., *Wear* 1988, *123*, 269–280.

[33] Madou, M. J., *Fundamentals of Microfabrication: The Science of Miniaturization*, 2nd Edn., CRC Press, Boca Raton, FL 2002.

18
Parallel mixing of photolithographically defined nanoliter volumes using elastomeric microvalve arrays*

*Nianzhen Li***, *Chia-Hsien Hsu***, *Albert Folch*
** *These authors contributed equally to the work.*

Portable microfluidic systems provide simple and effective solutions for low-cost point-of-care diagnostics and high-throughput biomedical assays. Robust flow control and precise fluidic volumes are two critical requirements for these applications. We have developed a monolithic polydimethylsiloxane (PDMS) microdevice that allows for storing and mixing subnanoliter volumes of aqueous solutions at various mixing ratios. Filling and mixing is controlled *via* two integrated PDMS microvalve arrays. The volumes of the microchambers are entirely defined by photolithography, hence volumes from picoliter to nanoliter can be fabricated with high precision. Because the microvalves do not require an energy input to stay closed, fluid can be stored in a highly portable fashion for several days. We have confirmed the mixing precision and predictability using fluorescence microscopy. We also demonstrate the application of the device for calibrating fluorescent calcium indicators. Due to the biocompatibility of PDMS, the device will have broad applications in miniaturized diagnostic assays as well as basic biological studies.

18.1
Introduction

Miniaturization of biochemical assays in “lab-on-a-chip” devices is revolutionizing the biomedical benchtop. Microfluidic systems, especially combined with soft lithography techniques (for a review, see [1]), provide simple, low-cost, and high-throughput implementations for these miniaturization platforms. Portable microfluidic systems with precise fluidic dispensing volumes and integrated flow control functions are of great interest in numerous biomedical assays.

* Originally published in Electrophoresis 2005, 26, 3758–3764

Microfluidic Applications in Biology. Edited by Niels Lion, Joël S. Rossier, and Hubert H. Girault

ISBN-10: 3-527-31761-9

Various methods have been reported to construct microvalves for controlling fluid flow. Beebe *et al.* [2] employed the expansion and contraction of pH-responsive hydrogel components to regulate fluid flow in microchannels. Quake and coworkers demonstrated the integration of pressure-driven polydimethylsiloxane (PDMS) "pinch-type" valves in microfluidic PDMS devices [3] for applications ranging from protein crystallization [4] and cell sorting [5]. The valve is constructed by overlapping two microchannels (a "control" channel atop a fluidic channel) orthogonally, separated by a thin (< 30 μm thick) PDMS membrane. The membrane deflects when the control channel is pressurized, which interrupts flow in the fluidic channel in a manner akin to pinching a flexible tube. Due to the hydrophobic nature of PDMS, the valves are leak-proof because water is naturally excluded from the spacing between the PDMS membrane and the bottom of the channel. A clever multiplexing scheme allows for controlling N channels using only $2 \log_2 N$ control channels (*e.g.*, 1000 microvalves using only 18 channels) [6].

However, microvalves in the aforementioned configurations require extra energy source to close the fluidic channel. As a result, the devices are not well suited for applications where the assay is not necessarily performed at the same location as the one where (some or all) reagents were loaded into the device. For valves to completely close, the pinch-type configuration [6] also requires the fluidic microchannels to be fabricated with a rounded cross-section (a square cross-section microchannel cannot be completely flattened at the walls). While satisfactory for many applications, the "photoresist reflow" method for fabricating rounded microchannels is not optimal, because the width of the features influences their final height; hence large heights (> 20 μm) and aspect ratios are not possible for every given feature width, which makes it nonideal for studies where a wide range of designs and heights on one device is critical (*e.g.*, cellular studies [7]).

We have adapted a previously reported PDMS microvalve design that is closed at rest [8–10] to demonstrate the parallel mixing of subnanoliter volumes. The main advantages of this microvalve design are that: (i) No extra energy source is required to close the fluidic path, hence the loaded device is highly portable; and (ii) the device can be built by PDMS replicas from photolithographically patterned SU-8 molds, allowing for microfabricating deep (up to 1 mm) channels with vertical sidewalls (*i.e.*, the height of the features can be specified independently of their width) [11] and resulting in very precise features. The prototype device features two arrays of photolithographically defined nanoliter fluidic chambers and two individually controlled sets of microvalves. We have assessed and verified the precision of the mixing with fluorescence microscopy and applied it to calibrate fluorescent calcium indicators using the micromixer. The device allowed for storage of subnanoliter sized fluid volumes for at least 7 days, which is highly desirable in portable point-of-care diagnostics.

18.2
Materials and methods

18.2.1
Fabrication of silicon masters

Standard SU-8 photolithography methods were used to create masters for the microfluidic layer and the valve control layer [12]. We designed the masks using CorelDraw software, and printed transparency masks at 8000 dots *per* inch (CAT/Art services, Poway, CA, USA). Negative photoresist (SU-8 2050; MicroChem, Newton, MA, USA) was spun on silicon wafers at 500 rpm for 10 s and 2300 rpm for an additional 30 s, prebaked by ramping up at 4°C/min to 95°C (15 min) with a programmable hot plate (Dataplate Hot Plate/Stirrer Series 730; Barnstead International, Dubuque, IA, USA), and exposed for 90 s with collimated UV light (Kaspar–Quintel Model 2001 aligner; the irradiances at 365, 405, and 436 nm are 0.2, 0.4, and 0.3 mW/cm^2, respectively). After exposure, the SU-8 substrate was heated by ramping up at 4°C/min to 95°C (5 min), allowed to cool at the same rate, and developed with SU-8 developer (MicroChem) at room temperature for 15 min. The heights of the SU-8 mold features were then measured with a Tencor P15 surface profiler (KLA-Tencor, San Jose, CA, USA). For these parameters, the typical feature height was ~42 μm. Other heights could be obtained by varying the spinning speed according to the SU-8 manufacturer's instructions.

18.2.2
Replica molding of PDMS from the master

To facilitate release, prior to PDMS replication the SU-8 master was silanized by exposure to a vapor of a fluorosilane ((tridecafluoro-1,1,2,2-tetrahydrooctyl)-1-trichlorosilane; United Chemical Technologies, Bristol, PA, USA) in house vacuum at room temperature for 30 min. A mixture of PDMS prepolymer and curing agent (10:1 w/w, Sylgard 184, Dow-Corning, Midland, MI, USA) was cast against the SU-8 master and cured at 65°C for 3 h. Pieces of silicone tubing were embedded into PDMS to create access inlets in the molds.

18.2.3
Volume measurement of each microchamber

The height of each microchamber was obtained by measuring the corresponding posts in the SU-8 master with the profilometer. The area of each microchamber was calculated by calibration from high-magnification images of the PDMS fluidic layer. The measured volumes were then compared with the expected values.

18.2.4
Thin PDMS membrane

In our microvalve devices, the middle layer consists of a ~12 μm thick PDMS membrane fabricated as in [13]. Briefly, a 10:1 weight ratio of PDMS prepolymer/curing agent mixture was mixed with hexane (3:1 weight ratio) and spun (inside a clean room) on a 3 in. diameter silicon wafer (derivatized with fluorosilane), resulting in a PDMS thin film. To cure the PDMS film, the wafer was heated to 85°C for 4 min on a hot plate. A dust-free environment is critical for ensuring that the PDMS membranes are free of defects; dust particles may result in membranes containing holes and/or defectively bound to the replica mold.

18.2.5
Device assembly and operation

The PDMS replica of control layer and the PDMS-coated wafer were brought into conformal contact for 10 min. To ensure bonding, the thin film and the replica mold were oxidized in oxygen plasma (Branson/IPC 2000 barrel etcher, 150 W, 1 Torr) for 30 s prior to bringing the two surfaces into contact [14]. The PDMS mold was then peeled from the silicon wafer, with the thin film chemically bound to the replica mold. This two-layer set was then visually aligned and sealed to the fluidic layer under a Nikon SMZ1500 stereoscope; this microscope was equipped with all the accessories necessary for fluorescence imaging. For opening and closing valves, pressures were controlled by a vacuum line (−30 kPa) and an air pressure line (10 kPa) connected through two pressure regulators to an array of miniature three-way solenoid valves (LHDA0511111H; Lee Company, Westbrook, CT, USA). The solenoid valves were connected to National Instruments data acquisition hardware (PCI 6025E, CB-50LP; National Instruments, Austin, TX, USA) controlled *via* Labview software (National Instruments).

18.2.6
Image acquisition and analysis

Fluorescein and Fluo-4 potassium salt were purchased from Molecular Probes (Eugene, OR, USA). For fluorescence imaging, light from a Mercury lamp (100 W) was filtered at 480 nm (480DF10; Omega Optical, Brattleboro, VT, USA). Fluorescent emission was collected through a dichroic mirror (510DRLP; Omega Optical) and filtered with a 515EFLP filter (Omega Optical). Images were taken with a SPOT RT CCD camera (Diagnostic Instruments, Sterling Heights, MI, USA) and analyzed with MetaMorph software (Universal Imaging, West Chester, PA, USA). To obtain the intensity profiles, the raw images were first corrected for background light and the unevenness of field illumination using a standard homogeneously doped fluorescent glass slide (Meridian Instruments, Kent, WA, USA). Three 20×20 pixel squares were then marked on each chamber pair and the average regional intensity was measured. We noticed that in the areas where the fluidic

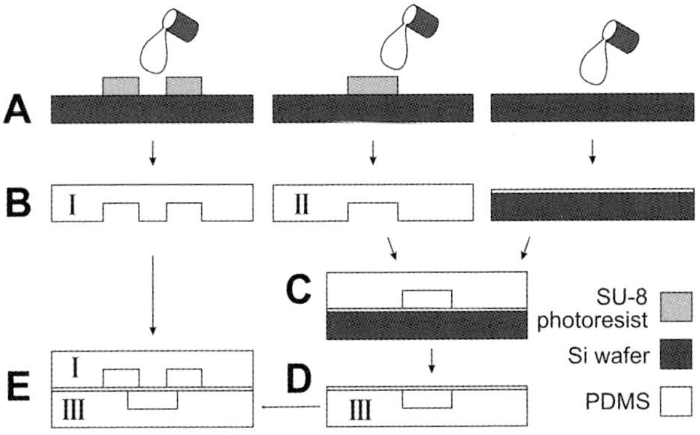

Fig. 1 Device fabrication and assembly. Schematics of soft lithography procedures to fabricate the fluidic layer ("I", left column), control layer ("II", middle column), and the thin PDMS membrane (right column). (A) PDMS prepolymer is poured onto a fluorosilanized, photolithographically patterned SU-8 master to mold fluidic and control layers; the thin PDMS membrane is made by spinning PDMS on a fluorosilanized silicon wafer. (B) Cross-section view of the resulting PDMS replicas. (C) Bonding of the control layer to the PDMS membrane after oxygen plasma oxidation of both. (D) All-PDMS control-layer assembly after release from silicon wafer (III). (E) Sealing of the fluidic layer (I) on top of the PDMS control layer (III) so that valve seats overlap two fluidic paths.

chambers overlapped with the control layer channels and valves, fluorescence intensity was slightly higher than in other areas, possibly due to the small deflection of thin membrane by pressure built up in the fluidic chambers; therefore the measured squares were picked outside of these overlapping areas.

18.3 Results and discussion

18.3.1 Fabrication and operation of device

We adapted a previously reported PDMS microvalve design [8–10], with the device consisting of three layers: a fluidic layer containing microchambers of various sizes, a "control layer" containing the microchannels necessary to actuate the microvalves, and a middle thin PDMS membrane that is bound to the control layer. Fig. 1 illustrates the modified fabrication procedures to create the three layers and assemble them to a final device (details in Section 2). In the original report from Hosokawa and Maeda [15], the authors had inlets to the fluidic layer on the top side of the device and inlets to the control layer from the bottom side of the device. This inlet placement makes devices difficult to visualize under the microscope. In our device, inlet regions are designed on the masks of both the fluidic and the control

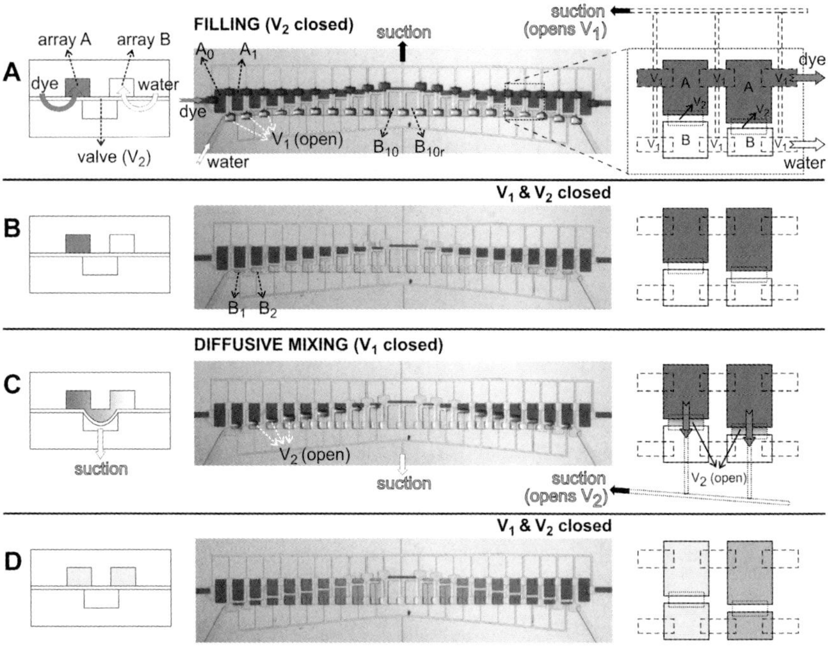

Fig. 2 Device operation. Operation of the device consists of four steps (A–D) shown in cross-sectional schematics (left column), top-view optical micrographs (middle column), and top-view schematics (right column). Left schematics correspond to a line orthogonal to both arrays and the right-column schematics correspond to an area comprising 2 × 2 microchambers, as indicated by the dashed-line square in the top-view optical micrograph in (A). For clarity, the pneumatic lines are shown only when suction is applied. (A) Initial filling of microchambers. Suction is applied to the top pneumatic lines (black arrow) in order to open valve set $\{V_1\}$ while valve set $\{V_2\}$ is kept closed; then array A is filled with colored dye (dark gray fluid, dark gray arrows) and array B is filled with de-ionized water (white arrows). Note the curvature of the $\{V_1\}$ membranes in the optical micrograph. (B) Isolation of each fluidic chamber after $\{V_1\}$ is closed. (C) Diffusive mixing between each A–B pairs after suction is applied to the bottom pneumatic lines (which open valve set $\{V_2\}$). Note the gradient forming in the fluidic connection after mixing starts. (D) State of the device at the end of a mixing experiment (after closing valve set $\{V_2\}$) where a sequence of 11 dilutions, one on each pair of isolated microchambers, is created due to differences in microchamber volumes between arrays A and B; the right-column schematics depict two of these 11 dilutions (dark and light shades of gray).

layers, but silicone tubing inlets are molded [16] only into one layer (for example, fluidic layer) of the device. To create inlets to the control layer, we manually remove or puncture the few sections of the membrane that cover the inlet regions. Therefore, after alignment and assembly, all microchannels (those that carry flow as well as those that control the valves) are accessible from the top of the device so that the bottom surface is planar, enabling imaging of the device on a conventional microscope stage.

In our prototype micromixer, the fluidic layer contains two arrays of microchambers (termed "array A" and "array B," see Fig. 2A). Along array A, the size of the microchambers (labeled A_0, A_1, ..., A_9) decreases, starting from the left, from 200 μm × 400 μm (A_0) to 200 μm × 40 μm (A_9); A_{10} is a 500 μm × 40 μm chamber and is used just for fluidic connection in Array A; to the right of chamber A_{10} is a set of chambers (A_{9r}, A_{8r}, ..., A_{0r}) symmetrically increasing in sizes from A_9's size to A_0's size. The chambers in array B are designed such that the added volume of any two adjacent chambers in different rows always equals to the volume of A_0, *i.e.*, volume (A_0) = volume (B_{10}) = volume (A_1) + volume (B_1) = volume (A_2) + volume (B_2) = ..., *etc.* A_0, A_{0r} and B_{10}, B_{10r} are designed as respective controls for solutions A and B without mixing. The heights of microchambers measured from the SU-8 master were ~40–42 μm, resulting in a microchamber volume progression from 0.32 to 3.36 nL. The control layer has two independently controlled sets of valves (Figs. 2A and C). A set of valves $\{V_1\}$ is used to connect arrays A and B with their respective inlets, whereas a second set of valves $\{V_2\}$ is used to connect each pair of chambers in the two arrays (*i.e.*, between A_1 and B_1, A_2 and B_2, ..., A_{1r} and B_{1r}, *etc.*).

The operation principle and sequences are illustrated in Fig. 2. The microvalve seat is the portion of the PDMS membrane that overlaps the bottom surface of the walls between two fluidic compartments (such as chamber A_0 and A_1 in the case of microvalve set $\{V_1\}$). At rest, due to the compliance and hydrophobicity of PDMS, the membrane seals (reversibly) against its seat, therefore the chambers remain isolated from each other without energy input. Valves can be opened by applying negative pressure (*e.g.*, house vacuum), so the PDMS membrane deflects down and separates from the surface that supports the wall between A_0 and A_1, thus connecting A_0 and A_1. Valve closure is then achieved by switching the pressure setting from vacuum to atmospheric pressure. (Additional positive pressure, *e.g.*, from pressurized air, accelerates valve closure.) Since the positive and negative pressures are adjustable, and opening and closing of the valves can be automated using pneumatic valves [4], the device is straightforwardly computer-operated with conventional scientific control software. To fill the microchambers, valve set $\{V_1\}$ is opened to allow the flow of dye and de-ionized water to arrays A and B, respectively (Fig. 2A). The valve set $\{V_1\}$ is then closed, thereby isolating each chamber in both arrays (Fig. 2B). Next, valve set $\{V_2\}$ is opened to allow fluid mixing between adjacent chambers in different arrays (*e.g.*, mixing between A_1 and B_1, *etc.*; see Fig. 2C). Due to the thinness of PDMS membrane (10–12 μm), upon vacuum/suction application the valve seats deflect down and touch the floor of the control layer (42 μm deep); the area of overlap between a microvalve and the chambers it connects ranges from 100 μm × 40 μm to 200 μm × 40 μm; therefore we estimate that the fluidic connection between two chambers is approximately 100 μm × 40 μm × 42 μm = 0.168 nL to 2 × 0.168 nL = 0.336 nL when the valve is open, 1/20 to 1/10 of the added volume of the pairing fluidic chambers (*i.e.*, the volume of A_0, 3.36 nL). We believe the fluidic chambers must have deformed to produce this new volume of fluidic connection. Mixing only takes ~1 min to complete for these volumes. Closing $\{V_2\}$ then pushes the fluid above back to each fluidic chamber and chambers deform back to their original shape.

Since the two fluidic arrays are designed with chambers of 11 different sizes, 11 different dilution ratios are produced (Fig. 2D). Because the time required for the membrane to reseal is very small compared to the mixing times, we have not tested the maximum switching frequency during operation; we anticipate that our valves are much slower than those based on the pinch-off design [4] because the area of the valve seat is much larger.

18.3.2 Accuracy of mixing using fluorescein

A key advantage of using photolithographically defined chambers is that the final concentrations after mixing can be predicted accurately from the designed chamber dimensions. To quantitate the accuracy of such predictions, we analyzed the fluorescence profiles of mixing fluorescein with water. (For this experiment the device was placed on the stage with the fluidic layer on top so as to avoid optical artifacts caused by emitted fluorescence light crossing features of the control layer; excitation light comes from above in this particular microscope.) Experimental procedures were the same as in the previous experiment with coloring dyes, except that arrays A and B were initially filled with de-ionized water and fluorescein (1 mg/mL), respectively (Fig. 3A). After mixing, the 11 dilution ratios from pure water to 1 mg/mL fluorescein are evident (Fig. 3B). We measured the after mixing fluorescence of each microchamber as described in Section 2. Since we have designed chamber A_0, A_{0r} and B_{10}, B_{10r} as internal calibration references (they have no connecting pairs in the opposite array, hence cannot be diluted with the other solution), the average intensity values in chambers A_0 and A_{0r} were subtracted from all the intensity values in the other chambers. These calibrated values were plotted against putative dilution factors (0, 0.1, . . ., 1) in Fig. 3C. We designed the left half and the right half to serve as internal mirror experiments in one chip, and found they yielded similar results (Fig. 3C). Since the R^2 values for left and right halves are both 0.992, the experimental values were very linear as we expected. To evaluate the precision of the experimental results, we compared the relative fluorescence intensity (pure water set as 0, undiluted 1 mg/mL fluorescein set as 1) from the left half to the expected values which should be a linear curve from 0 to 1 (Fig. 3D, "experimental fluorescence" and "expected fluorescence", respectively).

Due to the microfabrication process, each chamber in an SU-8 master may have a slightly different height and the actual areas of features may also deviate slightly from design, so the volume ratios of actual chamber pairs and thus the actual dilution factors might be slightly different from design (0, 0.1, . . ., 1). A systematic error could derive from this difference and reduce the precision of the mixing. This error was very small in our microdevice, as we found that the actual volume of each microchamber was very close to the designed volume, with only 0.1–1.0% deviation from expected values. The exceptions were the smallest (200 μm × 40 μm) chambers (5.8% deviation), due to their slightly (~2–3 μm) smaller heights than that of all the other chambers, possibly because they are located at the edge of the device. By using the actual volume ratios and compensating the fluorescence intensity

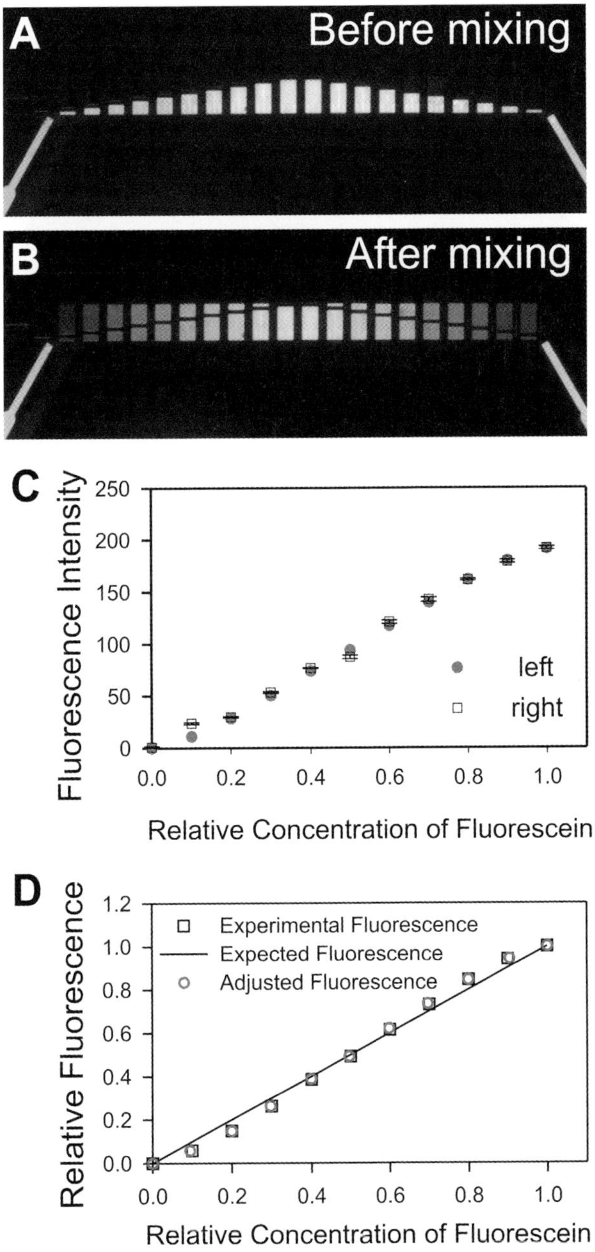

Fig. 3 Quantitative analysis of mixing precision with fluorescein. (A) Before mixing. Fluorescence image of isolated fluidic chambers, with array A containing de-ionized water and array B containing 1 mg/mL fluorescein. (B) After mixing. Fluorescence image showing results of 11 different dilutions. (C) Measured fluorescence intensities (after background correction) in the left (circles) and the right (squares) half of the device *versus* relative dye dilution. Error bars are SEM for three different regions from each chamber. (D) Comparison of the experimental results with predictions. Experimental fluorescence graph shows the relative intensity values (intensity from chambers of undiluted 1 mg/mL fluorescein solution is assigned the arbitrary value of 1) of the left half of the device as a function of the dilution factors (0, 1:10, 2:10, to 1:1); the expected fluorescence was calculated from the original design parameters; the adjusted fluorescence curve was obtained from the experimental fluorescence after factoring in the actual height and area of each microchamber (see text for details).

values with the measured heights (rather than those predicted from design), we plotted the "adjusted fluorescence" data in Fig. 3D. Note the data points in "adjusted fluorescence" were very close to the points in "experimental fluorescence" and both very close to the "expected fluorescence" curve, which confirmed that we can increase the precision in our prediction of mixing by factoring in the fabrication-resulted deviations, but due to the high precision of photolithography procedures, the deviations were very small.

18.3.3
Ca^{2+} indicator calibration using the microdevice

As a proof of the concept, we used the parallel mixing microdevice for calibrating a fluorescent calcium indicator, a key procedure in calcium imaging measurements. A standard method to calibrate the dissociation constant (K_d) of an ion indicator has been described by Tsien and Pozzan [17]. Briefly, the protocol requires preparing two dilute samples of the indicator, one sample in 10 mM K_2EGTA (zero Ca^{2+} sample) and the other in 10 mM CaEGTA (high Ca^{2+} sample). These two samples are then cross-diluted to produce a series of 11 solutions with the amount of total Ca^{2+} increasing by 1 mM CaEGTA with each dilution. The $[Ca^{2+}]_{free}$ in each dilution can be calculated from the K_d of CaEGTA. If indicator concentration, temperature, ionic strength, and pH are kept constant, the K_d of Ca^{2+} indicator can then be calculated from the fluorescence intensity curve against these known Ca^{2+} concentrations. Since a total of 11 different dilutions have to be carried out and 11 different images (or fluorometer measurements) have to be acquired, this method is time consuming (~1 h). In addition, for fluorescence imaging, care has to be taken to ensure that the thickness of each sample is constant. Our microdevice is ideal for this procedure, due to its parallel (automated) mixing capability and to the precise mixing profiles created from photolithographically fabricated chambers. The heights of the chambers are very close and can be measured in advance if more precise fluorescent intensity values are required.

Fig. 4 shows the fluorescence plot that serves to calibrate Fluo-4 (pentapotassium salt) in just one step using the microdevice. The operation procedures were similar as previous experiments with fluorescein. In this case, the chamber array A was filled with 10 mM K_2EGTA, and chamber array B was filled with 10 mM CaEGTA; both also contain 10 μM Fluo-4, 100 mM KCl, and 10 mM MOPS buffer (pH 7.3). After mixing the two solutions in different ratios, the free Ca^{2+} concentrations in these microchambers were estimated by the WinMaxC program (version 2.50, Chris Patton, Hopkins Marine Station, Stanford University, CA, USA) to correct estimates of free Ca^{2+} for temperature and ionic strength. We calculate the K_d of Fluo-4 to be 338 nM, within 2% of the value reported by Molecular Probes (345 nM at pH 7.2).

18.4
Concluding remarks

In conclusion, we have presented a prototype of a parallel subnanoliter micromixer. The device contains arrays of isolated microchambers of different sizes (ranging from 200 μm × 40 μm × 40 μm = 0.32 to 3.36 nL) and was controlled by automated

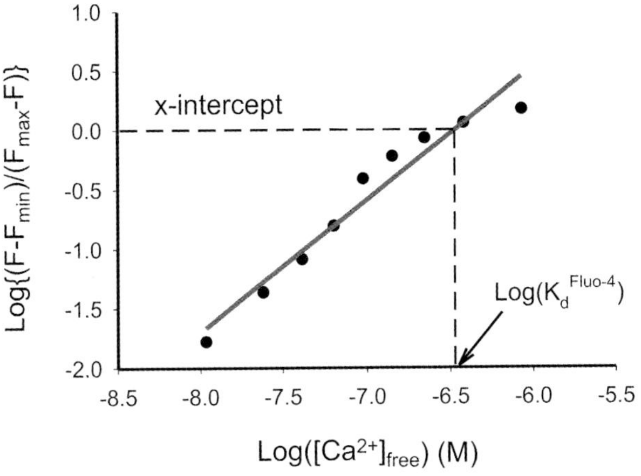

Fig. 4 Fluo-4 calibration. The graph shows the result of calibrating Fluo-4 (pentapotassium salt) from one experiment using the microdevice. *x*-Intercept of the log plot, *i.e.*, the K_d of Fluo-4, is extrapolated from a linear regression of the data, yielding a value of 338 nM.

individual elastomeric microvalves. The device is easy to fabricate and simple to control. The volumes are photolithographically defined and, thus, very precise, as demonstrated in a parallel dilution experiment. Fluid and reagents can be stored in the microdevice for several days, allowing for highly portable assays. Notably, PDMS is biocompatible, so the device has broad applicability in miniaturized diagnostic assays as well as in cell-based assays such as drug screening and enzyme-based biomolecule detection.

This work was supported by the National Science Foundation (CAREER Award to Albert Folch) and by NIH Grant No. 61-0295. Clean-room processing was done at the Washington Technology Center.

18.5 References

[1] Xia, Y. N., Whitesides, G. M., *Angew. Chem. Int. Ed. Engl.* 1998, *5*, 551–575.

[2] Beebe, D. J., Moore, J. S., Bauer, J. M., Yu, Q., *Nature* 2000, *6778*, 588–590.

[3] Unger, M. A., Chou, H. P., Thorsen, T., Scherer, A., Quake, S. R., *Science* 2000, *5463*, 113–116.

[4] Hansen, C. L., Skordalakes, E., Berger, J. M., Quake, S. R., *Proc. Natl. Acad. Sci. USA* 2002, *26*, 16531–16536.

[5] Fu, A. Y., Chou, H. P., Spence, C., Arnold, F. H., Quake, S. R., *Anal. Chem.* 2002, *11*, 2451–2457.

[6] Thorsen, T., Maerkl, S. J., Quake, S. R., *Science* 2002, *5593*, 580–584.

[7] Tourovskaia, A., Figueroa-Masot, X., Folch, A., *Lab Chip* 2005, *1*, 14–19.

[8] Hosokawa, K., Hanada, K., Maeda, R., *J. Micromech. Microeng.* 2002, *1*, 1–6.

[9] Jo, B. H., Moorthy, J., , Beebe, D. J., *Micro Total Analysis Systems 2000*, Kluver Academic Publishers, Dordrecht, The Netherlands 2000, pp. 335–338.

[10] Jo, B. H., Van Lerberghe, L. M., Motsegood, K. M., Beebe, D. J., *J. Microelectromech. Syst.* 2000, *1*, 76–81.

[11] Lee, K. Y., LaBianca, N., Rishton, S. A., Zolgharnain, S., *J. Vac. Sci. Technol. B* 1995, *6*, 3012–3016.

[12] Chen, C., Hirdes, D., Folch, A., *Proc. Natl. Acad. Sci. USA* 2003, *4*, 1499–1504.

[13] Hoffman, J., Shao, J., Hsu, C.-H., Folch, A., *Adv. Mater.* 2004, 23–24, 2201–2206.

[14] McDonald, J. C., Duffy, D. C., Anderson, J. R., Chiu, D. T., Wu, H., Schueller, O. J. A., Whitesides, G. M., *Electrophoresis* 2000, *21*, 27–40.

[15] Hosokawa, K., Maeda, R., *J. Micromech. Microeng.* 2000, *3*, 415–420.

[16] Folch, A., Toner, M., *Biotechnol. Prog.* 1998, *3*, 388–392.

[17] Tsien, R., Pozzan, T., *Methods Enzymol.* 1989, *172*, 230–262.

19
Method development and measurements of endogenous serine/threonine Akt phosphorylation using capillary electrophoresis for systems biology*

Suresh Babu C. V., Sung Gook Cho, Young Sook Yoo

Signal transduction studies have indicated that Akt is essential for transducing the signals originating from extracellular stimuli. An exploration of the Akt signal transduction mechanism depends on the ability to assay its activation states by determining the ability of Akt to phosphorylate various substrates. This paper describes a CE-based kinase assay for Akt using a UV detection method. The RPRAATF peptide was used as the specific substrate to determine the Akt activity. Under the CE separation conditions used, the phosphorylated and nonphosphorylated forms of the RPRAATF peptide were rapidly resolved in the Akt reaction mixture within 20 min. Using this method for measuring the Akt activity, the incubation time for the Akt reactions as well as the kinetic parameters (K_M) were examined. Furthermore, the developed method was applied to a PC12 cell system to assess the dynamics of the Akt activity by examining the effectiveness of the RPRAATF peptide substrate under various cytokine-stimulated environments. These results highlight the feasibility of the CE method, which is a simple and reliable technique for determining and characterizing various enzyme reactions particularly kinase enzymes.

19.1
Introduction

Signaling proteins, which are component members of different cascades, are currently the focus of intense interest, and are of prime importance in understanding the biological processes. Protein kinases are the key signaling proteins of a complex signaling network involving tightly regulated dynamic processes [1]. The main mechanism by which these protein molecules transduce their signals is *via* protein and/or kinase phosphorylation and dephosphorylation. Phosphorylation and dephosphorylation renders the protein and/or kinase either active or inactive, and

* Originally published in Electrophoresis 2005, 26, 3765–3772

Microfluidic Applications in Biology. Edited by Niels Lion, Joël S. Rossier, and Hubert H. Girault

ISBN-10: 3-527-31761-9

the activity of various cell surfacer poteins is responsible for diverse signaling processes [2]. Among the variety of kinases, a serine/threonine Akt, which is also known as protein kinase B (PKB), has been reported to be a crucial regulator of wide range of cellular processes including glucose uptake, apoptosis, metabolism, proliferation, differentiation, cell cycle progression, and transcriptional regulation [3]. Akt is recruited to the plasma membrane following the activation of phosphatidylinositol 3-kinase (PI3K) [4–6], whose activity is tightly regulated in response to various extracellular cues as well as intracellular signals, and they mediate the phosphorylation of the specific substrates. Determining the Akt activities is important in research as well as in identifying the lead compounds in drug discovery.

To date, many studies have demonstrated the kinase assay methods. An in-gel kinase assay and Western blot analysis are two traditional methods used to perform kinase assays [2]. The in-gel kinase assay is an activity staining technique in which the substrate protein is immobilized on a gel and phosphorylated by the protein kinases, which is followed by SDS-PAGE separation and visualization of the incorporated [^{32}P]phosphate by autoradiography. While this method is sensitive, it is also cumbersome and requires special precautions and equipment for handling the radioactivity, which makes it unsafe to work and unsuitable for high-throughput applications. Western blot analyses are performed using antibodies that recognize only the phosphorylated version of the protein of interest. Although Western blotting is less tedious than in-gel kinase assays, it requires the preparation of nuclear or whole-cell extracts and separation by SDS-PAGE. Furthermore, this process is expensive due to the large quantity of phosphospecific antibodies required. Although these techniques provide reliable methods for measuring the kinase activities, they require substantial manipulation and are not readily extended to multi-kinase formats. Owing to the limitations of these methods, researchers are moving toward nonradioactive and nonantibody-based assay approaches. Fluorescence based assay techniques are another active area of development for cell analyzing the kinase activity [7–12]. This method utilizes fluorophore-labeled peptides and/or antibodies, and the activity is determined by a direct measurement of the fluorescent signals of the fluorophore-labeled peptides and/or antibodies. The technology provides a universal method but requires specialized equipment and assays dependent on the fluorophore-labeled reporter. Besides, the changes in the fluorescent signals in these methods are still too small to be used for sensitive monitoring.

A prerequisite for bioanalysis is the ability to obtain accurate, reliable, and reproducible data across a broad enough dynamic range to efficiently measure the analyzed biological systems of interest. The assay sensitivity, signal linearity, and dynamic range are the state-of-the-art in developing and accessing the kinase assay methodology. Bioanalytical methods are at the forefront of emerging science and blends traditional kinase assay methods with new tools to help answer the questions in the systems biology field. Recent innovations in bioanalytical methods have significantly improved its application in the study of biological systems. Among the wide array of analytical technologies, the evolving field of proteomics and systems biology has increased the interest in using CE as a protein analyzing tool [25]. Although CE technology has been applied to many areas of bioresearch, it gained its importance in the study of the development of new assay approaches that have traditionally been difficult [13]. Several laboratories are actively investigating the use of CE for evaluating the protein kinase activity through the substrate preference, and have reported success in establishing high-throughput quantitative kinase assay systems [14, 15]. CE-based measurements of assay signal stability, dynamic range, and system reproducibility explain the importance of CE for bioassays when compared with traditional assay methods. The most valuable asset of CE is its extremely high efficiency, which allows separations to be obtained in a short analysis time. Method development is rapid in many cases, requires only small sample volumes, and the separations are often complementary to those obtained by LC. Overall, CE is a promising system for displaying innovative information for solving various biological questions.

Because CE shows greater promise for meeting the pioneering information in biological science, understanding the cellular responses to external stimuli, intracellular and intercellular communication, disease states, and drug specificity will be enhanced as the CE sampling, detection, and throughput improves. Previously, a CE method was established for detecting extracellular signal-regulated kinase2 (ERK2) localization using a green fluorescent protein and for monitoring the phosphorylation and translocation of the ERK2 proteins in PC12 cell systems [16–18]. Current studies in our laboratory have been focusing on analyzing the cellular signaling proteins for systems biology studies, which employ a systematic study [19, 20, 25], and a part of this study is presented here. This study aims to develop a CE based kinase assay method for monitoring and measuring protein kinase. Our initial efforts in developing a method involved determining the CE separation conditions for the phosphorylated and nonphosphorylated form of the substrate peptide, the phosphorylation reaction time, and the kinetics. Akt was chosen to show the feasibility of the kinase assays using CE in order to demonstrate the potential application of the developed CE method. CE was used to measure the dynamics of the Akt activity status in various cytokine treated PC12 cells samples.

19.2 Materials and methods

19.2.1 Chemicals

The recombinant Akt and Akt substrate peptide Arg-Pro-Arg-Ala-Ala-Thr-Phe-NH_2 (RPRAATF) was obtained from Calbiochem (La Jolla, CA, USA). Trizma hydrochloride, β-glycerol phosphate, EGTA, EDTA, sodium orthovanadate, DTT, PMSF, pepstatin A, aprotinin, leupeptin, magnesium chloride ($MgCl_2$), adenosine triphosphate (ATP) and cytokines, tumor necrosis factor α (TNFα), and interferon γ (IFNγ) were purchased from Sigma (St. Louis, MO, USA). The water was obtained from a Milli-Q plus water purification system from Millipore (Millipore, Bedford, MA, USA). All the samples were passed through a 0.45 μm membrane filter unit and carefully degassed prior to use.

19.2.2
Preparations of buffer solutions

The kinase buffer was 20 mM Tris-HCl buffer, pH 7.2, containing 25 mM β-glycerol phosphate, 5 mM EGTA, 1 mM DTT, and 1 mM sodium orthovanadate (Na_3VO_4). A stock solution of an ATP (500 μM) and $MgCl_2$ (75 mM) cocktail was prepared by dissolving the appropriate amount of each compound in a kinase buffer. The cell lysis buffer A consisted of 50 mM β-glycerophosphate, pH 7.3, 1.5 mM EGTA, 1 mM EDTA, 0.1 mM Na_3VO_4, 1 mM DTT, and a protein inhibitor cocktail (1 mM PMSF, 2 μg/mL pepstatin A, 10 μg/mL aprotinin, and 20 μM leupeptin). The PBS consisted of 137 mM NaCl, 27 mM KCl, 8.1 mM Na_2HPO_4, and 1.5 mM KH_2PO_4. The CE electrolyte was prepared by dissolving the appropriate amount of sodium phosphate (100 mM) in distilled deionized water while adjusting the pH to 2.8 with phosphoric acid.

19.2.3
Cell culture and cytokine stimulation

Rat pheochromocytoma PC12 cells (ATCC, Rockville, MD, USA) were maintained in DMEM supplemented with 10% horse serum 5% fetal bovine serum (Invitrogen, Grand Land, NY, USA), 100 IU/mL penicillin, and 50 μg/mL streptomycin. The cells were maintained in T-75 cm^2 culture flasks at 37°C in a humidified incubator with 5% CO_2 until they were harvested for the experiments. For biochemical analysis, the culture dishes were treated with 0.1 μg/mL collagen and rinsed with the medium. The cells were then plated and grown on a 100 mm dish until they were 70–80% confluent. After the cytokine treatment, the cells were washed twice with ice-cold PBS and harvested with a cell lysis buffer A, which was followed by homogenization and centrifugation (15 000 rpm, 15 min, 4°C). The collected supernatants were analyzed for their protein content using a MicroBCA Protein Assay kit (Pierce, Rockford, IL, USA USA), and stored frozen at −20°C until use. The cytokines (TNFα, IFNγ, and a mixture of TNFα and IFNγ—TNFα + IFNγ) were treated as indicated in the legend of each figure.

19.2.4
Kinase reactions

All the Akt reactions were performed in a kinase buffer. After the optimal conditions had been established, the Akt reactions were carried out in a final volume of 10 μL by adding aliquots of an RPRAATF substrate peptide solution (30 μM) to Eppendorf tubes containing 5 μL of the ATP/$MgCl_2$ cocktail (500 μM ATP and 75 mM $MgCl_2$ in kinase buffer). The reaction mixtures were briefly vortexed and equilibrated at 30°C for 15 min prior to adding the Akt. The assay was initiated by adding Akt (2 ng/mL) to the reaction mixture. Alternatively, the whole-cell extracts (10 μg) were directly assayed for the Akt activity as described. All the reaction components were combined on ice. The reaction was terminated by placing the tubes in boiling water. The reaction mixture was then frozen until needed for the CE analysis, and the reaction products were filtered prior to analysis.

19.2.5
Akt reaction progression

Reaction mixtures containing 30 μM of the RPRAATF substrate peptide and 2 ng/mL of the recombinant Akt along with the ATP/$MgCl_2$ cocktail in a kinase buffer were incubated at different time intervals (as indicated in Fig. 2). The reactions were quenched after a specified time and subjected to CE analysis. The peak area of the phosphorylated form of RPRAATF (hereafter referred to as RPRAATF-P) was divided by the RPRAATF substrate peptide peak area, and the product peak area percentage ratio generated was plotted as a function of the incubation time in order to generate the Akt reaction progression plot.

19.2.6
Akt kinetics experiment

Reaction mixtures containing different amounts of the RPRAATF substrate peptide (10–950 μM) and 2 ng/mL of the recombinant Akt along with ATP/$MgCl_2$ cocktail in kinase buffer were incubated at 30°C. After 20 min, the reactions were quenched and subjected to CE analysis. A kinetic plot was generated by plotting the generated product peak area percentage ratio *versus* the different RPRAATF substrate peptide concentrations. The K_M value was determined by performing a nonlinear least-square best fit to the Michaelis–Menten equation using Microsoft Excel 2000 (version 5.0; Microsoft, Redmond, WA, USA).

19.2.7
Akt standard curve preparation

Five microliters of aliquots of the Akt stock solutions in the kinase buffer system were added to separate Eppendorf tubes containing 30 μM of the RPRAATF substrate peptide solution and 10 μL of ATP/$MgCl_2$ and kinase buffer solution. The resulting mixture was vortexed and incubated at 30°C for 20 min. The final concentration of the Akt enzyme in the reaction mixtures ranged from 0.5 to 1000 ng/mL. After incubation, the reaction was stopped and frozen. The resulting solutions were analyzed by CE in triplicate. A linear regression standard curve was generated by plotting the average product peak area percentage ratio for the generated product obtained for each Akt standard with the Akt concentration.

19.2.8
CE

The CE instrument used was a P/ACE™MDQ electrophoresis system equipped with a photodiode array detector that was obtained from Beckman Coulter (Fullerton, CA, USA). The separations were carried out in a 40.2 cm × 75 μm ID uncoated fused-silica capillary (Polymicro Technologies, Phoenix, AZ, USA) with an effective length of 35 cm to the detector window. The samples were injected by pressure (0.5 psi for

5 s) and the separation was conducted at 8 kV for 25 min (current: 68 μA). All the measurements were carried out at 25°C. A diode array detection was used over the range of 190–300 nm to obtain the spectral data and detection took place at 190 nm. Data acquisition was carried out using 32 Karat™ software (Beckman Coulter). Prior to using of the uncoated capillary, it was conditioned by flushing it with 0.1 M NaOH for 30–45 min, water for 15–20 min, and with a separation buffer consisting of 100 mM sodium phosphate pH 2.8 for 30–60 min. The capillary was rinsed with 0.1 M NaOH, water, and the buffer between each run.

19.3 Results and discussion

Since the establishment of a protein kinase assay entails somewhat difficult procedures, it is important to develop more convenient assays for discovering drug candidates. An increasing number of reports have highlighted the importance of the role of substrate peptides in CE based kinase assay systems. In this study, initial attempts were made to develop a CE-UV method for a kinase assay using RPRAATF as the specific substrate peptide and Akt as the model signaling transduction enzyme.

19.3.1 CE analysis of Akt assay reaction mixture

The RPRAATF substrate peptide used for the Akt reaction has a net positive charge, which tends to be absorbed by a silica capillary when analyzed by the CE method. In addition, since it has a high p*I* value (p*I* 12), it was not possible to optimize the CE separations with high pH buffers. CE in acidic buffer systems was investigated in order to overcome the analyte–wall interactions in the uncoated fused-silica columns. The choice of the acidic buffers in this study has distinct advantages. These include the high reproducible separations and at pH values well below the pK_a of the free silanols on the fused-silica wall (pK_a 6.3), their dissociation becomes significantly suppressed and the negative charge on the surface is removed. The EOF should be negligible, thus any interaction of substrate and product with the wall is suppressed. Citrate and phosphate were the first buffers employed because they are the most common buffers used for basic peptide or protein analysis with uncoated capillaries. The phosphate buffer system showed a sharp and more sensitive peak, whereas citrate had a poor peak shape (data not shown). Therefore, in order to obtain a good peak shape with a higher sensitivity, all further CE analyses of the Akt reactions were carried out using a phosphate buffer.

This study next evaluated the Akt reaction mixture components. Fig. 1 shows an electropherogram of the separation of the Akt assay reaction mixture components within 20 min. The phosphorylation reaction mixture was prepared by mixing 30 μM of the RPRAATF substrate peptide and 2 ng/mL Akt. Under the CE conditions used, the baseline resolution was obtained for both RPRAATF and RPRAATF-P, which migrated at 12.1 and 18.2 min, respectively. The extent of the phospho-

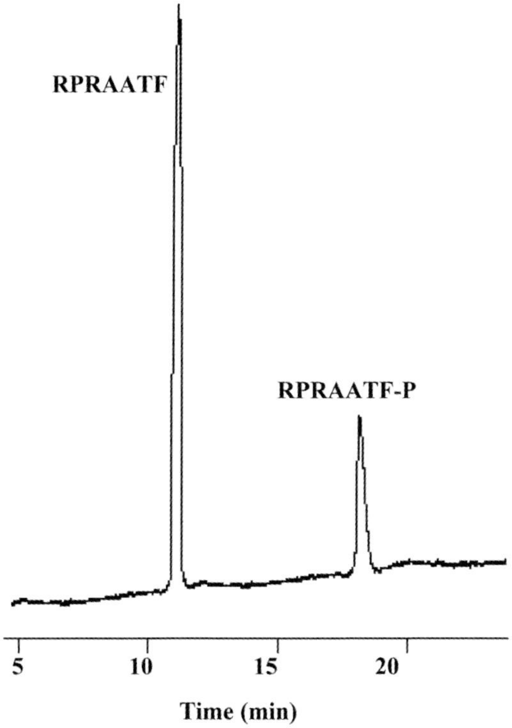

Fig. 1 CE separation of RPRAATF substrate peptide and RPRAATF-P product in Akt reaction mixture using 30 μM RPRAATF substrate peptide with 2 ng/mL Akt. CE conditions: uncoated fused-silica capillary 40.2 cm × 75 μM ID, 45 cm; 100 mM phosphate buffer pH 2.8; applied, voltage 8 kV; capillary temperature, 25°C.

rylation reaction by Akt was identified either from the decrease in the RPRAATF substrate peak or the increase in the RPRAATF-P peak. The migration time reproducibility and the peak area of RPRAATF and RPRAATF-P, which is given in terms of the percentage RSD, were <5%. The satisfactory separation of the reactants and products of the reaction revealed the suppression of the analyte–capillary wall interactions under the electrophoretic conditions used.

19.3.2
Effect of Akt incubation time

The developed CE method was also used in this study to determine the influence of the incubation time on the Akt reaction. Fig. 2 shows the correlation between the incubation time of the RPRAATF phosphorylation reaction and the product peak area percentage ratio using 30 μM RPRAATF substrate peptide and recombinant Akt concentration of 2 ng/mL. The RPRAATF-P level increased linearly with increasing incubation time. A linear rate of product formation was observed within 20 min, which represents the useful reaction time for the kinetic studies. Therefore, the incubation time for the Akt assay was set to 20 min in subsequent studies.

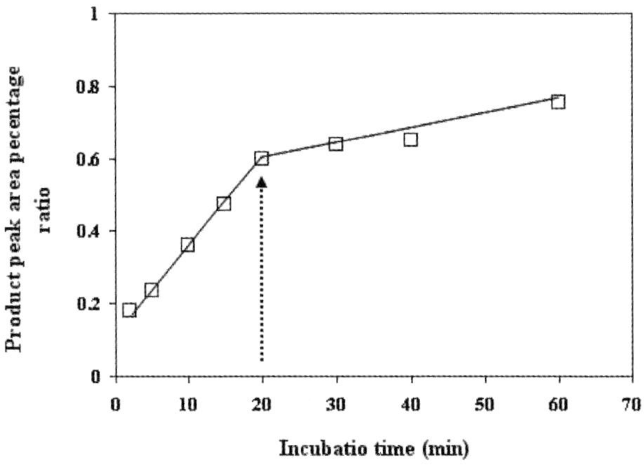

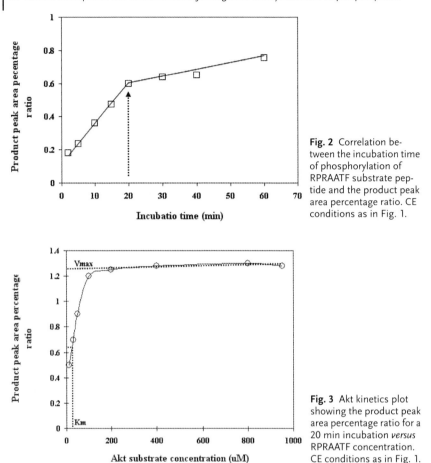

Fig. 2 Correlation between the incubation time of phosphorylation of RPRAATF substrate peptide and the product peak area percentage ratio. CE conditions as in Fig. 1.

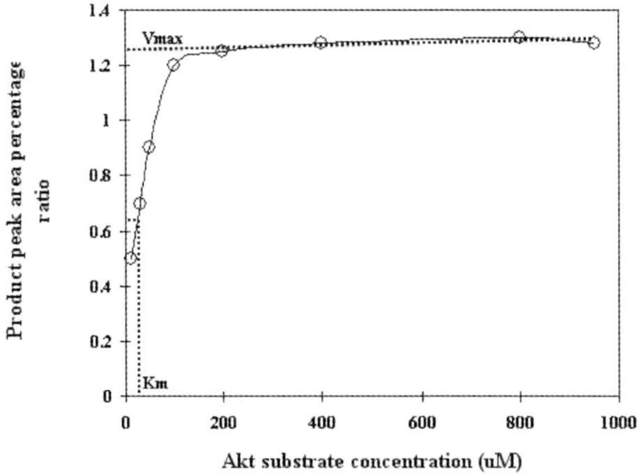

Fig. 3 Akt kinetics plot showing the product peak area percentage ratio for a 20 min incubation *versus* RPRAATF concentration. CE conditions as in Fig. 1.

19.3.3
Determination of RPRAATF substrate peptide concentration: Akt kinetics

In developing an Akt enzyme assay conditions, it is important to determine the optimal RPRAATF substrate peptide concentration for the Akt reaction. Ideally, the substrate concentration should be on the flat part of the Michaelis–Menten plot in order to avoid dramatic changes in the reaction rate due to small errors in the substrate concentration and to provide for the maximum reaction rate. The Michaelis–Menten constants, K_M, can be determined for the substrate by examining the reaction rate as a function of the substrate concentration. The Akt kinetics was determined by changing the amount of the RPRAATF substrate peptide, from 10 to 950 μM, in the presence of a fixed amount of Akt. The product peak area percentage ratio was used for the reaction rate because it reflects the extent of the reaction directly. The Michaelis–Menten curve (Fig. 3) was generated by plotting the determined product peak area percentage ratio *versus* the different RPRAATF substrate peptide concentrations. As shown in Fig. 3, the rate of the reaction increased as the RPRAATF substrate peptide concentration increased. However, the increase in the reaction rate was reduced as the RPRAATF substrate peptide concentration increased. Eventually it reached a plateau that remained constant regardless of the RPRAATF substrate peptide concentration, where the Akt was saturated and was working at its maximum reaction rate (V_{max}). A K_M value of 27.42 μM for the RPRAATF substrate peptide was obtained. The K_M is in good agreement with the value reported in the literature [21]. An RPRAATF substrate peptide concentration of 30 μM was chosen for further experiments.

19.3.4
Akt standard curve

Based on the above observations, a recombinant Akt standard curve was constructed. The average product peak area percentage ratio for the generated RPRAATF-P was obtained for each recombinant Akt standard and was plotted as a function of the recombinant Akt concentration to generate a linear regression standard curve. The linearity of the method was evaluated over a concentration range of 0.5–1000 ng/mL. Fig. 4 shows the standard curve for the recombinant Akt. The data for regression analysis, which was performed using the least-squares method, gave an equation $y = 0.0996x + 3.5346$. The correlation coefficient (0.981) confirms the linearity of the method. The assay was found to have a high linearity in the low concentration ranges of Akt. The RSD of the migration time, the peak area, and the peak height were lower which demonstrates the application of this developed method for analyzing the Akt levels in the ligand treated cell samples. The precision was evaluated by repeatedly measuring the peak areas of RPRAATF-P, which was <7%, indicating good precision without an internal or external standard. This method was reproducible for the Akt assay.

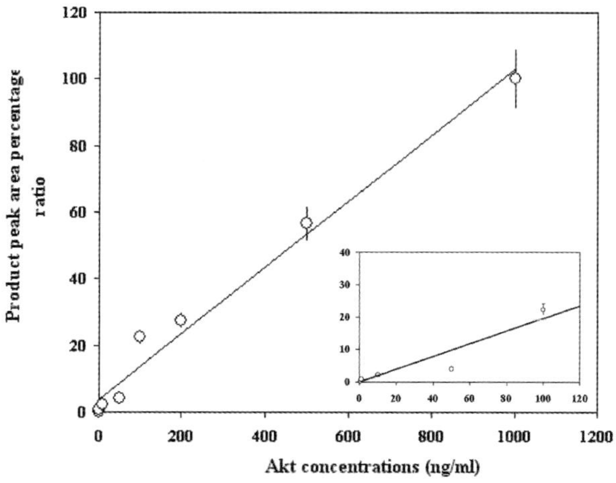

Fig. 4 Standard curve for Akt. Recombinant Akt (0.5–1000 ng/mL) was incubated with 30 μM substrate and ATP/$MgCl_2$ for 20 min. Each standard represents the average value obtained from three replicates. Figure inset corresponds to magnification of the standard curve in the range of 0.5–100 ng/mL. CE conditions as in Fig. 1.

19.3.5
Analysis of PC12 cells for Akt activation levels

A significant improvement in understanding the role of the Akt transducing signals has been made. Akt plays a key role in cancer progression by stimulating cell proliferation and inhibiting apoptosis, and has been suggested to be a key mediator. These findings indicate that Akt is likely to be a hot drug target for the treatment of cancer, diabetes, and stroke [22, 23]. However, experimental methods for assessing its activation and phosphorylation state are vital to the understanding of Akt signaling in cells. The findings from the increasing amount of the Akt study suggest that the phosphorylation mechanism and substrate are important pharmacological and therapeutic targets. Various *in vivo* and *in vitro* substrates for Akt have been suggested including BAD, GSK3, caspase-9, raf-1 kinase, and rac1. However, these substrates are phosphorylated by other kinases such as $p70^{s6k}$ and the mitogen-activated protein kinase (MAPK) kinase-activated protein kinases 1, and are not specific for Akt. Recently, Bozinovski *et al.* [24] reported the use of the RPRAATF substrate peptide for the specific assay of the Akt activity in cell extracts.

In order to illustrate the above-developed CE method, this study examined the use of the RPRAATF substrate peptide for assaying the Akt activities in PC12 cells in response to various cytokines. Cytokines are proteins secreted by a cell that signals itself or other cells in a paracrine manner. The binding of the cytokine to the receptors activates the cascade of proteins by phosphorylation including the activation of Akt. In this study, PC12 cells were stimulated using various cytokines, such as TNFα, IFNγ, and a mixture of TNFα and IFNγ (TNFα + IFNγ) at different time intervals, and the Akt levels were analyzed using the above-developed CE based Akt assay approach. The electropherograms obtained from the TNFα-stimulated PC12 cells are shown in Fig. 5. The TNFα concentration and stimulated time are indicated in the figure legends. These electropherograms showed the RPRAATF substrate peptide and RPRAATF-P peaks illustrating that the Akt was active and phosphorylated in the cell samples at the indicated time and cytokine concentration levels. However, the electropherograms of the IFNγ- and TNFα + IFNγ-stimulated sample (data not shown) showed only an RPRAATF substrate peptide peak and no product phosphorylated peptide peak was separated. This shows that the Akt was not active in these cytokine-stimulated samples. The concentrations and activation levels of the activated Akt in the cell extract stimulated with the different cytokines at various times were determined from the standard curve and are shown in Fig. 6. The activated Akt levels determined were 0.46 ng/mL (TNFα, IFNγ, TNFα + IFNγ – 0 h), 50 ng/mL (TNFα – 3 h), and 1.1 ng/mL (TNFα – 6 h). These experiments clearly demonstrate the ability of the CE assay developed in this study to measure the Akt activities in cell extract.

In the next course of experiments, an attempt was made to determine the effect of the different cytokine concentrations on the Akt activation status. In order to ascertain Akt activation levels, the PC12 cells were treated with different cytokine concentrations and subjected to an assay with the RPRAATF substrate peptide followed by CE separation. Fig. 7 shows the CE analysis of the dose-dependent effect

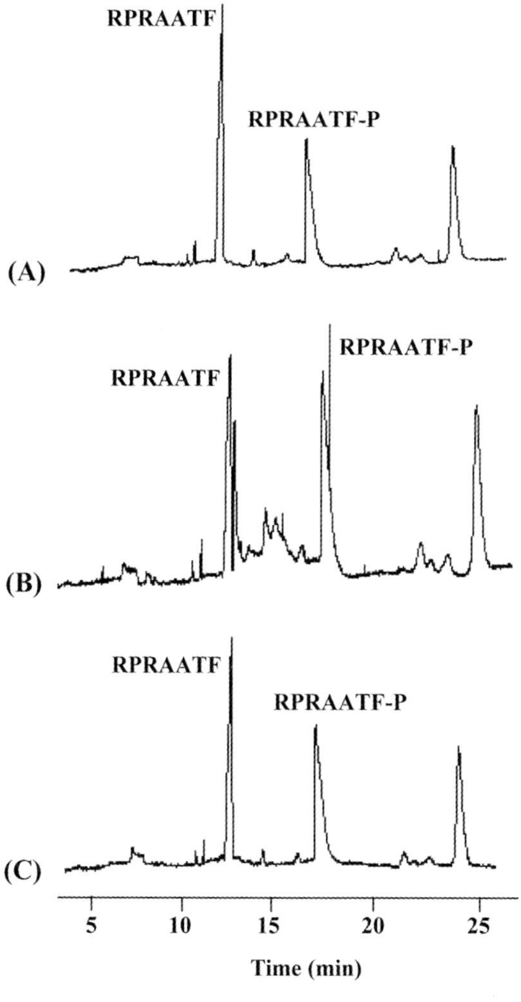

Fig. 5 Electropherograms showing the separation of RPRAATF and RPRAATF-P in 10 ng/mL TNFα-stimulated PC12 cell samples for (A) 0 h (B) 3 h, and (C) 6 h. CE conditions as in Fig. 1.

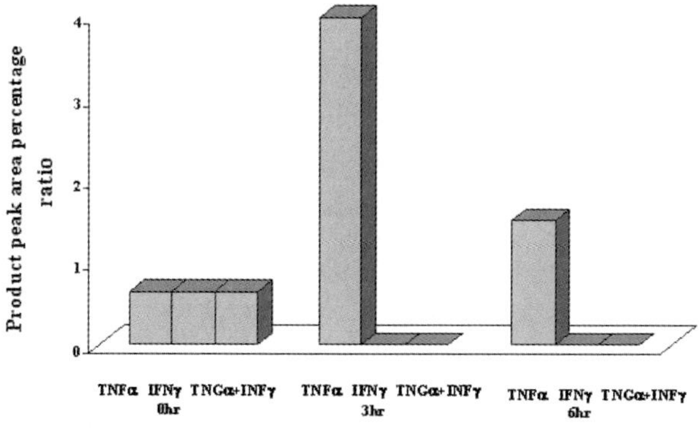

Fig. 6 CE experimental results on Akt activation levels upon cytokines (10 ng/mL) treated cell samples. CE conditions as in Fig. 1.

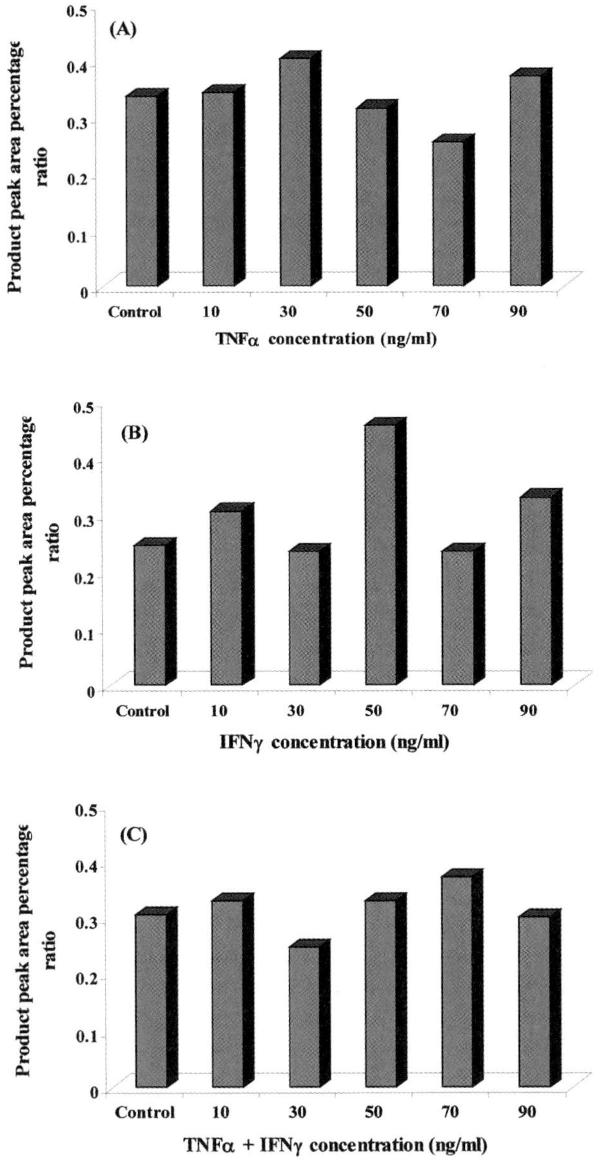

Fig. 7 CE experimental results on Akt activation levels upon different concentrations of cytokine (0, 10, 30, 50, 70, and 90 ng/mL) stimulated PC12 cell samples. CE conditions as in Fig. 1.

of the indicated cytokines on the Akt activation levels. These results demonstrate that the Akt was active at all the studied conditions. The concentrations of levels of activated Akt determined from the standard curve were within the range of 0.12–0.27 ng/mL. In these experiments, CE analysis of the cytokine-stimulated samples was difficult in some cases because the activated Akt concentrations were near the detection limit (0.05 ng/mL). Furthermore, this CE method is sensitive, because it can measure Akt activity in 10 μg of cell lysate. In addition to these experiments,

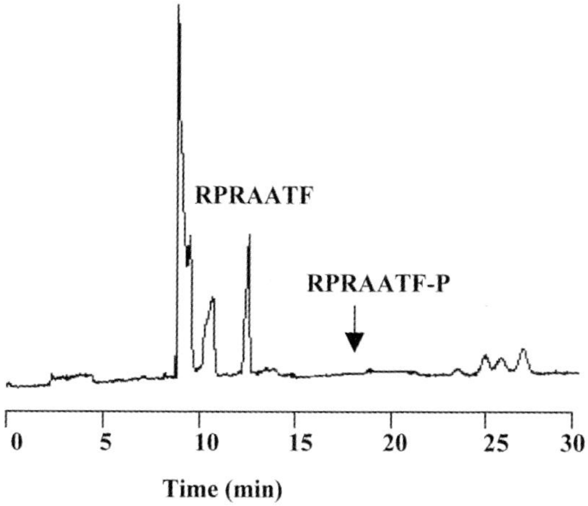

Fig. 8 CE experimental results on Akt activation levels upon TNFα (10 ng/mL, 15 min) stimulated PC12 cell. CE conditions as in Fig. 1.

CE analysis was undertaken to determine the effect of the cytokines at the early time intervals. The time-dependent cytokine-stimulated samples (0, 15, 30, 45, and 60 min) were subjected to a CE based Akt assay. These experiments indicated that Akt was not active at the early time points. Fig. 8 shows a typical electropherogram from these experiments. Concurrently, the cytokine-stimulated samples of this study were examined by Western blot experiments and the results were found to be consistent. A part of this work was discussed in [20, 25].

19.4
Concluding remarks

In summary, a CE-based method for assaying the Akt activities was developed. The assay is based on a specific peptide substrate RPRAATF, which is phosphorylated by Akt. The Akt activity could be estimated by monitoring the RPRAATF and RPRAATF-P peaks. The assay could be easily optimized for each kinase/substrate pair to produce rapid and quality data and did not require any antibodies, radioactive beads, radiolabeled ATP, or specifically modified substrate sequences. The above results demonstrate that the developed CE-based assay was specific, accurate, and sensitive for quantitating of phosphorylated Akt from the cell lysates, and that the assay provides a common platform that can be used to systematically measure the activity of other signaling proteins. The results of the CE-based Akt activity in this study is a systems based bioanalytical approach for interpreting the Akt/PKB signal transduction pathway mechanisms [25].

This work was supported in part by grants from the Korea Ministry of Science and Technology (Systems Biology Research Grant Nos.: 2N26170, 2N27680; and International Cooperation Research Grant No.: 2U02840).

19.5 References

[1] Hunter, T., *Cell* 2000, *100*, 113–127.
[2] Cohen, P., *Nat. Cell Biol.* 2002, *4*, E127–E130.
[3] Brazil, D. P., Yang, Z. Z., Hemmings, B. A., *Trends Biochem. Sci.* 2004, *29*, 233–242.
[4] Patel, R. K., Mohan, C., *Immunol. Res.* 2005, *31*, 47–56.
[5] Toker, A., Newton, A. C., *Cell* 2000, *103*, 185–188.
[6] Vara, J. A. F., Casado, E., de Castro, J., Cejas, P., Belda-Iniesta, C., Gonzales-Buron, M., *Cancer Treat. Rev.* 2004, *30*, 193–204.
[7] Sportsman, J., Daijo, J., Gaudet, E. A., *Comb. Chem. High Throughput Screen* 2003, *6*, 195–200.
[8] Gaudet, E., Huang, K. S., Zhang, Y., Huang, W., Mark, D., Sportsman, J. R., *J. Biomol. Screen* 2003, *8*, 164–175.
[9] McIlroy, B. K., Walters, J. D., Johnson, J. D., *Anal. Biochem.* 1991, *195*, 148–152.
[10] Higashi, H., Sato, K., Omori, A., Sekiguchi, M., Ohtake, A., Kudo, Y., *NeuroReport* 1996, *7*, 2695–2700.
[11] Higashi, H., Sato, K., Ohtake, A., Omori, A., Yoshida, S., Kudo, Y., *FEBS Lett.* 1997, *414*, 55–60.
[12] Meredith, G. D., Sims, C. E., Soughayer, J. S., Allbritton, N. L., *Nat. Biothechnol.* 2000, *18*, 309–312.
[13] Sims, C. E., Allbritton, N. L., *Curr. Opin. Biotechnol.* 2003, *14*, 23–28.
[14] Meredith, G. D., Sims, C. E., Soughayer, J. S., Allbritton, N. L., *Nat. Biothechnol.* 2000, *18*, 309–312.
[15] Gibbons, I., *Drug Discov. Today* 2000, *5*, 33–36.
[16] Yoon, S., Han, K. Y., Nam, H. S., Nga, L. V., Yoo, Y. S., *J. Chromatogr. A* 2004, *1056*, 237–242.
[17] Nam, H. S., Ban, E., Yoo, E., Yoo, Y. S., *J. Chromatogr. A* 2004, *976*, 79–85.
[18] Yoon, S., Ban, E., Yoo, Y. S., *J. Chromatogr. A* 2002, *976*, 87–93.
[19] Suresh Babu, C. V., Yoon, S., Nam, H. S., Yoo, Y. S., *Syst. Biol.* 2004, *1*, 213–221.
[20] Cho, S. G., Yi, S. Y., Yoo, Y. S., *Neurosci. Lett.* 2005, *378*, 49–54.
[21] Alessi, D. R., Caudwell, F. B., Andjelkovic, M., Hemmings, B. A., Cohen, P., *FEBS Lett.* 1996, *399*, 333–338.
[22] Vara, J. Á. F., Casado, E., Castro, J. D., Cejas, P., Belda-Iniesta, C., González-Barón, M., *Cancer Treat. Rev.* 2004, *30*, 193–204.
[23] Hill, M. H., Hemmings, B. A., *Pharmacol. Ther.* 2002, *93*, 243–251.
[24] Bozinovski, S., Cristiano, B. E., Marmy-Conus, N., Pearson, R. B., *Anal. Biochem.* 2002, *305*, 32–39.
[25] Suresh Babu, C. V., Cho, S. G., Wi, M. H., Yoo, Y. S., Systems Biology Challenges in the Capillary Electrophoresis of Cell Signaling Network, Proceedings in The Asia-Pacific International Symposium on Microscale Separation and Analysis (APCE 2004), Dec. 5–8, 2004, Seoul, Korea.

20
Comparison of a pump-around, a diffusion-driven, and a shear-driven system for the hybridization of mouse lung and testis total RNA on microarrays*

Johan Vanderhoeven, Kris Pappaert, Binita Dutta, Paul Van Hummelen, Gert Desmet

In the present study, we demonstrate the benefits of a shear-driven rotating microchamber system for the enhancement of microarray hybridizations, by comparing the system with two commonly used hybridization techniques: purely diffusion-driven hybridization under coverslip and hybridization using a fully automated hybridization station, in which the sample is pumped in an oscillating manner. Starting from the same amount of DNA for the three different methods, a series of hybridization experiments using mouse lung and testis DNA is presented to demonstrate these benefits. The gain observed using the rotating microchamber is large: both in terms of analysis speed (up to tenfold increase) and in final spot intensity (up to sixfold increase). The gain is due to the combined effect of the hybridization chamber miniaturization (leading to a sample concentration increase if comparing iso-mass conditions) and the transport enhancement originating from the rotational shear-driven flow induced by the rotation of the chamber bottom wall.

20.1
Introduction

With the completion of the human and other genome projects (and the identification of 30 000+ potential human genes), traditional models of examining one gene at a time are being supplemented by large-scale screening technologies. DNA hybridization arrays have become a common form of screening technology and allow the analysis of hundreds to thousands of genes in parallel [1]. The power and universality of DNA microarrays as experimental tools are derived from the exquisite specificity and affinity of complementary base-pairing. Researchers are thereby provided with an instant experimental handle on DNA or RNA unlike for any other biological molecules [2]. During the past decade DNA microarray assays became an

* Originally published in Electrophoresis 2005, 26, 3773–3779

Microfluidic Applications in Biology. Edited by Niels Lion, Joël S. Rossier, and Hubert H. Girault

ISBN-10: 3-527-31761-9

important biological tool for obtaining high-throughput genetic information in studies of human disease states [3], toxicological research [4], and gene expression profiling [5]. Microarrays have proven to have the potential to drastically influence biomedical research, diagnostics, and medicine. However, as with any relative young technology, a number of open questions still need to be answered and some protocols should be developed to circumvent a number of inherent shortcomings such as lack of uniformity, slow hybridization speeds, *etc.* One of the problems with conventional microarrays is that without active agitation, the number of target molecules available for hybridization is limited by molecular diffusion. For a typical DNA molecule, this distance has been estimated to be less than 1 mm during an overnight experiment [6, 7], as can be seen from Einstein's law of diffusion.

$$l^2 = 2D_{\mathrm{mol\,nolimits}}t \qquad (1)$$

using a value of $D_{\mathrm{mol}} = 10^{-11}\ \mathrm{m^2/s}$ to represent the rate of diffusion of DNA strands in a typical microarray analysis. This suggests that, for any given spot on an 18 mm × 54 mm array, less than 0.3% of the complementary targets present in the sample can reach the given spot during an overnight diffusion assay. If a target is in low abundance, it may become depleted near the complementary probe spot [7].

With microarray experiments gradually becoming a common tool in analytical laboratories, the enhancement of the speed and the detection sensitivity has become an important commercial factor and has attracted the interest of many research groups. Several methods and devices, in some cases already commercially available, have been described to enhance the hybridization rate on a microarray slide. van Beuningen *et al.* [8] and Benoit *et al.* [9] described the use of flow-through microarrays in which the reactive surface of the material is increased 500-fold by using long branched capillaries and moving the sample through these structures. Anthony *et al.* [10] successfully used these microarrays for the direct detection of *Staphylococcus aureus* mRNA with secondary probes. It is also believed that micromixing is required for the acceleration of hybridization kinetics and for improvement of the hybridization uniformity [11, 12]. Toegl *et al.* [13] proposed the use of a surface acoustic wave (SAW) microagitation chip to agitate sample volumes as small as 10 μL. A device in which fluid oscillation is achieved by an electrically driven heater was developed by Lenigk *et al.* [14]. The cooling and heating of an air pocket leads to this oscillation which is used to overcome the diffusion limitation. Other hybridization enhancement methods used include the use of electronical assistance methods [15, 16], pump-around systems (http://www1.amershambiosciences.com/aptrix/upp01077.nsf/Contect/Products?OpenDocument&ParentId=460778), chaotic mixing [17], *etc.* Although most of the above-mentioned devices for hybridization enhancement require larger volumes in comparison to hybridization under coverslip, the ideal hybridization environment would combine some form of active mixing with a volume, smaller or similar to the volume required for hybridization experiments under coverslip. Therefore, Bynum and Gordon [18] recently introduced a new and robust mixing protocol for microarray hybridization called "planetary centrifugal mixing" in which the small chamber sizes of coverslip systems are combined with the uniform mixing of large-volume hybridizers, leading to improved reaction kinetics, reduction of the required sample volumes, detection of weak markers, and shortening of the hybridization time.

Our group previously reported the use of a rotating disk flow system [7, 19] wherein convection of the sample is achieved by applying the shear-driven flow principle [20, 21] in a rotating circular microchamber. The shear-driven flow principle relies on the dragging effect of a moving channel or chamber wall and therefore does not exhibit any of the pressure-drop or Joule-heating limitations on the fluid velocity noticed in pressure and electrically driven transport enhancement methods. It can be demonstrated from hydrodynamic theory that the rotational velocities which can be imposed in a shear-driven flow are independent of the channel or chamber depth. This implies that a strong convective transport can be generated, even if the chamber has a sub-μm thickness. When dividing the rotating chamber in a series of concentric circles (Fig. 1), it can be shown that, when neglecting the small regions near the side walls of the chamber, a flow with a substantially linear radial velocity profile is established along the circumference of each circle, with a mean fluid velocity given by

$$\omega_{\text{mean}} = \frac{\omega_{\text{wall}}}{2} \qquad (2)$$

wherein ω_{wall} is the velocity of the rotating disk with a similar outline.

The study in [19] showed that, in comparison with diffusion-driven hybridizations under a coverslip, the combination of the shear-driven flow transport and a reduced volume microchamber significantly enhances the hybridization on microarrays. The gain in final hybridization was nearly inversely proportional to

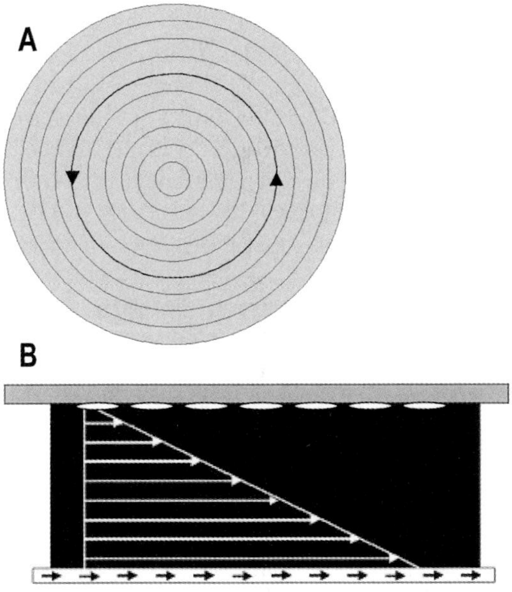

Fig. 1 Top view of rotating disk showing pathways of equal velocities (full lines) (A) and diagonal section along one pathway (B) demonstrating the sample being dragged on average at half the velocity of the moving wall (*i.e.*, rotating disk).

the applied sample volume reduction (keeping the amount of dry DNA mass constant). In agreement with one's physical expectations, this gain was independent of the transport method, *i.e.*, was the same for diffusion- and shear-driven conditions. Comparing hybridizations on the same microchamber, with and without rotational transport, it was shown that the advantage of the applied shear-driven convective transport method only relates to the rate with which the final equilibrium is achieved and not to its value. The final equilibrium spot intensity, which in diffusion-driven conditions is normally reached only after overnight incubation, was always reached must faster (typically in about 1–2 h for a 330 bp polA-probe [19]) due to the convective transport.

The present study has been set up to verify whether the DNA hybridization enhancement realized for a simplified sample (containing only three matching and three nonmatching probes) in [19] by a combination of a hybridization chamber reduction (allowing to increase the sample concentration under constant DNA mass conditions) and a convective shear-driven flow transport, can still be achieved for a complex sample of mouse lung and testis total RNA. Another aim of the study was to verify how the rotating microchamber system would perform in comparison with two conventional hybridization techniques (the diffusion-driven coverslip system and a commercial pump-around system), applying the same amount of genetic material to each system.

20.2
Materials and methods

20.2.1
Rotating microchamber setup for shear-driven experiments

The two main components of the employed microchamber/microarray slide system are schematically shown in Fig. 1. The circular microchamber (diameter = 2 cm and depth = 3.7 μm ± 0.2 μm) was etched in the center of a round, flatly polished (flatness = $\lambda/4$ at λ = 512 nm) borosilicate glass wafer (Radiometer Nederland, The Netherlands) with thickness 6 mm and diameter 5 cm. The etching was carried out using a 50% HF-solution (Fluka Chemie GmbH, Buchs, Switzerland) during 100 s. After the etching, the channel depth was measured using a Talystep™ step profiling apparatus (Rank Taylor Hobson, UK). A WYKO®-surface scan measurement showed that variations of maximally 0.2 μm (*i.e.*, approximately 5.4%) were noticeable over the surface of the 3.7 μm deep microchamber. The channel substrates were subsequently hydrophobically coated, according to the following procedure: 2 h in a 1 M KOH solution (E. Merck, Darmstadt, Germany), 1 h in a 0.03 M HCl solution (Fluka Chemie GmbH), and overnight in a methanol solution (Fluka Chemie GmbH) with [3-(heptafluoro-isopropoxy)propyl]-trichlorosilane, 97% (Aldrich, Steinheim, Germany).

The wafer was subsequently clamped in a home-built, circular stainless-steel holding piece (not represented in Fig. 1), having a stainless-steel rotation shaft attached to its bottom wall. The rotation shaft was in turn connected to an automated rotation stage equipped with a stepper motor (M-060, Physik Instrumente, Karlsruhe, Germany), and controlled with NetMove420 software (Physik Instrumente) running on an ME Windows controlled PC. With this automated rotation stage, the rotation rate can be varied between = 0 rpm and = 60 rpm. In the present study, convection of the sample was established by intermittent movement. It was noticed in a previous study that intermittent movement yielded the best hybridization results [22]. The chamber was rotated for 5 s at a rotation rate of = 3 rpm after each stop period of 3 min during 1 h, resulting in a total amount of five rotations. The rotation shaft was running through the bottom of a circular holding cup (also not represented in Fig. 1), machined in a piece of PVC, and surrounding the glass wafer such that only the top few millimeters of the glass wafer were emerged above the top level of the holding cup. The holding cup was arranged with an inlet- and overflow outlet-port to allow the circulation of a temperature controlled water flow. The temperature of the heating water was controlled using a Julabo® F32-MD heating (and refrigerated) circulator (JD Instruments, Houston, TX, USA), set at $T = 42°C$ during all the experiments.

To carry out the hybridization experiments, a freshly spotted microarray slide (see Section 2.2) was gently deposited (spotted surface facing downwards) on the top surface of the glass wafer. Since in the shear-driven rotating disk flow system more sample volume is needed than the nominal microchamber volume (1.16 μL), a volume of 5 μL was applied with a 20 μL pipette (Gilson, Middleton, WI, USA) in a 3.7 μm deep microchamber with diameter 2.0 cm. This was needed because of the possible formation of air bubbles caused by the nonperfect flatness of the microarray slide, leading to a higher effective microchamber volume than the calculated one, and due to a mechanical friction problem, caused by glass-on-glass contact, resulting in a possible evaporation of the sample. This problem would most probably be solved if microarray slides with a smaller degree of large-scale waviness would be used. For the time being, the problem is solved by adding a small volume excess forming a lubrication layer between the microarray and the substrate surface.

During the rotation of the glass wafer, the microarray slide was held in place by four small metal holding pins arranged around the circumference of the holding cup, one near every corner of the microarray slide. Intimate contact between the channel substrate and the microarray slide was maintained by putting a 500 g weight on top of the microarray slide. As the microarray slide and the microchamber are two independent, separate parts, the microarray slide can simply be picked up from the microchamber substrate after the experiment.

The outer ring is etched about 100 μm deep to minimize the contact surface between the microarray slide and the substrate surface. No sample is injected in the outer etched ring as this was found to lead to excessive evaporation problems.

20.2.2
Diffusion-driven experiments under coverslip

For the diffusion-driven hybridization under coverslip, the standard overnight hybridization procedure of the MicroArray Facility Laboratory [23] was followed. The procedure simply consisted of pipetting a 30 μL sample and putting it on the microarray slide, which was then topped by a thin glass coverslip (2 × 5 cm) and sealed off with rubber glue. The slides were subsequently kept overnight (16 h) in an incubator at 42°C.

20.2.3
Hybridization using the automated slide processor (ASP) system

Hybridizations using the ASP system (Amersham Biosciences, Buckinghamshire, UK) are performed in independently temperature-controlled chambers, allowing customized process parameters to be used for each slide. The system consists of up to five modules, each containing six chambers that clamp to slide surfaces. Labeled sample is injected *via* a septum injection port, and wash solutions are circulated from wash storage bottles *via* tubing to each chamber. Up to 200 μL can be injected into each chamber. Chambers are constructed with chemically resistant polyetheretherketone (PEEK) and have a patented o-ring design. Wash solutions are pumped in and out of each chamber using a 1000 μL syringe pump. Flow rate can be adjusted with the ability to toggle flow back and forth through the chamber (10–50 μL). Clean, dry air is used to dry slides as well as actively pump wash solutions back and forth across the slide surface (http://www.amershambiosciences.com).

20.2.4
Microarray procedures

All hybridization experiments were conducted using conventional microarray procedures and made use of the VIB_Mouse_5K_II chip (MicroArray Facility, Leuven, Belgium). As array elements, mouse testis and lung total RNA was used (BD Biosciences Clontech, Palo Alto, CA, USA). All different cDNA strands were obtained by PCR amplification and were arrayed on Amersham type 7 star slides (Amersham Biosciences) using a commercial Generation III Array Spotter (Amersham Biosciences). The array elements were all spotted at a concentration of 200 ng/μL. The spots (diameter 100 μm and spaced 20–30 μm apart), situated in the circular spotted area (see (4) in Fig. 2), were grouped in 2 × 5 different blocks of 12 × 32 spots.

Prior to each experiment, the concentration of the samples and the amount of incorporated Cy3 molecules was measured with a NanoDrop™ spectrophotometer (NanoDrop Technologies, Montchanin, DE, USA). To achieve maximal hybridization, the arrays were slightly dampened and UV-cross-linked at 50 mJ. The slides were then prehybridized in 2 × SSPE/0.2% SDS at 26°C for 30 min, rinsed with Milli-Q water and dried by centrifugation. After heat denaturation of the samples, the hybridization was carried out at 42°C, either on the microchamber system, under coverslip or using the ASP system.

After hybridization using the rotating disk flow system and under coverslip, the slides were washed three times for 10 min in different SDS/sodium chloride/sodium citrate (SSC) solutions (1 × SSC/0.2% SDS, 0.1 × SSC/0.2% SDS, and 0.1 × SSC/0.2% SDS, respectively) at 56°C and one time for 1 min in 0.1 × SSC at

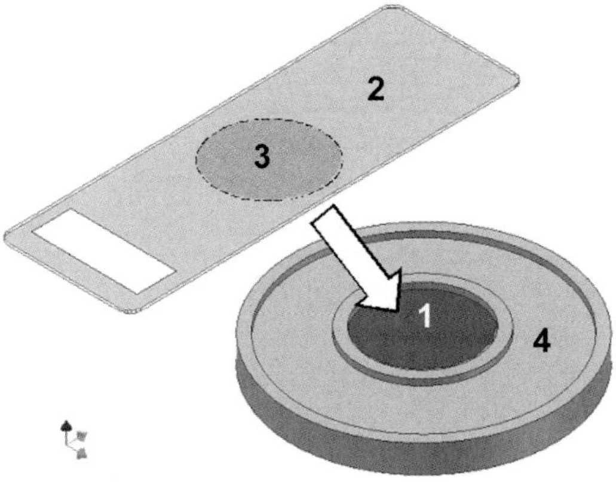

Fig. 2 Bird's eye view schematic of the rotating microchamber hybridization setup, showing the rotatable sample chamber (1, dark gray) and the DNA microarray (2) with spotted area (3) covering the sample chamber during the operation of the device. Outer etched ring (4) only serves to limit the contact area between the rotating bottom substrate and the microarray.

room temperature. After washing, all slides were rinsed with Milli-Q water and finally dried by centrifugation. The washing procedure after hybridization in the ASP system is a fully automated process: after hybridization, the microarray slides are washed two times for 10 min, one time for 4 min, and another time for 10 min in, respectively, 1 × SSC/0.2% SDS, 0.1 × SSC/0.2% SDS, 0.1 × SSC, and Milli-Q water. The slides were then scanned at 532 nm using a commercial Array Scanner Generation III (Amersham Biosciences). Image analysis was done using Array Vision 7.0 software (Imaging Research, ON, Canada) and Spotfire 8 software (Spotfire, Somerville, MA, USA).

20.2.5
Conducted experiments

All experiments were started from 5 μg testis total RNA and 5 μg lung total RNA *per* sample. This was first amplified by T7 *in vitro* transcription [23] and 5 μg tRNA was used for every labeling reaction. All samples originated from the same mouse species batch. After labeling with Cy3-dye, both samples were mixed and for every experiment identical amounts of sample, *i.e.*, 40 pmol Cy3-dye, were divided over different aliquots. The cDNA was further dissolved in a solution containing 50% formamide, 25% buffer, 15% Mouse COT, and 7.5% polydT35. This cDNA-mixture was then, respectively, denatured at 96°C for 3 min, set on ice for 2 min, heated to 42°C for 5 min, centrifugated at 12 000 rpm for another 5 min and subsequently held on ice before applying it on the microarray slide.

The hybridization efficiencies of three different kinds of hybridization assays were compared: (i) shear-driven hybridizations on a rotating disk system, (ii) conventional diffusion-driven hybridizations under coverslip, and (iii) hybridization experiments conducted in an ASP system. The sample volumes needed for the three different kinds of hybridization experiments were, 5, 30, and 210 μL, respectively.

Tab. 1 Comparison of different microarray hybridization systems for mouse lung and testis total RNA on the VIB_Mouse_5K_II microarray chip

	Shear-driven flow system	Diffusion under coverslip	ASP
Amount incorporated Cy3, pmol	40	40	40
Sample volume, μL	5	30	210
Sample concentration, pmol Cy3/μL	8.00	1.33	0.19
Hybridization time, min	60	960	720
Surface microchamber, mm^2	314	1200	1200
Hybridized spots (max. 1760)	1559 (89%)	1415 (80%)	1173 (67%)
Relative intensity	396%	189%	100%

20.3 Results and discussion

All experiments started from the same amount of genetic material (*i.e.*, 40 pmol Cy3-dye-labeled cDNA). The volume of the shear-driven flow system is approximately six times smaller than in the purely diffusion-driven coverslip system, which in turn requires approximately another factor of seven less volume than the fully automated ASP system (*cf.* Tab. 1). As a result, a sixfold and approximately 40-fold higher concentration, respectively, is obtained when using the rotating microchamber system instead of the diffusion-driven procedure under coverslip and the ASP system. Even at the highest concentration, the concentration is still sufficiently small (*i.e.*, well below the 100 μg/μL limit [24]) to justify the assumption that the diffusion rates of the DNA strands are independent of the applied concentration.

Fig. 3 compares the confocal laser scans of the third block (12 × 32 spots) after hybridization with the three different hybridization systems. The first row of spots consists of unknown genes which are not considered in the further analysis. Although the visual differences between the different scans already give a clear first indication of the gains in hybridization speed, the confocal laser scans were analyzed using an ArrayVision 7 software package to exactly measure the spot intensity and the local background. In this procedure, a spot was considered as "hybridized" if its density surpasses the sum of its local background and two times the SD on the average pixel intensity.

It was noticed that, in comparison with the diffusion-driven hybridization and hybridization using the ASP system, the rotating microchamber system resulted in a larger amount of hybridized spots in a shorter hybridization time. In 1 h, about 89% of the 1760 spots in the microchamber were hybridized using the rotating microchamber system while for the diffusion-driven hybridization under coverslip, an overnight hybridization only resulted in about 80% of hybridized spots. The comparison with the automated ASP system is even more favorable because only about 67% of the spots were hybridized after overnight hybridization. Overall, it was hence observed that, starting from a given amount of mouse lung and testis

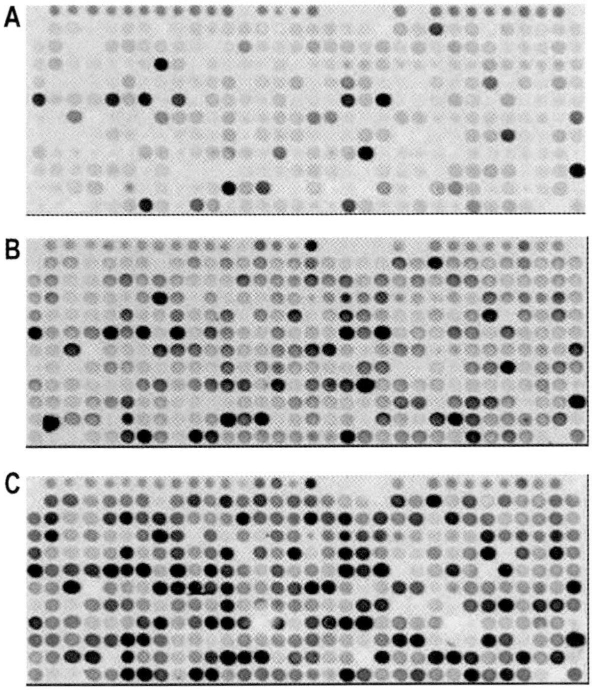

Fig. 3 Comparison of obtained scan images after hybridization of mouse lung and testis total RNA, using a fully automated hybridization station (A), using diffusion-driven hybridization under coverslip (B), and using the shear-driven flow principle (C). Average gain in final spot intensity using the shear-driven flow system was about 110 and 296%, respectively, for genes expressed in the three assays. The number of expressed genes was, respectively, about 9 and 22% higher using the shear-driven flow system while hybridization time was shortened from 16 and 12 h to 60 min, respectively.

total RNA, switching from the ASP system to the rotating disk hybridization system led to an approximately 22% increase in the amount of hybridized spots, emphasizing the benefits of a combined miniaturization of the system volume with a shear-driven convective flow for the screening of low-abundant genes.

A comprehensive comparison of the spot intensities obtained by diffusion-driven hybridization under coverslip and by shear-driven hybridization in the rotating microchamber is presented in Fig. 4A. The spots which are considered as hybridized in both systems (*i.e.*, 80% of the 1760 spots in the microchamber) are shown in Fig. 4B. The spots in the rotating microchamber system showed on average a more than twofold increase in intensity when compared to the coverslip system. A comparison of the spot intensities in the fully automated ASP system and in the rotating microchamber system is presented in Fig. 5A. Fig. 5B only shows the spots which are considered as hybridized in both systems, *i.e.*, 67% of the 1760 spots in the microchamber.

Averaging the intensities of all the spots shown in Figs. 4B, 5B, we noticed an average gain in hybridization intensity with a factor of approximately two between the 1 h rotating microchamber system and the overnight diffusion-driven hybridization under coverslip, and a factor 3.96 if compared to the overnight ASP analysis. These increases in final spot intensity explain the larger number of spots which was denoted by the ArrayVision 7 software as being hybridized in the rotating microchamber system than in the ASP system or in the coverslip system.

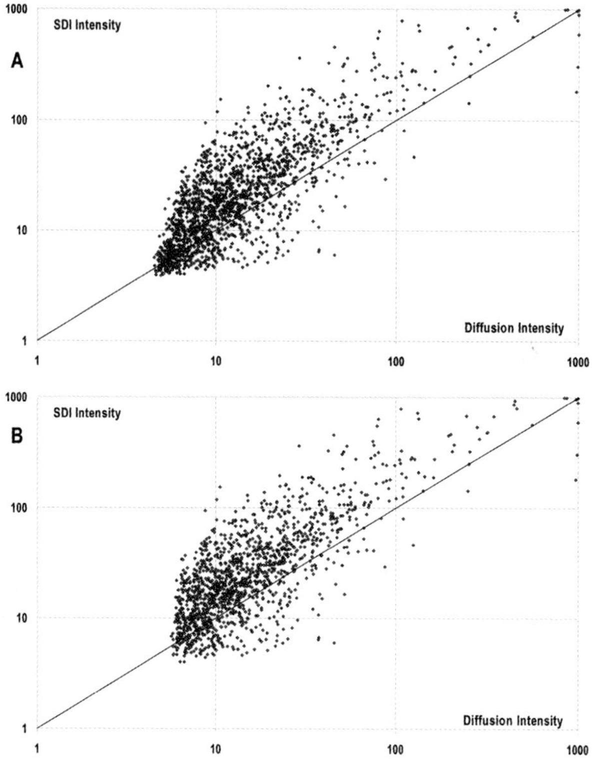

Fig. 4 Comparison of all spot intensities obtained after hybridization using the shear-driven flow system and diffusion-driven hybridization under coverslip (A) and comparison of all spots which are considered hybridized in both systems (B).

20.4 Concluding remarks

Three hybridization methods were compared for the hybridization of mouse lung and testis total RNA: a shear-driven hybridization system, diffusion-driven hybridization under a coverslip, and hybridization in an ASP system. In only 1 h, the shear-driven system, due to a combination of the applied sample volume reduction and the induced convective sample transport, can yield hybridization intensities which are about twofold larger than a traditional overnight experiment under a coverslip while the amount of hybridized spots increased about 9%. Compared to an overnight hybridization in a commercial ASP hybridization station, requiring larger sample volumes and hence inducing a larger dilution factor if starting from a fixed amount of DNA, the mean average hybridization intensity in the 1 h shear-driven system was almost fourfold larger while an increase of about 22% in the amount of hybridized spots was achieved. Of all the three considered hybridization systems, the rotating microchamber system is likely to be the most successful approach if screening for low abundant genes, even in a 16 times shorter hybridization period.

The present results show that DNA microarray analysis systems require both a high sample concentration (which has to be ensured by designing highly miniaturized systems requiring a minimal sample volume) and a strong convective

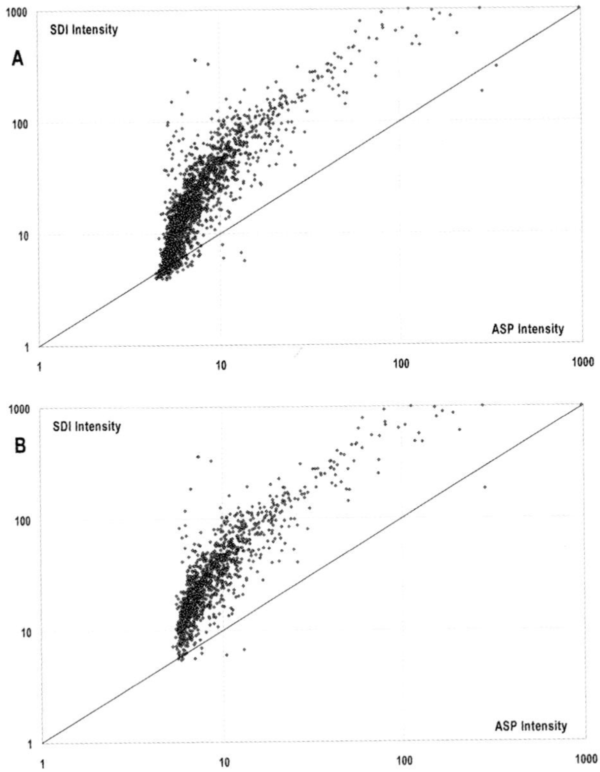

Fig. 5 Comparison of all spot intensities obtained after hybridization using the shear-driven flow system and hybridization using the ASP system (A) and comparison of all spots which are considered hybridized in both systems (B).

transport. For most fluid transport methods, the requirements of a strong degree of mixing and the use of small fluid volumes are very often contradictive. The presently proposed shear-driven flow approach does not suffer from this incompatibility and offers the possibility to induce large lateral convective transport rates in highly miniaturized systems with ultrathin fluid layers (order of 1 μm and less).

Obviously, the currently applied miniaturization approach cannot be continued without end, and will find its limits in the dissolution limit of the DNA samples. With the typical surface waviness of commercial microarray slides, it also seems unfeasible to think of fluid layers which are thinner than 1 or 2 μm and still have a sufficiently uniform fluid layer thickness.

20.5 References

[1] Vrana, K. E., Freeman, W. M., Aschner, M., *NeuroToxicology* 2003, *24*, 321–332.

[2] Brown, P. O., Botstein, D., *Nature Genet. Suppl.* 1999, *21*, 33–37.

[3] Cole, K. A., Krizman, D. B., Emmert-Buck, M. R., *Nature Genet. Suppl.* 1999, *21*, 38–41.

[4] Nuwaysir, E. F., Bittner, M., Trent, J., Barrett, J. C., Afshari, C. A., *Mol. Carcinog.* 2002, *24*, 153–159.

[5] Debouck, C., Goodfellow, P. N., *Nature Genet. Suppl.* 1999, *21*, 48–50.

[6] Adey, N. B., Lei, M., Howard, M. T., Jensen, J. D., Mayo, D. A., Butel, D. L., Coffin, S. C., *et al.*, *Anal. Chem.* 2002, *74*, 6413–6417.

[7] Pappaert, K., Vanderhoeven, J., Dutta, B., Van Hummelen, P., Clicq, D., Baron, G. V., Desmet, G., *J. Chromatogr. A* 2003, *1014*, 1–9.

[8] van Beuningen, R., van Damme, H., Boender, P., Bastiaensen, N., Chan, A., Kievits, T., *Clin. Chem.* 2001, *47*, 1931–1933.

[9] Benoit, V., Steel, A., Torres, M., Yu, Y.-Y., Yang, H., Cooper, J., *Anal. Chem.* 2001, *73*, 2412–2420.

[10] Anthony, R. M., Schuitema, A. R. J., Oskam, L., Klatser, P. R., *J. Microbiol. Methods* 2005, *60*, 47–54.

[11] Liu, R. H., Lenigk, R., Druyor-Sanchez, R. L., Yang, J., Grodzinski, P., *Anal. Chem.* 2003, *75*, 1911–1917.

[12] Liu, Y., Rauch, C. B., *Anal. Biochem.* 2003, *317*, 76–84.

[13] Toegl, A., Kirchner, R., Gauer, C., Wixforth, A., *J. Biomol. Tech.* 2003, *14*, 197–204.

[14] Lenigk, R., Liu, R. H., Athavale, M., Chen, Z., Ganser, D., Yang, J., Rauch, C., *et al.*, *Anal. Biochem.* 2002, *311*, 40–49.

[15] Erdman, C. F., Raymond, D. E., Wu, D. J., Tu, E., Sosnowski, R. G., Butler, W. F., Nereberg, M., Heller, M. J., *Nucleic Acids Res.* 1997, *25*, 4907–4914.

[16] Feng, L., Nerenberg, M., *Gene Therapy Mol. Biol.* 1999, *4*, 183–191.

[17] McQuain, M. K., Seale, K., Peek, J., Fisher, T. S., Levy, S., Stremler, M. A., Haselton, F. R., *Anal. Biochem.* 2004, *325*, 215–226.

[18] Bynum, M. A., Gordon, G. B., *Anal. Chem.* 2004, *76*, 7039–7044.

[19] Vanderhoeven, J., Pappaert, K., Dutta, B., Vanhummelen, P., Baron, G. V., Desmet, G., *Electrophoresis* 2004, *25*, 3677–3686.

[20] Desmet, G., Baron, G. V., *Anal. Chem.* 2000, *72*, 2160–2165.

[21] Desmet, G., Vervoort, N., Clicq, D., Gzil, P., Huau, A., Baron, G. V., *J. Chromatogr. A* 2002, *948*, 19–34.

[22] Vanderhoeven, J., Pappaert, K., Dutta, B., Vanhummelen, P., Desmet, G., *Anal. Chem.* 2005, *77*, 4474–4480.

[23] Puskás, L. G., Zvara, Á., Hackler, Jr. L., Van Hummelen, P., *BioTechniques* 2002, *32*, 1330–1341.

[24] Nkodo, A. E., Garnier, J. M., Tinland, B., Ren, H., Desruisseaux, C., McCormick, L. C., Drouin, G., Slater, G. W., *Electrophoresis* 2001, *22*, 2424–2432.

21
Microfluidic devices for the analysis of apoptosis*

Jianhua Qin, Nannan Ye, Xin Liu, Bingcheng Lin

Apoptosis is the outcome of a metabolic cascade that results in cell death in a controlled manner. Due to its important role in maintaining balance in organisms, in mechanisms of diseases, and tissue homeostasis, apoptosis is of great interest in the emerging fields of systems biology. Research into cell death regulation and efforts to model apoptosis processes have become powerful drivers for new technologies to acquire ever more comprehensive information from cells and cell populations. The microfluidic technology promises to integrate and miniaturize many bioanalytical processes, which offers an alternative platform for the analysis of apoptosis. This review aims to highlight the recent developments of microfluidic devices in measuring the hallmarks as well as the dynamic process of cellular apoptosis. The potential capability and an outlook of microfluidic devices for the study of apoptosis are addressed.

21.1
Introduction

Apoptosis (programmed cell death) is a complex and tightly regulated process by which a cell orchestrates its own destruction in response to specific internal or external triggers [1, 2]. The apoptotic processes are crucially involved in the development of cells, maintenance of tissue homeostasis, regulation of immune system, and elimination of damaged cells [3, 4]. Enhanced or inhibited apoptotic cell death can result in severe pathological alterations, including developmental defects, autoimmune diseases, neurodegeneration, or cancer [5]. Due to its importance in such biological processes, there has been an explosive progress toward dissecting its molecular basis, with an exponential increase in PubMed citations. Recently, the

* Originally published in Electrophoresis 2005, 26, 3780–3788

Microfluidic Applications in Biology. Edited by Niels Lion, Joël S. Rossier, and Hubert H. Girault

ISBN-10: 3-527-31761-9

system-levels' understanding of the biological system has been pursued. Research into system-levels' characteristics by discovering cell response (die or live) to certain chemicals, cell death regulation, and efforts to model cell processes are of great interest in the emerging field of systems biology. These have become powerful drivers for the new technologies to acquire ever more comprehensive information from single cells and cell populations.

Apoptotic cells are characterized by a distinct set of morphological and biochemical characteristics observed exclusively with this mode of cell death [6]. These apoptotic features mainly include the loss of cell volume or cell shrinkage, chromatic condensation, internucleosomal DNA fragmentation, and the formation of "apoptotic bodies" that are then phagocytized macrophages or neighboring cells to rid the body of the dying cells, thus preventing the induction of inflammatory response. Differential expression patterns or alternate mechanisms in the apoptosis pathways among tissues and cell lines make it necessary to have two or even three parallel experiments to confirm and compare the kinetic events on cell death. Techniques designed to identify, quantify, and characterize apoptosis are numerous. Examples are fluorescence microscope, flow cytometry, and microplate analysis [1, 2, 7]. Among these, flow cytometry remains the methodology of choice in studying the apoptotic cascade in relation to cell type, trigger, and time [8–10]. However, these methods still have some limitations: (i) a large number of cells are required for each flow cytometric assay ($>10^4$/mL); (ii) cell preparation prior to analysis is time-consuming, requiring some additional time for cell culture, cell staining, and washing steps; thus, real-time monitoring of cell death kinetics is problematic; (iii) cells are often fixed and destroyed so that intact single living cells cannot be analyzed. These restrictions make current assays of apoptosis cumbersome and impractical for clinical use.

The development of miniaturized (microfabricated, microfluidic-based) analytical devices and their integration to create micro total analysis system (μTAS) [11] is one of the directions to address the future analytical needs. Microfluidic technologies allow for the miniaturization of various functional units, such as pumps, valves, reactors, *etc.*, and make it possible to build novel integrated microsystems for various biological applications [12, 13]. It has been well recognized as a foreseeable "enabling technology" that may, possibly, play important roles in future biology. This observation is, partly, demonstrated in the efforts in building functional microsystems for various cellular analysis [14, 15]. Some typical work in this field [16, 17] showed that novel scheme could be performed on the microfluidics platform to investigate cellular behaviors, which otherwise cannot be possible with a conventional analytical system. It is said that the contemporary microfluidic technologies provide a bridge between biological materials and systems biology through large-scale multiparameter analysis, with applications ranging from molecular dissections of single cells and very small cell populations to multiparameter disease diagnostics from cell and blood [18].

Microfluidic devices incorporated with living cell manipulation and comprehensive cellular analysis capabilities make it an alternative platform for a wide range of cellular applications, including apoptosis analysis. Though, for now, there are

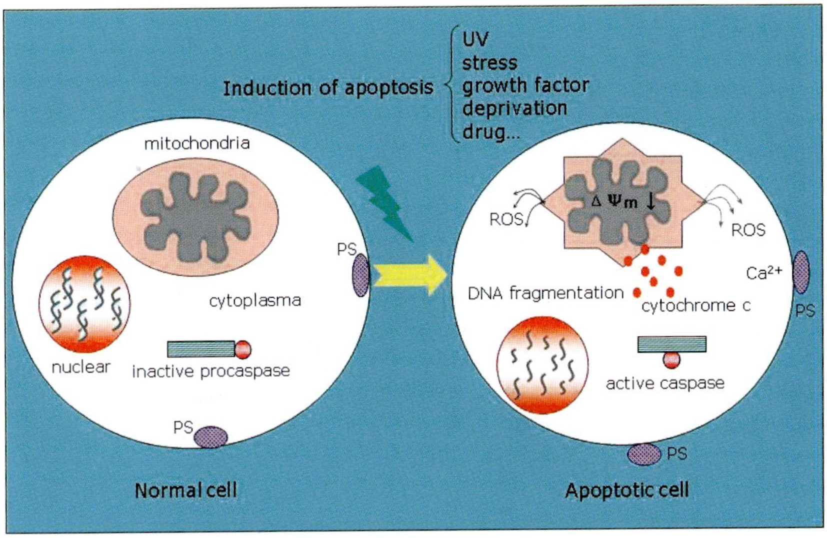

Fig. 1 An outline of the apoptotic processes mostly mentioned in this review. PS, phosphatidylserine; $\Delta\Psi_m$, mitochondrial transmembrane potential.

fewer works dealing specifically on characterizing cell apoptosis on microfabricated devices, the importance of apoptosis in biology as well as the extensive works on cellular analysis by microfluidics urged us to write this review to highlight the present development of microfluidic devices in the unique area of apoptosis research. The potential capabilities and an outlook of this technology for further advances in apoptosis studies are presented.

21.2 Characterizing the hallmarks of apoptosis on microfluidic devices

Apoptosis is a complex biological process, which is regulated by several cell signal transduction pathways. The study of apoptosis is involved in the measurement of biological and morphological hallmarks at a variety of stages in the apoptotic cascade. A simplified outline of the apoptotic processes that are mostly mentioned in this review is depicted in Fig. 1. In Section 2.1, some examples of microfluidic devices that have been developed for the analysis of hallmarks of ongoing apoptotic processes are presented.

21.2.1 Alterations of mitochondria integrity and function

A variety of key events in apoptosis are related to mitochondria, such as the release of cytochrome *c*, the changes in electron transport, the loss of mitochondrial transmembrane potential, the altered cellular oxidation reduction, and the partici-

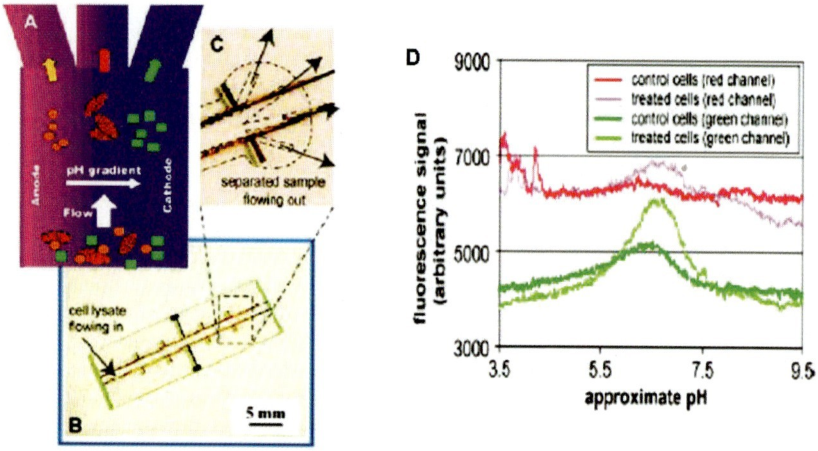

Fig. 2 (A) Schematic of field-flow IEF of mitochondria. (B) Photograph of the microfabricated device before final assembly with (C) enlarged view of the fractionation end of the device. The device consists of electroplated gold electrodes and microfluidic channels formed in a photopatternable epoxy. (D) IEF of JC-1-labeled mitochondria from apoptotic Hela cells showing larger signals (in both red and green channels near neutral pH), which suggests that a number of mitochondrial from the apoptotic cell population have lost the transmembrane potential. Reprinted from [20], with permission.

pation of pro- and antiapoptotic bcl-2 family proteins, *etc.* [19]. Among these, the disruption of the mitochondrial transmembrane potential is one of the earliest intracellular events that occur following the induction of apoptosis.

Lu *et al.* [20] presented a unique microfabricated device for continuous separation of intact and comprised membrane potential mitochondria by microIEF in apoptotic HeLa cells. The schematic of field-flow IEF and the photograph of the microfabricated device are shown in Figs. 2A–C. The device is fabricated using a combination of photolithography, thin-film metal deposition/patterning, and electroplating techniques. In the microIEF experiment using JC-1-stained whole-cell lysate, HeLa cells were treated with TNF and cycloheximide to induce apoptosis. The different p*I*s for intact and compromised membrane potential mitochondria could potentially be used to assay fractions of mitochondria that lose membrane potential when cells undergo apoptosis (Fig. 2D). This microdevice provides fast separation in very small samples while avoiding large voltages and heating effects typically associated with conventional electrophoresis-based device.

21.2.2
Cell membrane alterations (surface lipid translocation)

One of the earliest events in programmed cell death is the externalization of phosphatidylserine, a membrane phospholipids normally restricted to the inner leaflet of the lipid bilayer [21, 22]. Annexin V, an endogenous human protein with a high affinity for membrane-bound phosphatidylserine, can be used *in vitro* to detect apoptosis before other well described morphologic or nuclear changes associated with programmed cell death [23].

Calipter technologies/Agilent has developed an annexin-V assay on microfluidic system, which allows flow cytometric analysis of apoptosis with a minimum number of cells [24, 25]. In this setup, the cells are moved by pressure-driven flow inside a network of microfluidic channels and are analyzed individually by two channel fluorescence detection. As a small number of cells (as few as 50–100 cells) are consumed *per* assay, it is particularly suitable for working with cells of limited availability, for example, primary cells. The system has been applied to evaluate staurosporine-induced apoptosis and annexin-V binding in human umbilical vein endothelial cells (HUVECs) and normal human dermal fibroblasts (NHDFs). Results are in good correlation with the results obtained using a conventional flow cytometry but demonstrate new dimensions in low reagent and cell consumption.

21.2.3
Shift in cellular redox state

Cellular reduction/oxidation (redox) state is found to be involved in the regulation of cell proliferation, differentiation, and apoptosis [26]. Generally, cellular redox state is balanced by the relative quantities of the intracellular oxidative (reactive oxygen species, ROS) and reductive substances (reduced glutathione, GSH). Conventionally, intracellular ROS and GSH can only be determined by flow cytometry, HPLC, or enzymatic methods, respectively [27–29]. Qin *et al.* [30] described a microfluidic device method with LIF detection for simultaneous and rapid determination of intracellular ROS and GSH in apoptotic leukemia cells on a glass chip. Both quantitative and qualitative information of the compounds investigated can be obtained and the two analytes can be detected in just 27 s by using this method. The depletion of GSH and the generation of ROS were observed in apoptotic NB4 cells after arsenic trioxide (As_2O_3) trigger (Fig. 3), showing that the proposed microchip method can readily evaluate the production of ROS, which occurs simultaneously with the consumption of the inherent antioxidant.

21.2.4
DNA fragmentation

The most familiar biochemical feature of apoptosis is endogenous endonuclease activation. It results in the production of domain-sized large (50–300 kb) DNA fragments and oligonucleosomal cleavage products commonly referred to as "DNA ladders" [31]. Traditional gel electrophoresis method is labor-intensive, requiring an additional time for cell lysing, and extraction of genomic DNA. Kleparnik and Horky [32] performed the detection of doxorubicin-induced apoptosis in individual cardiomyocytes on a microfluidic device. Microstructures integrated on a CD-like plastic disk were adapted for the selection of individual cells, the lysis of cells in an alkaline environment, and the separation of released apoptotic DNA fragments.

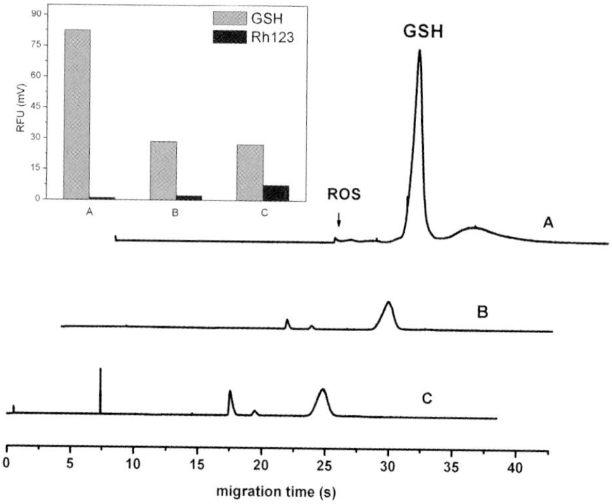

Fig. 3 Microchip electropherograms of ROS and GSH in apoptotic NB4 cells. (A) Control; (B) 1.0 μM As_2O_3; (C) 1.5 μM As_2O_3. The cells were exposed to As_2O_3 for 72 h.

The fragments with typical 180 bp ladder pattern were electrophoretically resolved in a 2% solution of linear polyacrylamide on a polycarbonate chip (Fig. 4). The causal relation between the enhanced doxorubicin concentration and the extent of DNA fragmentation in a single cell was confirmed.

In addition to electrophoresis for the identification of the characteristic DNA ladders, DNA content analysis by flow cytometry is often employed to identify cells undergoing DNA fragmentation. This method is based on the fact that permeablized apoptotic cells release fragmented DNA, resulting in DNA content that is less than that in live diploid cells. Sohn *et al.* [33] developed a novel microfluidic technique – capacitance cytometry – that can be used to quantify the polarization response of DNA within the nucleus of single eukaryotic cells and to analyze the cell-cycle kinetics of populations of cells. The fundamental basis of capacitance cytometry is an alternating current (AC) capacitance measurement. It is found that there is a linear relationship between the DNA content of eukaryotic cells and the change in capacitance that is evoked by the passage of individual cells across a 1 kHz electric field. By monitoring the DNA content of populations of cells with the technique, one can produce a profile of their cell-cycle kinetics. Thus, this might be advantageous in tumor cell detection and in real-time monitoring of the effects of a pharmacological agent on cell cycle and cell death.

21.2.5
Changes of caspase activity

A number of specific changes in enzyme activity occur during apoptotic programmed cell death. The protease, caspase-3, increases in activity during apoptosis and participates in the degradation of several cellular proteins essential for DNA repair. Tabuchi and Baba [34] developed a self-contained on-chip cell culture and pretreatment microdevice system, which was used for screening of caspase-3

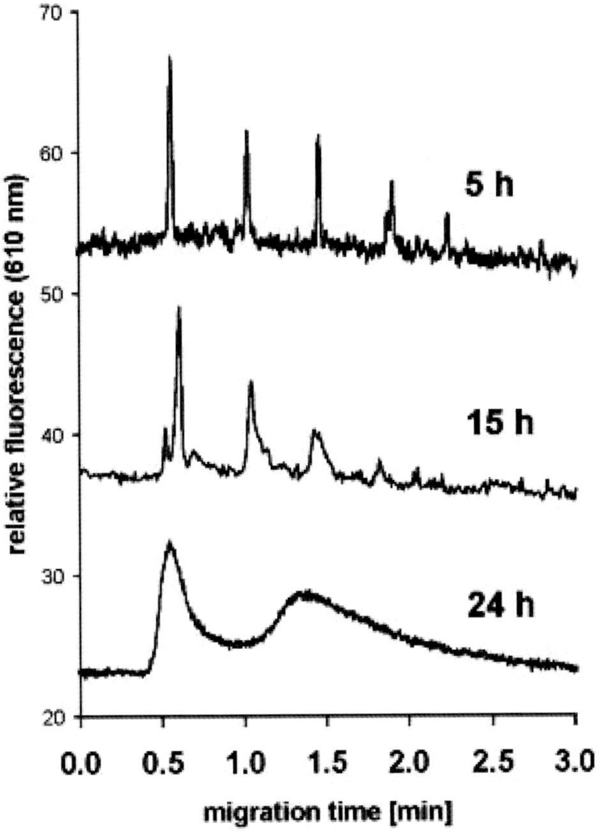

Fig. 4 Comparison of typical records of DNA fragmentation in cells treated by doxorubicin at a concentration of 30 μM for 5, 15, and 24 h [32].

expression in stress-shock triggered apoptosis of Jurkat cells. The pretreatment microdevice includes hand-made 96 wells and 12-well inner cup, requiring no external pipetting and centrifugation steps for cell culture and washing. The cells are placed into the inner cup and inserted into wells that are filled with culture medium. During the media exchange, the inserted cups containing cells are moved into the next row of wells, which are filled with new buffer or reagent. By connecting all procedures to an electrophoretic separation chip, cell culture, cell washing, stimulation, protein extraction, and electrophoretic separation processes could be integrated. The microdevice can manipulate 12 samples simultaneously and the total analysis time is around 1/8 shorter as compared to the conventional analytical method. Using this microdevice, the annexin-V assay in camptothecin or heat-shock treated cells was also presented.

21.2.6
Protein release from cell

Release of cytochrome *c* from mitochondria to cytosol is thought to be a key event for signal transductions in the apoptotic process. It is suggested to induce a series of biochemical reactions that result in caspase activation and subsequent cell death [35]. Tamaki *et al.* [36] developed a microchip-based system for direct monitoring of cytochrome *c* distribution during the apoptosis process. This system incorporated microscale cell culture, chemical stimulation, and a scanning thermal lens microscope detection system under a microscope, which results in secure cell stimulation and coincident *in vivo* observation of the cell responses. Monitoring of cytochrome *c* distribution in a neuroblastoma-glioma hybrid cell was performed in the microflask fabricated in a glass microchip. Cytochrome *c* was detected with an extremely high sensitivity (~10 zmol) without using any labeling substrate.

21.2.7
Intracellular Ca^{2+} concentration changes

Several lines of evidence indicate that alterations in the cytosolic Ca^{2+} concentration and/or intracellular Ca^{2+} compartmentalization are involved in the regulation of apoptosis [37]. These apoptosis-associated Ca^{2+} fluxes are critical targets of their evolutionarily conserved actions. Weeler *et al.* [38] developed a microfluidic device constructed from poly(dimethylsiloxane) (PDMS) for ionophore-mediated intracellular Ca^{2+} flux measurements, and multistep receptor-mediated Ca^{2+} measurements at single-cell level. The schematic diagram of the microfluidic device and the results showing the lonomycin-mediated Ca^{2+} flux measurements on Jurket T-cells are presented in Fig. 5. The microfluidic network enables the passive and gentle separation of a single cell from the bulk cell suspension, and integrated valves and pumps enable the precise delivery of nanoliter volumes of reagents to that cell. The cell trapping and perfusion method with focusing and shield buffers provide an innate "washing", minimizing reagent consumption, analysis time, and stress on the cell. Thus, it might be useful for measuring intracellular Ca^{2+} concentration in individual cells during apoptosis.

In addition, a microfluidic device for monitoring of cellular Ca^{2+} uptake reactions has been developed by Yang *et al.* [39]. The device allows cell transport and immobilization, and dilution of an analyte solution to generate a concentration gradient. The ATP-dependent calcium uptake reaction of HL-60 cells was monitored, demonstrating the feasibility of using the microchip for real-time monitoring of cellular Ca^{2+} flux during apoptosis upon the treatment of a concentration gradient of test solution.

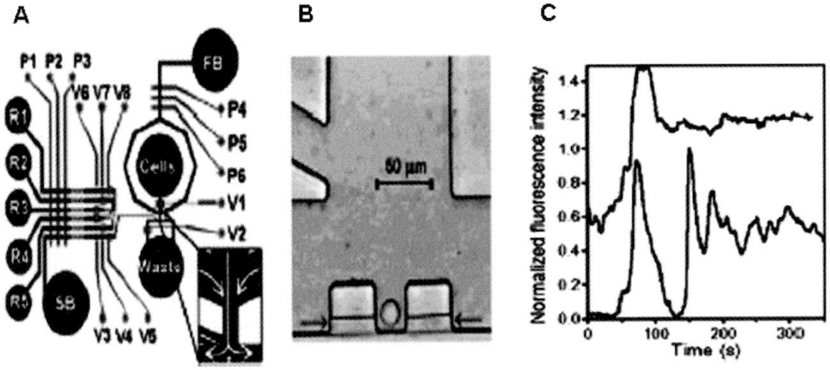

Fig. 5 (A) Single cell analysis device schematic of device fluidic channels are dark. Control channels are light. R1–R5 are reactant inlets, SB and FB are shield and focusing buffer inlets, respectively. Valves are actuated by applying pressure to control inlets V1-Vb. Pumps are activated by actuating P1–P3 or P4–P6 in series. Inset: closeup of cell trapping region. Cells are pushed by hydrostatic pressure from cell inlet to waste. They are focused to the center of the stream by the focusing buffer. (B) CCD image of an individual Jurkat T-cell trapped in cell dock. (C) Ionomycin-mediated calcium flux measurements on Jurkat T-cells. After perfusion with ionomycin, cells responded typically with ringing (low trace) or with sustained $[Ca^{2+}]_i$ elevation (upper trace). Traces have been normalized and offset for comparison. Reprinted from [38], with permisson.

21.3
Perspectives on microfluidic devices for apoptosis studies

The central component of apoptosis is a genetically programmed network within cells, which drives the new technology to be developed for a system-levels study. As mentioned above, microfluidic devices have been employed to characterize most of the key hallmarks involved in apoptotic processes. Also, there has been an attempt to integrate the ability to make multiple measurements on one microfluidic chip. However, in view of technical points, there is still enough space for this technology to be further explored for apoptosis study in the following aspects.

21.3.1
Single-cell analysis

Apoptosis is found to be a process that proceeds in different cells at different times after an initial stress. However, the current experimental measurements on particular cell apoptotic processes are typically performed on large cell populations. Such experiments necessarily lead to measurements of an inhomogeneous distribution of cell responses. Therefore, it is particularly attractive to be able to carry out some of the external perturbations at single-cell level.

The microfluidic technology appears to offer an effective platform for single-cell analysis and many efforts are underway to mimic these properties of single cells and to design chips that are as efficient as cells. Various techniques such as hydrostatic pressure heads in device inlets [39], electrokinetic pumping [40] with the usage of multiple integrated off-chip pump [41], and optical or dielectrophoretic (DEP) traps [42–44] have been tried to control liquid or manipulate individual cells in microdevice. The development of multilayer soft lithography fabrication [45–47] has greatly improved control of fluids and cells in microfluidic devices with the advantage of forming near-zero dead volume valves. An integrated microfluidic process on a single chip using mechanical microvalves to achieve parallel, nanofluid processing in a highly integrated system has been demonstrated in Quake's group. The system is capable of trapping mammalian cells, lysis, and purifying measurable amounts of mRNA from as little as a single cell [48]. These works lie the ground for further study of cell death behavior from single cells that deviate from that of the bulk population, which otherwise would not be possible due to system complexity.

21.3.2 *In situ* dynamic apoptosis analysis

Apoptosis is a dynamic and physiologic process in normal development and homeostasis of multicellular organisms. The ability to monitor cell death events overtime provides a unique tool for the evaluation of therapeutic efficacy of experimental therapeutic agents as well as for studies on the role of apoptosis in various disease processes. Conventionally, apoptosis in the population is determined at a specific time by microscopic examination or flow cytometry. However, apoptotic cells exist *in vitro* for a limited time and occur only in a relatively narrow temporal window, after which they promptly disintegrate [49]. Thus, an accurate determination of apoptosis in a population of cells requires either frequent determinations or continuous monitoring.

Microfluidic devices enable to incorporate cell culture, living cell manipulation, or cell lysing into a microchip-based platform, making it possible to investigate the cell death events from the same cell or a population of cells after stimuli overtime. Recently, there has been some work dealing with dynamic apoptosis analysis in real-time on microfluidic devices. Muñoz-Pinedo *et al.* [50] designed a microfluidic chip with filter structure so that a certain number of cells could be retained in the microchamber, meanwhile, the well-defined height of the chamber (25 μm) could confine the cells within the focal plane of the confocal microscope (Fig. 6). This technology was used to monitor and image cell death events in human leukemic U937 cells over time. Multiple apoptosis events such as loss of mitochondrial transmembrane potential, exposure of phosphatidyl-serine, membrane blebbing, and permeabilization of the cell membrane have been observed, in real-time, in individual cells.

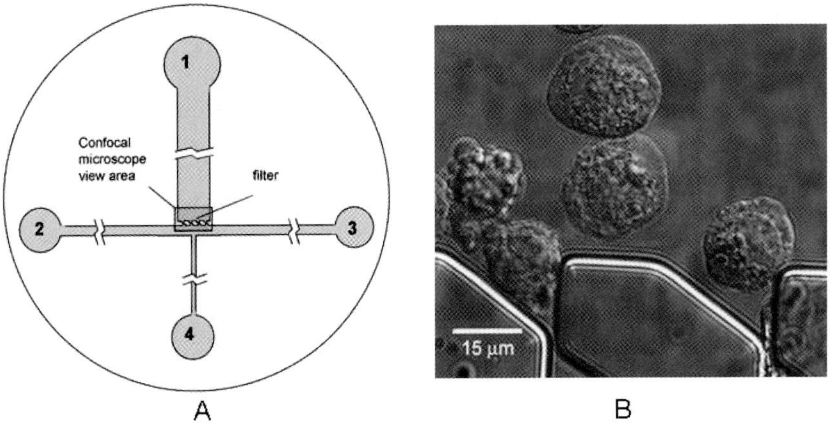

Fig. 6 (A) Schematic layout of the PDMS chip with filter culture and (B) a differential image contrast (DIC) picture of U937 cells in filter chamber. Reprinted from [50], with permission.

21.3.3
High-throughput apoptosis assay

Moving apoptosis and toxicity testing to an earlier stage in pharmacological screening for drug candidates has become a major focus in drug discovery. In the case of cancer cells, the goal is to develop chemotherapeutics that promote differential cell death. This search is often a quest to screen the triggers of apoptosis in high-throughput assay. In addition, the investigation on the comparative effectiveness of antiviral and antimicrobial compounds and examination of the general cytoxicity of new drugs are also important reasons to analyze cell death in high-throughput assay.

Microfluidic technology enables the sorting of cells upon the basis of dynamic functional responses to stimuli, and the manipulation of smaller volumes and cells while performing more complex biological experiment. Recently, Valero *et al.* [51] have presented the design and fabrication of a microfluidic cell trap device for potential high-throughput analysis of apoptosis. The microfluidic cell trapping device consists of two channels, which join together in a crossway. The cross-flow devices introduced in this study offer the tools to characterize single cells in a format that allows integration of high-throughput techniques with unique assay functionality. The microfluidic silicon-glass chip enables the immobilization of cells and real-time monitoring of apoptosis processes. The tumor necrosis factor (TNF-α), in combination with cycloheximide-induced cell death dynamics of HL-60 cells was studied by using this chip.

21.4
Conclusions

Microfluidics is a fast growing field of technology that holds the promises to overcome many limitations of today's biological research. As some reviewed work here, microfluidic technologies are being developed and integrated into a system biology laboratory for various biological experiments including apoptosis analysis. There has been an attempt to integrate the ability to make multiple interrelated measurements on one microdevice, which can potentially analyze some key hallmarks of apoptosis and monitor the cell death dynamic events, in real-time, and at the level of a single cell or small cell population. Though it is in many ways still in its infancy, the field is currently moving from exploratory demonstration to utility, where the unique properties of microfluidic devices can be leveraged to provide new or enhanced functionality.

Currently, apoptosis research is focused on understanding the complex regulation of this machinery and on utilizing this understanding to develop new therapies for a variety of human diseases. Despite extensive progress, however, fundamental questions remain unanswered. Hopefully, with the advent of microfluidic technology, in combination with other technologies (*e.g.*, nanotechnology) and the endeavors from the researchers with different experiences, it will provide improved understanding of the overall process of cell death. At that time, the resulting technologies will achieve first success in translating this understanding into beneficial therapies.

This work was supported by the Fund of the National Science Council of the People's Republic of China (20275039, 20035010, 20299030), the Incentive Fund of Dalian Institute of Chemical Physics, and The Fund of National Gymnasium Committee for Olympic Project (2002BA904B04-04).

21.5
References

[1] Jacobson, M. D., Weil, M., Raff, M. C., *Cell* 1997, *88*, 347–354.
[2] Rathmell, J. C., Thompson, C. B., *Cell* 2002, *109* (*Suppl*), S97–107.
[3] Hanahan, D., Weinberg, R. A., *Cell* 2000, *100*, 57–70.
[4] Thompson, C. B., *Science* 1995, *267*, 1456–1462.
[5] Fadeel, B., Orrenius, S., Zhivotovsky, B., *Biochem. Biophys. Res. Commun.* 1999, *266*, 699–717.
[6] Saraste, A., Pulkki, K., *Cardiol. Res.* 2000, *45*, 528–537.
[7] Vermes, I., Haanen, C., Steffens-Nakken, H., Reutelingsperger, C., *J. Immunol. Methods* 1995, *184*, 39–51.
[8] Gaforio, J. J., Serrano, M. J., Algarra, I., Ortega, E., Alvarez de Cienfuegos, G., *Cytometry* 2002, *49*, 8–11.
[9] Komoriya, A., Packard, B. Z., Brown, M. J., Wu, M. L., Henkart, P. A., *J. Exp. Med.* 2000, *191*, 1819–1828.
[10] Lecoeur, H., de Oliveira-Pinto, L. M., *J. Immunol. Methods* 2002, *265*, 81–96.
[11] Manz, A., Graber, N., Widmer, H. M., *Sens. Actuators B* 1990, *1*, 244–248.
[12] Reyes, D. R., Iossifidis, D., Auroux, P. A., Manz, A., *Anal. Chem.* 2002, *74*, 2623–2636.
[13] Reyes, D. R., Iossifidis, D., Auroux, P. A., Manz, A., *Anal. Chem.* 2002, *74*, 2637–2652.

[14] Andersson, H., van den Berg, A., *Curr. Opin. Biotechnol.* 2004, *15*, 44–49.

[15] Andersson, H., van den Berg, A., *Sens. Actuators B* 2003, *92*, 315–325.

[16] Takayama, S., Ostuni, E., Le Duc, P., Naruse, K., Ingber, D. E., Whitesides, G. M., *Nature* 2001, *411*, 1016–1016.

[17] Lucchetta, E. M., Lee, J. H., Fu, L. A., Patel, N. H., Ismagilov, R. F., *Nature* 2005, *434*, 1134–1138.

[18] Hood, L., Heath, J. R., Phelps, M. E., Lin, B. Y., *Science* 2004, *306*, 640–643.

[19] Green, D. R., Reed, J. C., *Science* 1998, *281*, 1309–1312.

[20] Lu, H., Gaudet, S., Schmidt, M. A., Jensen, K. F., *Anal. Chem.* 2004, *76*, 5705–5712.

[21] Martin, S. J., Reutelingsperger, C. P. M., McGahon, A. J., Rader, J. A., van Schie, R. C. A. A., La Face, D. M., Green, D. R., *J. Exp. Med.* 1995, *182*, 1545–1556.

[22] Zwaal, R. F. A., Schroit., A., *Blood* 1997, *89*, 1121–1132.

[23] Wood, B. L., Gibson, D. F., Tait, J. F., *Blood* 1996, *88*, 1873–1880.

[24] Buhlmann, C., Valer, M., Mueller, O., *Tissue Antigens* 2004, *64*, 392–392.

[25] Chan, S. D. H., Luedke, G., Valer, M., Buhlmann, C., Preckel, T., *Cytometry Part A* 2003, *55A*, 119–125.

[26] Sauer, H., Wartenberg, M., Hescheler, J., *Cell Physiol. Biochem.* 2001, *11*, 173–186.

[27] Coolen, S. A. J., Huf, F. A., Reijenga, J. C., *J. Chromatogr. B* 1998, *717*, 119–124.

[28] Yi, J., Gao, F., Shi, G., Li, H., Wang, Z., Shi, X., Tang, X., *Apoptosis* 2002, *7*, 209–215.

[29] Stamler, J. S., Loscalzo, J., *Anal. Chem.* 1992, *64*, 779–785.

[30] Qin, J. H., Ye, N. N., Yu, L. F., Liu, D. Y., Wang, W., Ma, X. J., Lin, B. C., *Electrophoresis* 2005, *26*, 1155–1162.

[31] Enari, M., Sakahira, H., Yokoyama, H., *Nature* 1998, *391*, 43–50.

[32] Kleparnik, K., Horky, M., *Electrophoresis* 2003, *24*, 3778–3783.

[33] Sohn, L. L., Saleh, O. A., Facer, G. R., Beavis, A. J., Allan, R. S., Notterman, D. A., *Proc. Natl. Acad. Sci. USA* 2000, *97*, 10687–10690.

[34] Tabuchi, M., Baba, Y., *J. Proteome Res.* 2004, *3*, 871–877.

[35] Liu, X., Kim, C. N., Yang, J., Jemmerson, R., Wang, X., *Cell* 1996, *86*, 147–157.

[36] Tamaki, E., Sato, K., Tokeshi, M., Sato, K., Aihara, M., Kitamori, T., *Anal. Chem.* 2002, *74*, 1560–1564.

[37] McConkey, D. J., Orrenius, S., *J. Leukoc. Biol.* 1996, *59*, 775–783.

[38] Weeler, A. R., Throndset, W. R., Whelan, R. J., Leach, A. M., Zare, R. N., Liao, Y. H., Manger, I. D., Daridon, A., *Anal. Chem.* 2003, *75*, 3581–3586.

[39] Yang, M., Li, C. W., Yang, J., *Anal. Chem.* 2002, *74*, 3991–4001.

[40] Li, P. C., Harrison, D. J., *Anal. Chem.* 1997, *69*, 1564–1568.

[41] Schilling, E. A., Kamholz, A. E., Yager, P., *Anal. Chem.* 2002, *74*, 1798–1804.

[42] Arai, F., Ichikawa, A., Ogawa, M., Fukuda, T., Horio, K., Itoigawa, K., *Electrophoresis* 2001, *22*, 113–116.

[43] Muller, T., Gradl, G., Howitz, S., Shirley, S., Schnelle, T., Fuhr, G., *Biosens. Bioelectron.* 1999, *14*, 247–256.

[44] Wang, M. M., Tu, E., Raymond, D. E., Yang, J. M., Zhang, H., Hagen, N., Dees, B., Mercer, E. M., Forster, A. H., Kariv, I., Marchand, P. J., Butler, W., *Nature Biotechnol.* 2004, *23*, 83–87.

[45] Unger, M. A., Chou, H. P., Thorsen, T., Scherer, A., Quake, S. R., *Science* 2000, *288*, 113–116.

[46] Quake, S. R., Scherer, A., *Science* 2000, *290*, 1536–1540.

[47] Liu, L., Enzelberger, M., Quake, S. R., *Electrophoresis* 2002, *23*, 1531–1536.

[48] Hong, W. J., Studer, V., Hang, G., Anderson, F., Quake, S. R., *Nature Biotechnol.* 2004, *22*, 435–439.

[49] Kravtsov, V. D., Greer, J. P., Whitlock, J. A., Koury, M. J., *Blood* 1998, *97*, 968–980.

[50] Muñoz-Pinedo, C., Green, D. R., van den Berg, A., *Lab Chip* 2005, *5*, 628–633.

[51] Valero, A., Merino, F., Wolbers, F., Luttge, R., Vermes, I., Andersson, H., van den Berg, A., *Lab Chip* 2005, *5*, 49–55.

22
Effect of iron restriction on outer membrane protein composition of *Pseudomonas* strains studied by conventional and microchip electrophoresis*

Ildikó Kustos, Márton Andrásfalvy, Tamás Kustos, Béla Kocsis, Ferenc Kilár

In this study the outer membrane protein (OMP) composition of six *Pseudomonas aeruginosa* strains had been analyzed by conventional CE and microchip electrophoresis. Bacterial OMPs are important virulence factors and play a significant role in the pathogenesis of infectious diseases. Changes in their composition might refer to the altered pathogenic properties and antibiotic sensitivity of a certain strain. Pathogenic bacteria invading the human host have to multiplicate under iron-restricted conditions that induce changes in the OMP composition. High-molecular-weight OMPs have to be expressed, which serve as receptors for the iron–siderophore complexes. OMP patterns of bacteria obtained by the two different methods in this study were similar, all major proteins could be detected by both techniques, and the molecular weights showed good correlations, although direct comparison of the peak areas is not straightforward due to the different detection methods (UV and LIF). Changes in OMP composition under iron restriction could be detected, and appearance of a 92 kDa protein in all six *P. aeruginosa* strains and a 94 kDa protein in the KT 2 strain could be demonstrated. Besides that up- and downregulation of certain proteins could be also detected. The increased separation speed, picoliters of sample consumption, baseline separation achieved more frequently by this method – especially in the high-molecular-weight region – showed the advantages of microchip electrophoresis in the analysis of clinical samples.

22.1
Introduction

Since the introduction of the first microanalytic system in 1990 [1], the interest in chip-based devices has increased significantly [2]. The lab-on-a-chip technology has undergone tremendous growth over the last decade [3], and the application field of

* Originally published in Electrophoresis 2005, 26, 3789–3795

Microfluidic Applications in Biology. Edited by Niels Lion, Joël S. Rossier, and Hubert H. Girault

ISBN-10: 3-527-31761-9

this miniaturized analysis system is still increasing [4]. This technology allows the analysis of complex biological samples [5]. Detection, identification, and also quantitation of all proteins presented in biological samples might be performed [6].

Adaptation of CE to microchips has many advantages. Miniaturized flow systems create the opportunity of faster, automated analysis (10-, 100-fold increase in separation speed) [7, 8]. Furthermore, miniaturized dimensions have the potential to simultaneously analyze multiple samples within minutes or less [3], in addition to this the dead volume is significantly reduced.

Chip-based separation techniques are based on the knowledge of CE, which is itself a recent and developing technology [9]. Because of its simplicity of implementation, zone electrophoresis remains the most frequently used method for chip-based separation [9]. Besides that the techniques of MEKC [10], capillary gel electrophoresis [11], and IEF [12] have also been adapted to the microchip format. Bouesse *et al.* [13] established a miniaturized capillary gel electrophoretic system that included gel loading, separation, labeling, and also detection by LIF. This device is now commercially available for protein separation (Bioanalyzer, Agilent Technologies).

While in the beginning chips were created from glass, quartz, or silicon, nowadays the use of polymers and plastics is preferable, *e.g.*, poly(dimethylsiloxane), poly-(methyl methacrylate) [14, 15]. CE on microchips is based upon microfabricating techniques developed in the semiconductor industry.

A sensitive detection system is essential in microfabricated devices due to the extremely small size of the detection cell. LIF is the most popular detection method for CE chips, so far, because of its sensitivity [16]. Electrochemical [17] and conductivity detectors are also frequently used [18, 19]. MS has also been combined with microfluidic CE systems [20]. Most microfluidic devices are coupled with ESI interfaces, since the fluidics of nanoESI is more compatible with the fluidics of the microchannels [21].

Electrophoresis on microchips has now been applied for the analysis of a wide variety of samples [3, 22], in investigation of cell cultures, in clinical diagnostics, protein, and DNA analysis [23, 24]. Advantages of microfabricated techniques might be useful first of all in situations with a need of ultrahigh throughput, *e.g.*, in emerging application fields of life sciences, especially in clinical diagnostics.

Pseudomonas aeruginosa is a threatening bacterial pathogen, which is known for its antibiotic resistance. It has the ability to adapt and thrive in many ecological environments, from water and soil to plant and animal tissues. In humans it is primarily an opportunistic pathogen causing a variety of infections in immunocompromized individuals, and in patients with severe burns, cancer, cystic fibrosis, and patients requiring extensive stays in intensive care units. It might cause urinary tract, respiratory system, soft tissue, bone and joint infections, or generalized sepsis.

Outer membrane proteins (OMPs) are very important components of bacterial cells, participating in several bacterial functions and in several steps of the infectious process. The best known OMPs are porins that produce pores and promote the transport of different substances into the bacterial cells. Furthermore, they serve as receptors for various bacteriophages and colicins. As surface antigens they induce immunological response in the host. They are important virulence factors, they play a role in the attachment process to host cells, specific interactions can occur between the OMPs and the specific receptors of the host tissues. Several factors can modify the OMP composition of bacteria, *e.g.*, mutations, antibiotic treatment, and culturing conditions. The effect of iron depletion on the OMP composition has been studied thorougly earlier [25].

Under iron-restricted conditions bacteria are forced to excrete high-affinity, low-molecular-weight iron-chelating molecules, known as siderophores [26]. These are able to compete with the endogenous iron-chelating compounds of the human host, like transferrin and lactoferrin [27]. Chelated iron is taken into the bacterial cells by iron uptake systems [28]: high-molecular-weight OMPs are important parts of this system, they function as receptors for the iron–siderophore complexes. Expression of these high-molecular-weight OMPs might influence the bacterial surface and related biological properties. Their appearance on the surface of bacterial cells might refer to the pathogenic properties of a certain strain.

The aim of our study was twofold, (i) we wanted to follow the changes in the OMP composition upon iron restriction, and (ii) to convert the conventional CE method applied by us earlier for the analysis of bacterial proteins [29] into microchip format.

22.2 Materials and methods

22.2.1 Chemicals

Tris, glycine, SDS, sucrose, lysozyme, EDTA, benzoic acid, 2,2'-dipyridyl, and deferoxamine mesylate salt were purchased from Sigma (St. Louis, MO, USA). 2-Mercaptoethanol and the run buffer of CE-SDS protein kit were supplied by BioRad (Richmond, CA, USA). The low-molecular-weight calibration kit used in capillary electrophoretic studies was obtained from Pharmacia (Uppsala, Sweden). The Agilent 2100 Bioanalyzer, Protein 200 LabChip kits, Protein 200 ladder, and upper marker were from Agilent Technologies GmbH (Waldbronn, Germany).

22.2.2 Bacterial growth conditions

Five *P. aeruginosa* strains (KT 2, KT 7, KT 25 Ps, KT 28, KT 39) isolated from clinical sources and a Hungarian standard (*P. aeruginosa* NIH Hungary 170000) have been examined in our study. Clinical strains were isolated from the wounds of orthopedic patients operated in 2000/2001 in the Department of Orthopaedics, University of Pécs, Hungary. Clinical strains were identified by standard microbiological methods [30].

Bacteria were cultivated in 2000 mL tryptic soy broth, under shaking (50 rpm) at 37°C for 24 h. Under iron-restricted conditions cultivation media was supplemented by 250 μmol 2,2'-dipyridyl and 10 μmol deferoxamine mesylate salt. Bacteria were harvested by centrifugation at 5000 × g for 15 min at 4°C, and then washed with physiological saline.

22.2.3
Preparation of OMPs

A detailed preparation method of the OMPs has been described earlier [29]. Briefly, the peptidoglycan layer of the bacterial cell wall was digested by lysozyme treatment (10 mg/mL) in the presence of EDTA (1.5 mM, pH 7.5), which made the outer membrane permeable for lysozyme, and converted the originally rod-shaped bacteria to spheroplasts. Bacterial cells were lysed by sonication (400 W, 4 × 2 min in ice bath) and the membrane fractions were collected by ultracentrifugation (100 000 × *g*, 1 h, 4°C). Cytoplasmic and outer membrane were separated by sucrose gradient ultracentrifugation (100 000 × *g*, 21 h, 4°C). Washed pellets were resuspended in sample buffer (0.125 M Tris-HCl (pH 6.8), containing 4% w/v SDS and 10% v/v 2-mercaptoethanol).

22.2.4
CE

Capillary electrophoretic measurements were performed in a Biofocus 3000 system (BioRad, Hercules, CA, USA) by dynamic sieving technique as described earlier [29]. Briefly, uncoated fused-silica (24 cm long, 50 µm ID) capillary was used. The commercially available run buffer of the CE-SDS kit (BioRad, catalog No. 148-5032), containing a hydrophil polymer, was applied as a background buffer. Samples were injected hydrodynamically (50 psi × s). The voltage applied was 15 kV and the temperature was 20°C. UV detection was used at 220 nm. Molecular weights of the individual proteins were determined by means of the low-molecular-weight calibration kit of Pharmacia.

22.2.5
Lab-on-a-chip technology

Chip-based separations were performed in the commercially available Agilent 2100 Bioanalyzer system (Agilent Technologies, Palo Alto, CA, USA). Protein 200 Plus LabChip kits (catalog No. 5065-4480) were used in the study. These chips contained 11 sample wells and a single separation channel. Chips were treated according to the protocol provided with the Protein 200 LabChip kit. The microcapillaries of the chips were filled with running buffer containing a linear polymer as a sieving matrix. The composition, length, and concentration of the polymer were adjusted to obtain optimal resolution within the 14 and 200 kDa molecular mass range. Samples were injected electrokinetically. LIF detection was applied. Staining dye was mixed in the sieving matrix, so SDS – protein complexes were labeled on-chip less than 100 ms after entering the separation channel. A postcolumn SDS dilution step was included in the chip to minimize background fluorescence. Reproducibility of measurement was ensured by an in-built thermostat.

Storage, evaluation, and interpretation of data were performed by the Protein 200 assay software. For determination of the molecular weights of proteins, the Protein 200 ladder (molecular weight markers) of the LabChip kit was applied. This assay uses internal lower and upper markers (6 and 210 kDa, respectively) within each lane, which ensure accurate sizing. Concentration of the proteins was determined by a one-point calibration, comparing the peak area of the analyzed protein to the peak area of the upper marker (myosin).

22.2.6 Calculation

To compare the changes (relative changes) in the protein composition upon iron-restricted conditions, we used the 19.4 kDa protein peak as an internal standard. The peak areas of the control and treated bacterial proteins were normalized and compared with the help of the area of this peak, *i.e.*, the peak area was considered to be 1 in the normalization.

22.3 Results

Bacteria were cultivated under control conditions and in iron-restricted media. The iron was removed by the addition of 250 μmol 2,2'-dipyridyl and 10 μmol deferoxamine mesylate salt. This treatment resulted in changes in the protein profiles. OMPs were prepared from a standard *Pseudomonas* strain and five clinical isolates, and were sent to conventional CE and chip-based separation. Electropherograms under control conditions (continuous line) and iron-depleted conditions (dashed line) obtained by microchip electrophoresis are presented in Fig. 1. Electrophoretic separations of bacterial proteins were completed within 45 s. Because of the complexity of the samples baseline separation of each protein could not be achieved, but the dominating bacterial proteins were separated clearly. The OMP composition of the six *Pseudomonas* strains is characterized by the presence of nine dominating peaks, with molecular masses of 11, 20, 22, 28, 40, 45, 47, 59, and 77 kDa, respectively. See Tab. 1 for details. Up-and downregulation of proteins relative to peak 2 are marked with arrowhead. An additional protein peak appeared in the profile of all *Pseudomonas* strains under iron restriction with a molecular mass of 92 kDa (marked by 10 in the figure). In the OMP pattern of KT 2 strain, a further peak with a molecular mass of 94 kDa could be seen (marked by 11 in the figure). Capillary electrophoretic patterns obtained by the Biofocus 3000 machine are presented in Fig. 2. CE separation of the OMPs was obtained within 10 min in each case. All dominating peaks mentioned at chip-based separations were presented in the capillary electropherograms. The relative area of the proteins was determined from the electropherograms recorded at 220 nm. Molecular weights of the dominating proteins showed good correlation with the chip-based results. Relative peak area (compared to the area of the 20 kDa peak) in control cases and under iron restriction are summarized in Tab. 1 for both microchip experiments and conventional CE. All measurements were repeated three times, changes in protein content were considered to be significant, if the differences of the increase or decrease were

Tab. 1 Relative peak areas of selected major protein peaks in the outer membrane profiles of *Pseudomonas* strains obtained by microchip electrophoresis and conventional CE

Peak number	MW ± SD	Microchip		Conventional CE	
		Relative area ± SD Control	Relative area ± SD Iron restriction	Relative area ± SD Control	Relative area ± SD Iron restriction
P. aeruginosa NIH Hungary 170000					
1	11.2 ± 0.1	0.74 ± 0.01	0.82 ± 0.02↑	0.04 ± 0.01	0.12 ± 0.01↑
2	19.4 ± 0.1	1	1	1	1
5	39.6 ± 0.2	1.07 ± 0.03	1.11 ± 0.03	1.65 ± 0.04	1.67 ± 0.03
9	76.8 ± 0.5	0.42 ± 0.04	0.41 ± 0.03	0.17 ± 0.04	0.30 ± 0.05
10	91.8 ± 0.4		0.40 ± 0.04↑		0.12 ± 0.03↑
KT 2					
1	11.3 ± 0.1	0.55 ± 0.02	0.27 ± 0.02↓	0.56 ± 0.02	0.35 ± 0.01↓
2	19.8 ± 0.1	1	1	1	1
5	39.5 ± 0.2	0.76 ± 0.08	0.55 ± 0.08	0.67 ± 0.04	0.79 ± 0.05
9	77.4 ± 0.8	0.45 ± 0.04	0.68 ± 0.03↑	0.14 ± 0.02	0.22 ± 0.02↑
10	92.7 ± 0.5		0.54 ± 0.05↑		0.75 ± 0.03↑
11	94.3 ± 0.4		0.57 ± 0.03↑		0.83 ± 0.04↑
KT 7					
1	11.1 ± 0.1	0.30 ± 0.01	0.39 ± 0.02↑	0.41 ± 0.01	0.43 ± 0.02
2	19.4 ± 0.08	1	1	1	1
5	39.4 ± 0.2	0.65 ± 0.08	0.88 ± 0.07	0.90 ± 0.06	0.74 ± 0.06
9	76.7 ± 0.4	0.57 ± 0.05	0.60 ± 0.04	0.24 ±0.02	0.29 ± 0.03
10	92.0 ± 0.4		0.25 ± 0.02↑		0.15 ± 0.02↑
KT 25 Ps					
1	11.0 ± 0.1	0.23 ± 0.02	0.22 ± 0.02	0.64 ± 0.06	0.74 ± 0.07
2	19.1 ± 0.2	1	1	1	1
5	40.8 ± 0.8	0.33 ± 0.03	0.41 ± 0.05	0.75 ± 0.04	0.64 ± 0.04
9	76.7 ± 0.6	0.38 ± 0.04	0.42 ± 0.05	0.20 ± 0.03	0.27 ± 0.04
10	92.4 ± 0.5		0.29 ± 0.06↑		0.21 ± 0.02↑
KT 28					
1	11.1 ± 0.1	0.36 ± 0.01	0.36 ± 0.02	0.34 ± 0.01	0.36 ± 0.02
2	18.9 ± 0.1	1	1	1	1
5	39.8 ± 0.3	1.62 ± 0.04	1.54 ± 0.05	0.87 ± 0.06	1.04 ± 0.07
9	76.8 ± 0.4	0.33 ± 0.05	0.39 ± 0.04	0.21 ± 0.02	0.22 ± 0.03
10	92.3 ± 0.3		0.21 ± 0.03↑		0.20 ± 0.03↑
KT 39					
1	11.4 ± 0.2	0.15 ± 0.01	0.11 ± 0.01↓	0.41 ± 0.03	0.21 ± 0.02↓
2	20.2 ± 0.3	1	1	1	1
5	40.7 ± 0.6	0.77 ± 0.02	0.84 ± 0.04	0.65 ± 0.03	0.74 ± 0.02↑
9	77.5 ± 0.7	0.38 ± 0.02	0.33 ± 0.03↓	0.26 ± 0.02	0.16 ± 0.02↓
10	92.6 ± 0.3		0.16 ± 0.04↑		0.15 ± 0.02↑

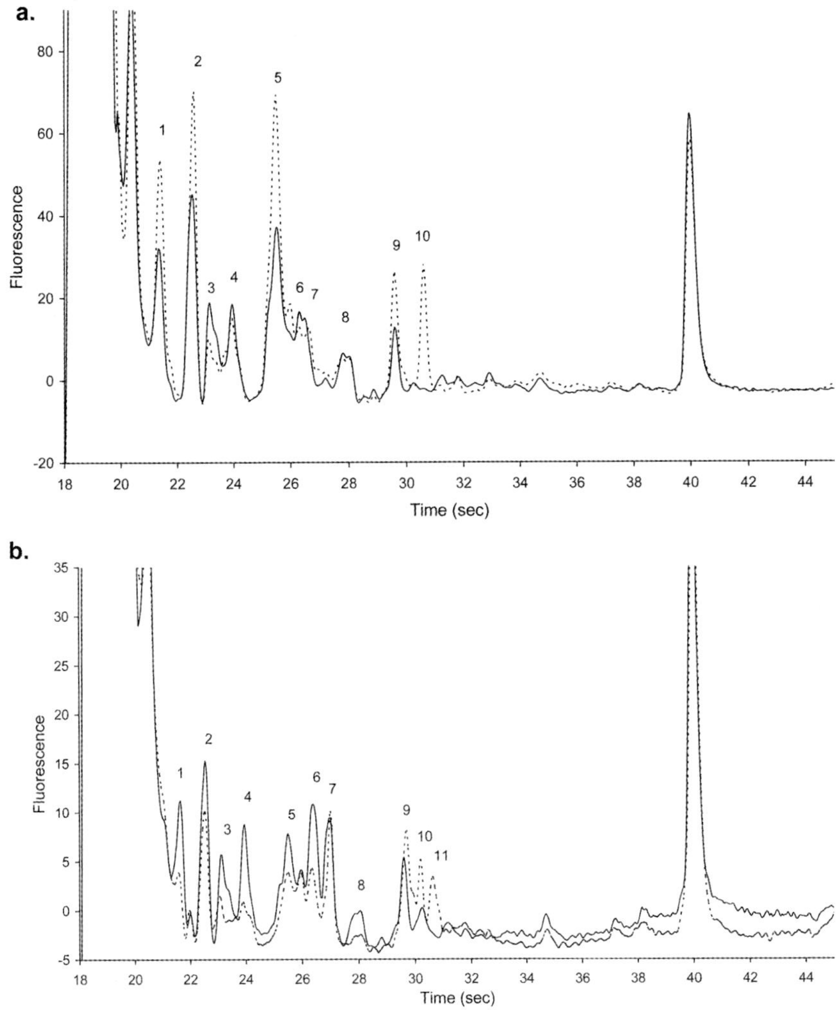

Fig. 1 Chip-based electropherograms of OMPs from (a) standard *P. aeruginosa* NIH Hungary 170000 strain and (b) KT 2 clinical isolate. Dominating peaks of the patterns are numbered (1–9). The molecular weights of these proteins are given in Tab. 1. Profiles obtained in normal condition are drawn by continuous lines, while the patterns obtained under iron-restricted conditions are shown by dashed lines. 92 and 94 kDa proteins appearing under iron restriction are marked by 10 and 11. Peaks related to the low- and high-molecular-weight internal standards can be seen at *ca.* 20 and 40 s. Experiments were performed on the Bioanalyzer 2100 system. Experimental conditions are described in Section 2.

Peak area of the 20 kDa protein was used as internal standard for normalization. Up- and down-regulation of protein expression (larger difference than three times the SD) is signed by arrowhead.

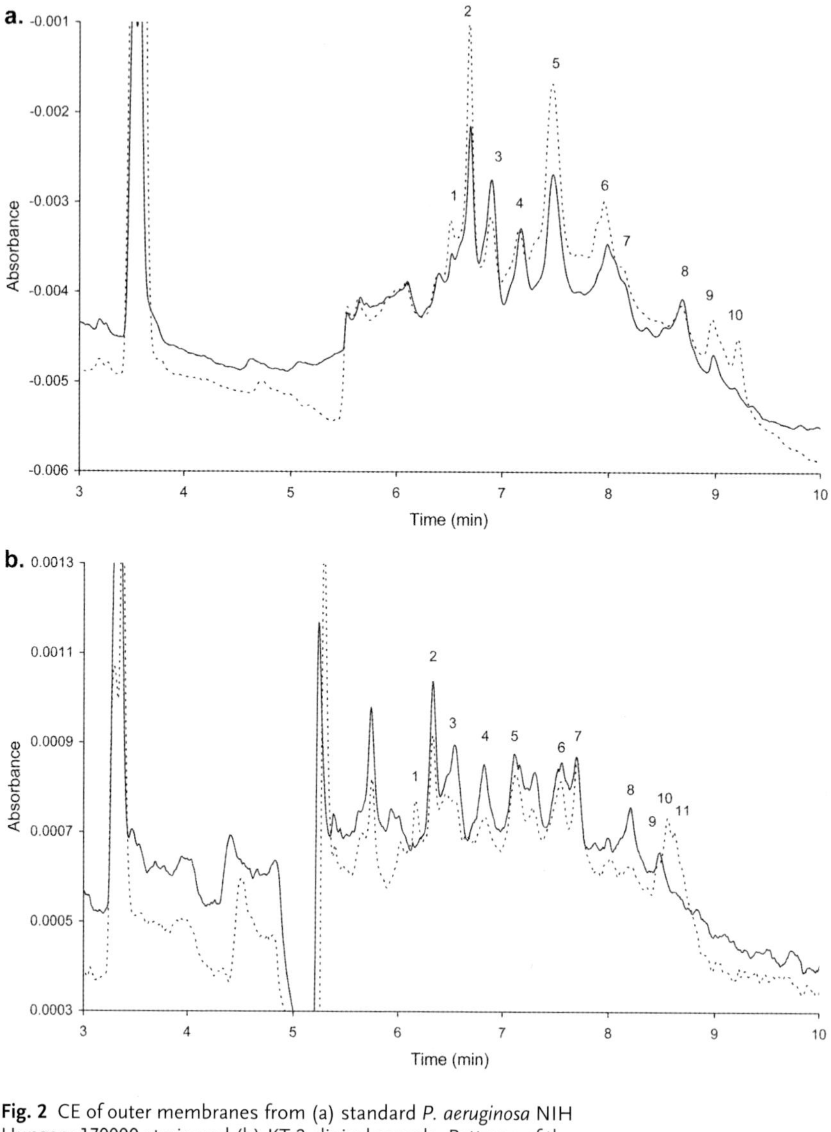

Fig. 2 CE of outer membranes from (a) standard *P. aeruginosa* NIH Hungary 170000 strain and (b) KT 2 clinical sample. Patterns of the control cells (continuous lines) and the iron-restricted cells (dashed lines) are shown. Numbers correspond to those in Fig. 1. Proteins appearing in higher amounts in iron-restricted media are signed by 10 and 11. The peak of the internal standard (benzoic acid) can be seen at *ca.* 3.5 min in the electropherograms. Experimental conditions: buffer, run buffer of CE-SDS protein kit of BioRad; injection, 50 psi × s; voltage, 15 kV; current, 19 μA; detection, 220 nm; temperature, 20°C; capillary, 24 cm × 50 μm ID, uncoated. Equipment: Biofocus 3000 system (BioRad).

higher than three times the SD. Number of plates and the plate height of each peak was also determined. Separation performance of the two methods was similar to each other; the plate number was between 1 and 7×10^4 with both techniques, while the plate height was between 0.1 and1.0 μm.

22.4 Discussion

Pathogenic bacteria invading the human host have to multiplicate under iron-depleted conditions [31]. Iron is an important nutrient required for the growth of bacteria, but it is not freely available in human tissues, it is firmly bound to endogenous chelating molecules such as transferrin in serum and lactoferrin on mucosal surfaces. Iron restriction induces the expression of high-molecular-weight OMPs, which function as receptors for iron–siderophore complexes [32]. The ability of bacteria to acquire iron is an important factor in their pathogenesis [33]. Expression of high-molecular-weight OMPs might influence the bacterial surface. Detection of these proteins in *in vitro* iron-restricted systems might refer to the pathogenic ability of a certain strain, and might provide data about its pathogenesis. CE was proven to be suitable for the analysis of complex bacterial protein samples [29]. It provided fast and automated quantitative data about the protein composition. OMP profiles were characteristic for the *Pseudomonas* genus. Its pattern could be distinguished from the profiles of other genera (*Escherichia coli, Shigella, Proteus, etc.*) [29] examined by us earlier. OMP composition of all six *Pseudomonas* strains could be characterized by the presence of nine dominating peaks. These dominating proteins could be detected in the profiles of all *Pseudomonas* strains, but the quantity and ratio of these proteins differed in the individual patterns (*e.g.*, the concentration of the 40 kDa protein was similar in the standard and in KT 39 strain, while in KT 2 and KT 25 strains this protein was presented in lower quantity).

High-molecular-weight OMPs are essential for the iron uptake of bacterial cells, therefore detection of the appearance of these proteins was our main purpose. Considerable changes in the OMP composition under iron-free conditions could be detected by CE: appearance of a 92 kDa protein could be observed in the profiles of all *Pseudomonas* strains, and in KT 2 strain the overexpression of 92 and 94 kDa proteins could be demonstrated. Besides the overexpression of these high-molecular-weight OMPs, several other changes could be detected in the electropherograms that were not mentioned earlier in literature. Down- and upregulation of certain OMPs could be observed under iron-restricted conditions. (See more details in Tab. 1.) Changes in the quantity of the proteins were considered significant, if it was higher than three times the SD (of the peak areas). The follow up of up- and downregulation of proteins always needs a reference point, which is now the 19 kDa protein peak. The comparison of relative areas of the peaks based on normalization is obviously correct, but it should be kept in mind that the reference protein can also be up- and downregulated and on applying another reference peak one might observe other changes, too.

CE in microchips is a rapidly emerging, new analytical technology. Applicability of the technique is now well established, and numerous applications have been demonstrated. In our measurements microchip-based method ensured tenfold increase in separation speed compared to conventional CE (bacterial OMP profiles were obtained within 45 s instead of 10 min) and picoliters of samples were enough for successful separation. Comparing the electropherograms developed by the two techniques the separation performance was similar. All major proteins (mentioned earlier by CE) could be distinguished by both techniques and the molecular weights showed good correlations, so the profiles obtained by the two methods were comparable with each other. The ratio (quantity) of the dominating protein peaks, obviously, showed differences with the two methods, especially in the high-molecular-weight region (*e.g.*, 77 kDa peak). This should be explained by the different detection methods of proteins. In conventional CE the UV absorbance is detected at 220 nm, while at microchip-based method LIF is applied. The baseline drift (which is usual in the cases of real samples) seen in conventional CE measurements that might make complications in the evaluation, was much less pronounced in the microchip electropherograms. Therefore, we might note that the microchip technique seems to be more effective in separation, and more selective (baseline) separation of high-molecular-weight proteins could be achieved with the latter method. In the chip electrophoretic system the upregulation of the 92 and 94 kDa proteins was also more clearly observed than in the conventional system.

The significance of these microchip-based measurements is that they are automated, and fast results about the bacterial protein composition might contribute to the improved speed of microbiological diagnostics and allow clinicians to make faster drug therapy decisions and earlier treatment.

22.5 References

[1] Manz, A., Graber, N., Widmer, H. M., *Sens. Actuators B* 1990, *1*, 244–248.

[2] Dolník, V., Liu, S., Jovanovich, S., *Electrophoresis* 2000, *21*, 41–54.

[3] Gawron, A. J., Martin, R. S., Lunte, S. M., *Eur. J. Pharm. Sci.* 2001, *14*, 1–12.

[4] Reyes, D. R., Iossifidis, D., Auroux, P. A., Manz, A., *Anal. Chem.* 2002, *74*, 2623–2636.

[5] Figeys, D., *Proteomics* 2002, *2*, 373–382.

[6] Lion, N., Rohner, T. C., Dayon, L., Arnaud, I. L., Damoc, E., Youhnovski, N., Wu, Z. Y., *et al.*, *Electrophoresis* 2003, *24*, 3533–3562.

[7] Lion, N., Reymond, F., Girault, H. H., Rossier, J. S., *Curr. Opin. Biotechnol.* 2004, *15*, 31–37.

[8] Bruin, G. J. M., *Electrophoresis* 2000, *21*, 3931–3951.

[9] Figeys, D., Pinto, D., *Electrophoresis* 2001, *22*, 208–216.

[10] Suljak, S. W., Thomson, L. A., Ewing, A. G., *J. Sep. Sci.* 2004, *27*, 13–20.

[11] Herr, A. E., Throckmorton, D. J., Davenport, A. A., Singh, A. K., *Anal. Chem.* 2005, *77*, 585–590.

[12] Zilberstein, G. V., Baskin, E. M., Bukshpan, S., Korol, L. E., *Electrophoresis* 2004, *25*, 3643–3651.

[13] Bouesse, L., Mouradian, S., Minalla, A., Yee, H., Williams, K., Dubrow, R., *Anal. Chem.* 2001, *73*, 1207–1212.

[14] McDonald, J. C., Duffy, D. C., Anderson, J. R., Chiu, D. T., Wu, H., Schueller, O. J. A., Whitesides, G. M., *Electrophoresis* 2000, *21*, 27–40.

[15] Wang, S. C., Perso, C. E., Morris, M. D., *Anal. Chem.* 2000, *72*, 1704–1706.

[16] Flanagan, J. H., Owens, C. V., Romero, S. E., Waddell, L., Khan, S. H., Hammer, R. P., Soper, S. A., *Anal. Chem.* 1998, *70*, 2676–2684.

[17] Woolley, A. T., Lao, K., Glazer, A. N., Mathies, R. A., *Anal. Chem.* 1998, *70*, 684–688.

[18] Deyl, Z., Miksik, I., Eckhardt, A., *J. Chromatogr. A* 2003, *990*, 153–158.

[19] Zuborova, M., Demianova, Z., Kaniansky, D., Masar, M., Stanislawsky, B., *J. Chromatogr. A* 2003, *990*, 179–188.

[20] Lazar, I. M., Ramsey, R. S., Sundberg, S., Ramsey, J. M., *Anal. Chem.* 1999, *71*, 3627–3631.

[21] Sung, W. C., Makamba, H., Chen, S. H., *Electrophoresis* 2005, *26*, 1783–1791.

[22] Verpoorte, E., *Electrophoresis* 2002, *23*, 677–712.

[23] Auroux, P. A., Iossifidis, D., Reyes, D. R., Manz, A., *Anal. Chem.* 2002, *74*, 2637–2652.

[24] Vilkner, T., Janasek, D., Manz, A., *Anal. Chem.* 2004, *76*, 3373—3386.

[25] Rocha, E. R., Smith, A., Smith, C. J., Brock, J. H., *FEMS Microbiol. Lett.* 2001, *199*, 73–78.

[26] Matinaho, S., von Bonsdorff, L, Rouhiainen, A., Lönnroth, M., Parkkinen, J., *FEMS Microbiol. Lett.* 2001, *196*, 177–182.

[27] Anwar, H., Dasgupta, M., Lam, K., Costerton, J. W., *J. Antimicrob. Chemother.* 1989, *24*, 647–655.

[28] Kadurugamuwa, J. L., Anwar, H., Brown, M. R. W., Zak, O., *Antimicrob. Agents Chemother.* 1985, *27*, 220–223.

[29] Kustos, I., Kocsis, B., Kerepesi, I., Kilár, F., *Electrophoresis* 1998, *19*, 2324–2330.

[30] Kiska, D. L., Gilligan, P. H., in: Murray, P. R., Baron, E. J., Pfaller, M. A., Jorgensen, J. H., Yolken, R. H. (Eds.), *Manual of Clinical Microbiology*, 8th edn., *Vol. 1*, American Society for Microbiology, Washington, DC 2003, pp. 719–728.

[31] Kadurugamuwa, J. L., Anwar, H., Brown, M. R. W., Shand, G. H., Ward, K. H., *J. Clin. Microbiol.* 1987, *25*, 849–855.

[32] Deneer, H. G., Potter, A. A., *J. Gen. Microbiol.* 1989, *135*, 435–443.

[33] Leung, K.-P., Folk, S. P., *FEMS Microbiol. Lett.* 2002, *209*, 15–21.

Index

Microfluidic Applications in Biology. Edited by Niels Lion, Joël S. Rossier, and Hubert H. Girault

ISBN-10: 3-527-31761-9